'This is the best and clearest treatment available that explains biodiversity conservation, agriculture, landscapes, agroecology and movement politics. And it does so in an easy to read and easy to understand style. A must read!'

Peter Rosset, Professor of Agroecology, ECOSUR, Mexico

'As today's threats of catastrophic biodiversity loss and increased human suffering mount, we have seen ever-more proposals in scientific literature and popular media seeking to superficially solve these problems. In contrast, Perfecto, Vandermeer, and Wright follow in the footsteps of all great revolutionary thinkers by asking us to confront these problems' deepest roots in all their complexity. This second edition of their classic work therefore does one better than the old saw, "Those who do not learn history are destined to repeat it." *Nature's Matrix* shows us as well that those who do not understand agroecology and food sovereignty are destined to recreate injustices against both people and nature. Luckily, *Nature's Matrix* shows us that, by taking allyship with social movements seriously, we can move the arc of history towards justice and sustainability for all of earth's inhabitants.'

M. Jahi Chappell, Centre for Agroecology, Water & Resilience, Coventry University, UK

'*Nature's Matrix* delves deep into the ecological, social, and political factors that drive biodiversity conservation in the tropics. The authors use both theory and case studies to successfully argue for the integration of agroecology, landscape ecology, and social justice for peasant farmers as necessary components for meeting conservation goals.'

Stacy Philpott, University of California, Santa Cruz, USA

'This powerfully written and accessible book is a must-read for anyone concerned with the future of our global food system and its capacity to support both food sovereignty and biodiversity. This new edition provides critical evidence that landscapes that support an agricultural matrix based on principles of social and ecological justice are the only way to sustainably feed the future.'

Hannah Wittman, Professor and Director, Centre for Sustainable Food Systems, University of British Columbia, Canada

Praise for the first edition

'This well written book is informed by sophisticated ecological theory applied to the complexities of modern tropical development in a dazzling critique of conventional thinking.'

Susanna Hecht, Professor of Urban Planning at the University of California, Berkeley, USA, and author of Fate of the Forest

'Greens of every stripe – agro-ecologists, conservationists, regional planners – recognize the need for ecologically farmed areas, an empowered farm population, preserved areas in any sustainable, just and productive mosaic landscape. We all advocate this integration. This book does it.'

Richard Levins, Professor of Population Sciences at Harvard University, USA, and author of The Dialectical Biologist *and* Biology Under the Influence

'As the authors demonstrate in this ground-breaking book, traditional agroecosystems not only offer promising models for other areas as they promote biodiversity, thrive without agrochemicals, and sustain year-round yields but are key for food sovereignty and the conservation of millions of wild species as they promote high quality matrices.'

Miguel Altieri, Professor at the University of California, Berkeley, USA, and author of Agroecology: The Science of Sustainable Agriculture

Nature's Matrix

When first published in 2009, *Nature's Matrix* set out a radical new approach to the conservation of biodiversity. This new edition pushes the frontier of the biodiversity/agriculture debate further, making an even stronger case for the need to transform agriculture and support small- and medium-scale agroecology and food sovereignty.

In the first edition, the authors set out a radical new approach to the conservation of biodiversity. This is based on the concept of a landscape as a matrix of diverse, small-scale agricultural ecosystems, providing opportunities to enhance conservation under the stewardship of local farmers. This contrasts with the alternative view of industrial-scale farms and large protected areas which exclude local people. However, since then the debate around conservation and agriculture has developed significantly and this is reflected in this updated second edition. The text is thoroughly revised, including:

- a reorganization of chapters with new and timely topics introduced,
- updates to the discussion of agroecology and food sovereignty, bringing it in line with the current debates,
- greater coverage of the role of agroecology, in particular agroforestry, as an important component of climate change adaptation and mitigation, highlighting recent studies on the role of intensive agriculture in climate change and loss of biodiversity, and
- more attention given to the discussion of land sparing versus land sharing.

By integrating the ecological aspects of agriculture and conservation biology, with a political and social analysis as well as historical perspective, the book continues to set a progressive agenda and appeals to a wide range of students and professionals.

Ivette Perfecto is George W. Pack Professor of Natural Resources and Environment, University of Michigan, USA.

John Vandermeer is Asa Gray Distinguished University Professor of Ecology and Evolutionary Biology, University of Michigan, USA.

Angus Wright is Professor Emeritus of Environmental Studies, California State University, USA.

Nature's Matrix

Linking Agriculture, Biodiversity
Conservation and Food Sovereignty

Second Edition

**Ivette Perfecto, John Vandermeer
and Angus Wright**

Routledge
Taylor & Francis Group

LONDON AND NEW YORK

earthscan
from Routledge

Second edition published 2019
by Routledge
2 Park Square, Milton Park, Abingdon, Oxon, OX14 4RN

and by Routledge
52 Vanderbilt Avenue, New York, NY 10017

Routledge is an imprint of the Taylor & Francis Group, an informa business

First edition published by Earthscan 2009

British Library Cataloguing-in-Publication Data
A catalogue record for this book is available from the British Library

Library of Congress Cataloging-in-Publication Data
Names: Perfecto, Ivette, author. | Vandermeer, John H., author. | Wright, Angus Lindsay, author.
Title: Nature's matrix : linking agriculture, biodiversity conservation and food sovereignty / Ivette Perfecto, John Vandermeer and Angus Wright.
Description: Second edition. | Milton Park, Abingdon, Oxon ; New York, NY : Routledge, 2019. | Revised edition of: Nature's matrix : linking agriculture, conservation and food sovereignty. 2009. | Includes bibliographical references and index.
Identifiers: LCCN 2019003033 (print) | LCCN 2019004236 (ebook) | ISBN 9780429028557 (eBook) | ISBN 9780367137779 (hbk) | ISBN 9780367137816 (pbk) | ISBN 9780429028557 (ebk)
Subjects: LCSH: Agricultural ecology. | Agrobiodiversity conservation. | Food sovereignty.
Classification: LCC S589.7 (ebook) | LCC S589.7 .P47 2019 (print) | DDC 577.5/5–dc23
LC record available at https://lccn.loc.gov/2019003033

ISBN: 978-0-367-13777-9 (hbk)
ISBN: 978-0-367-13781-6 (pbk)
ISBN: 978-0-429-02855-7 (ebk)

Typeset in Bembo
by Wearset Ltd, Boldon, Tyne and Wear

Contents

Figures

Boxes

Foreword: To be or not to be?

Rethinking the conservation question through inhabited environments

Susanna B. Hetch

To be or not to be? This is no longer a lead in to a joke, or even to one of the most famous soliloquies in the English language. Now as we trudge through the corpses of the sixth extinction, the question of what will remain as Earth catapults into an increase of more than 3.5 degrees centigrade, and we have transformed much of the worlds tropical forests – key ecosystems for CO_2 absorption – into car fuel, cosmetics, and animal flesh, the issue is no longer an idle question, or a poetic inquiry. It is an existential issue: what remains of the Anthropocene when we and this now unruly planet are through with each other. In another realm of great poetic lines from epic tragedies, is that lament of Faust's: 'linger a while, you are so beautiful', a yearning we might soon be echoing about our ravaged Earth. Faust, as one recalls, is the person who sold his soul to the devil. Wikipedia tries to answer the innocent questions posed by the Faust narrative: 'Who was Faust? An Alchemist, which is a kind of scholar of natural philosophy.' Alchemy had at its practical heart the idea of turning base materials into gold. Next question: 'What does selling your soul mean?' The answer is a bit more ambiguous than 'sacrificing everything for power or knowledge' the Wiki's tweetable response. If we add nature into the interrogations, the questions are much more profound because of transformations that imply the very annulling of the version of the planet on which we had learned to live until the last 200 years, and where, now, turning everything into gold seems to be the central ambition. We always expected some kinds of continuity of the world we have known: we'll always have elephants, *and* Paris, insects will annoy us to the end of time, now perhaps by outliving us, although perhaps not, since an insect apocalypse also seems to be underway. It's useful to juxtapose Hamlet and Faust, because Hamlet remains paralysed by and wracked by his inability to act even as he invites his own doom by inaction. Faust, the alchemist dreams of domination. There are a few endings to the Faust tale, both in the original conceptualizations of the poet Goethe, and the subsequent elabora- tions by other writers such as Thomas Mann. In one version, Goethe is carried off to hell, by that poodle who turns out to be Mephistopheles, the fallen angel. This is the declensionist tale of knowledge, greed, and power leading to destruction. But there is a redemptive version as well. In that version, Faust's striving for knowledge actually saves him – his understanding is in the end part of his salvation. It is a bit more complicated than I have made out, but the point is that there is more, can be more, than one ending, and this is perhaps how we should begin to think about the dynamics of conservation in the ever rougher Anthropocene that now confronts us. The recent IPCC report (The Intergovernmental Panel on Climate Change 'Global Warming of 1.3 degrees Centigrade') and from the US National Climate Assessment Report in 2018[1] are not bearing good news about the magnitudes and speed of change that climate dynamics are causing. The recent World Wildlife Fund's *Living Planet Report*[2] further describes levels of population decline by more than 60 per cent globally for mammals, birds, fish, reptiles, and amphibians, and levels greater than 80 per cent for the tropics, all this since 1970. We are in a new climate regime on a planet we seem intent on killing. So, we can remain wracked by our indecisions, thrust in to hell because we refuse to change, or possibly we can transform the trajectory, attenuate or at least modify its processes. These are not issues that have a solution per se. They are the outcome of long social processes and how we live with them and how to transform them involves a great deal of rethinking.

As Mike Hulme, a famous analyst of climate change has suggested, climate change is an environmental, political, and cultural process that is re-casting the way we think about ourselves, our societies, and humanity's place on earth.[3] We have some examples of profound and interesting responses of people to major historical climate events, such as the US Dustbowl,[4] and the Tambora volcanic explosion in 1815,[5] this last seems to have stimulated Mary Shelley to compose 'Frankenstein' – the parable of a destroying and desperate hybrid nature/culture monster. Since our planet is now a suffering hybrid, it is useful to take some time to rethink and recalibrate so that we don't slump off in howling despair at the ruin we created. But what we confront now requires a quite different set of approaches which have principles within them that can address the diversity and power of the transformations ahead of us. In this context of unrelenting challenges, we do have a few tools, but also, we need to reimagine what our relation to the world will be. We are no longer masters of our own destiny, but maybe we can be partners in it. This will require reviewing our paradigms of conservation which is exactly what matrix ecology does.

The question of the matrix – which is thoroughly explained in this book – thus moves into one of the central issues for climate, conservation and rural livelihoods. The Matrix idea and theory, which is a broader ecosystem and landscape approach to conservation, is especially relevant as we need to share our habitats with wildlife of animals and plants, both for protected areas and for the habits of daily life. The idea of domesticated landscapes has deep roots everywhere – but has recently been most strongly articulated for the tropics in terms of the basic planetary infrastructure of environmental services and rural livelihoods.[6] What this means is that matrices between old growth systems, in the world of fragmented ecologies – the ecologies that define our globe now – rather than being a kind of stigmatized degraded territory between valued 'primary landscapes', become the beloved fulcrum on which conservation depends. Furthermore, survival of rural populations are likewise dependent, relying on an array of benefits, from firewood, commercial and food crops to pollinators, that derive from such agroecologies and ecosystems; and even we, who reside in cities, whose profuse carbon output is rapidly sucked up by these secondary and complex systems whose rate of uptake is far more rapid than those of mature forests, we would be in far worse climatic shape without the active management of the inhabited environments by their landscape stewards – small farmers.

The matrix also invites us to rethink some other sets of questions and issues that will also be necessary: (1) We will need to change our values and epistemes (that is frameworks for understanding) from one that views nature as simply a platform for our enterprises to one of a more coevolutionary understanding of our relationship, and ourselves, as nature and part of a society of nature. These epistemes are widely found in native, indigenous, and rural populations which makes the survival of these societies all the more crucial;[7] (2) We need a view that sees nature and non-humans as, in the words of Bruno Latour, *actants*, that is, the non-human elements that shape our world, construct our time and lives as much as we do.[8] This decentres the questions of power over and power in the world to one in which we come into agreement with nature rather than insisting, incorrectly, that at the end of the day, we have mastery over it; (3) We need to think of ecological justice as well as environmental justice. Environmental Justice asserts that environmental burdens are borne more by the poor who suffer the brunt of climate disruption, pollution toxicities, and human-initiated catastrophic events. Ecological justice involves the recognition that, if the data from the World Wildlife Fund hold true, it is the world's organisms, large and small – from the rhinos to the beetles just to name the most obvious ones, that are paying for our interactions on the planet with their lives and evolutionary lines and futures. Extinction after all, is forever, and right now we are deciding which evolutionary trajectories will remain open and which are lost for all time.

The matrix ecology model presented in this book provides us with a way forward in both conservation and production landscapes and addresses the environmental as well as ecological justice issues. Biodiversity and the complex chains that support it – the sinews and infrastructures of the biosphere – also require some kinds of justice and, at the very least, a feeling of kinship and engagement. Ecological justice without rural environmental justice – this is the classic conservation set aside, the people less parks – will result in the sacrifice of both. At a moment when de-gazetting of protected areas by formal law or by fiat is rapidly underway everywhere in the tropics, the defensive bulwark of inhabited landscapes still resides with local populations, even if such places remain under deadly assault for their resources and as a prelude to land transformation through political, climatic, and structural transformations.

Currently in play in the conservation realm is the Borlaug hypothesis, the idea that intensification permits more land areas to be conserved, the so-called land-sparing model that would hive off conservation areas, even up to half the planet.[9] Elsewhere this has been summarized as a world of 'sacred groves and sacrifice zones',[10] which, if it would ever be implemented, would constitute the largest land grab perpetrated on almost half of the world's population, since the greatest conservation 'reward' is the diversity that resides in the tropics. There are several problems with this model, besides the injustice of its implied land theft from among the most marginalized populations in the world.[11] This is a useful moral argument, but there are other issues as well.

One issue, in the context of avid demand for tropical products, is that of Jeavon's paradox. Jeavon's paradox, which is currently being materialized in Latin America's soy zones, comes into play when technological progress or government policy increases the efficiency with which a resource is used but the rate of consumption of that resource rises due to increasing demand. Thus, while one is doing more with less, the efficient land use expands nullifying the purported conservation gains from efficiency. Therefore, in spite of significant technical change and intensifications in Latin America, soy production has expanded relentlessly to areas with less regulation, cheaper land prices or infrastructural and political benefits, as well as actively predatory land grabs of conservation and indigenous lands. The Borlaug hypothesis, unfortunately, has very little empirical basis.[12] The land-sparing model turns out to incarnate the Jeavon's paradox as it has applied to the South American soy system, and has condemned 50 million hectares of diverse forests, from Amazonian rain forests, to the Cerrado, the dry Caatinga, and Chaco to monocultures that can only be maintained through massive chemical inputs in lands increasingly vulnerable to climate change.[13]

The land-sharing model, which is predicated on the matrix idea, involves the idea of the multiple landscapes with intensive management through ecosystem-based local knowledge. This model has been the most durable idea for constructing highly diverse forest has produced a reality when aligned with a politics of rights, markets, and justice, which has conserved over 56 per cent of Amazonia,[14] as can be seen in the map (Figure P.1).

What matrix ecology, proposed in this book, helps us do is reconfigure our relationship to the living world. It's a powerful theoretical and epistemic system, as you will discover, but also an immensely practical way of being in the world of mutuality between people and nature. We are coming up into a bad planetary time, and, although I over-quote him, I think its relevant to repeat the phrase of the great Amazon explorer, Brazilian writer and humanist, Euclides da Cunha, who described the tropic like this: 'it is the last unfinished page of Genesis, and it is still to be written.' What remains to be seen is whether the text is redemptive or a simple obituary.

Figure P.1 Map of the Amazonian region in 2016 showing protected areas and indigenous territories.

Source: Amazonian Network of Geo-Referenced Socio-Environmental Information.

Notes

1 United States Global Change Research Program (2018).
2 World Wildlife Fund (WWF) (2018).
3 Hulme (2009).
4 Worster (1979).
5 Wood (2015).
6 Erickson (2006); Levis *et al.* (2012); Hecht (2014); Hecht *et al.* (2014); Maezumi *et al.* (2018).
7 Latour (2004); Descola (2013); Castro (2016).
8 Latour (2012).
9 Wilson (2016).
10 Oliveira and Hecht (2016).
11 Wisborg (2013); Davis *et al.* (2014); Balehegn (2015); Goyes and South (2016).
12 Rudel *et al.* (2009).
13 Oliveira and Hecht (2018).
14 Soares-Filho *et al.* (2006, 2010); Ricketts *et al.* (2010); Hecht (2012, 2014); Walker *et al.* (2013).

Preface

The earth's environment entered its current crisis state with two distinct miner's canaries – increasing greenhouse gases and decreasing biodiversity. These dual catastrophes have many interrelated parts, both causative and consequential, but they intersect at a critical nexus – the industrial agricultural system. While vested interests urge us to acknowledge the need for more intensive agriculture in face of the expected new wave of climate refugees needing food, and other well-meaning actors urge more intensive agriculture so that more land can be spared for biodiversity preserves, the elephant sneers from the middle of the room. Some political actors today urge that more greenhouse gases be burned to salvage what they apparently see as a crumbling world politico-economic system. Their cousins lobby for pesticide and fertilizer manufacturers, for large-scale animal feeding operations, for international economic concentration of lands for their export crops. All want yet more, and more, and more. More subsidies to support what they cannot support in the legitimate market. More support for land takeovers, especially in tropical lands. More investment in high profit technologies like genetic modification and robotics. Can we see something of the definition of chutzpah in asking us to support more of the same agricultural intensification that caused much of the problem, in order to solve the problem?

Beginning with a remarkably perceptive foreword, Susanna Hecht lays out the overall structure of the issues we face, in all their interdisciplinary complexity. Indeed, our coverage of the issue, based on the idea that conservation of biodiversity requires a focus not on the areas where that biodiversity is imagined to be restricted, but on the larger landscape, both physical/biological and cultural/political. While we begin with the incontrovertible ecological logic that biodiversity survival in the long run depends on a physical landscape that includes both their homes and the area between their homes, the larger intellectual landscape includes all the historical and political and cultural elements of that landscape, a central feature of our analysis.

This new edition adds nothing to the basic arguments in the first edition. What we have done is fill in some of the gaps that either we inadvertently left in the first edition, or that we now have evidence for as having emerged from events unfolding between the first edition and now. Having added considerable material, we also re-collated much of the text into 11 chapters from the original six. But, as the reader will no doubt recognize, the bulk of the text remains from the original edition and the basic argument remains untouched.

Acknowledgements

In this new edition we want to thank the millions of farmers in the world who consider biodiversity as an integral part of their system and do not make a distinction between nature and their farms. In particular we thank Edgardo and Annette Alvarado, and Lotty and Bernardo Morales form Puerto Rico and Walter Peters from Mexico for opening up their homes and farms to us and our students and sharing with us their love of the land. Our acknowledgements have not changed since the first edition, and we offer renewed thanks. Eschewing the risk of omitting someone who we value, we decline to list all who have contributed in some way to the ideas herein. You all know who you are and we trust you already know that we sincerely appreciate the continued interactions and support you have given us during the intervening years between the two editions of this work.

1   Matrix matters

An overview

The birds of New York and the coffee of Mesoamerica

Bird-watching requires a morning buzz. The best sightings occur with the rising sun, and even the most enthusiastic bird-watcher normally has to have an all-important cup of Joe before setting out to increment that ever-expanding list of species 'gotten' (the word bird-watchers normally use for 'having seen'). But the association of birds with coffee turns out to be far more important than the casual need for stimulants to get going in the morning.

It all started when someone noticed a striking coincidence: populations of songbirds in the eastern US were declining at the same time as some Latin American countries were engaged in the 'technification' of their coffee production systems in the never-ending quest for higher production. Technification[1] meant elimination of the shade trees that normally covered the coffee bushes in the more traditional form of growing coffee. Those traditional coffee farms were known to be forest-like habitats for various kinds of birds (Figures 1.1 and 1.2). Many of the songbirds that experienced population declines were migratory species that flew south for the winter. Putting a couple of evident observations together, it was not a gigantic leap in logic to suggest that eliminating the shade trees that constituted most of their wintering habitat would have a significant impact on those birds that sought the tropical climates to avoid the northern winter.

It remains debatable as to how much of the decline in populations of North American songbirds can be directly attributed to the technification (or intensification) of coffee production in Latin America. Regardless, the logic of the situation was enough to mobilize many people – bird-watchers, conservationists, environmentalists – to become apprehensive about this transformation of an agroecosystem. And that apprehension was a watershed. Previously, there had always been a kind of knee-jerk attitude that agriculture automatically meant degradation of natural habitat, which obviously translated into a loss of biodiversity, bird and otherwise. But here we have a case where it was arguably the type of agriculture – not the simple existence of agriculture, rather its particular form – that was having an effect on biodiversity.

Figure 1.1 Songbirds that are commonly found in coffee plantations in the Neotropics.
Source: Gerhard Hofmann.

Figure 1.2 A shaded coffee plantation that serves as habitat for Neotropical migratory birds.

Source: Hsun-Yi Hsieh.

What emerged from the coffee–biodiversity connection was a new discourse about nature. Previously, conservationists had clearly separated agricultural systems from natural ones in terms of their value for biodiversity conservation. But the discovery that many birds apparently could not tell the difference between a diverse canopy of shade trees in a traditional coffee plantation and a diverse canopy of trees in a natural forest caused a reassessment of the very idea of 'natural'. Was it legitimate to be concerned with keeping the remaining small tracts of untrammelled tropical forest forever untrammelled? If we just ignored the coffee technification were we not insisting that the birds adopt our particular interpretation of what a natural forest was? If the birds were mainly concerned with the existence of shade canopy cover and not so much with where it came from or who managed it (nature or a farmer), should we maybe rethink our peculiar notion of natural as something managed only by 'nature' and never by a farmer?

The debate still rages. Indeed, this book is in part intended to inform one piece of that debate. On the one hand, it is tacitly assumed that the proper goal is to intensify agriculture so as to produce as much as possible on as little land as possible, thus leaving the maximum possible amount of land under 'natural' conditions. On the other hand – and this is our position – agroecosystems are important components of the natural world, intricate to biodiversity conservation. Consequently, their thoughtful management should be part of both a rational production system and a worldwide plan for biodiversity conservation. To fully explain our position requires focusing on three areas:

1   the nature of biodiversity itself;
2   how we have arrived at our current state of agroecosystem management; and
3   the current way that humans relate to tropical landscapes.

These three areas are the topics of Chapters 2–7. In Chapters 8–10 we present some examples of how these three areas intersect in the real world. We conclude, in Chapter 11, with our full argument of what we regard as a new practical programme that derives from the underlying social and biological theory.

In the present chapter we provide an overview of the three areas, followed by a glimpse at what we regard as a new paradigm of conservation.

The argument

The ecological argument

In the last few years we have flown over areas that used to be continuous tropical forest in eastern Nicaragua, southern Brazil, southern Mexico, south-western India, and south-western China. In a general way, all areas look very similar: a patchwork of forest fragments in a matrix of agriculture (Figure 1.3). With few exceptions (mainly in the Amazon and Congo basins), the terrestrial surface of the tropics looks like this. And we always remain mindful that this is the location of the overwhelming majority of the world's biodiversity. So, if we are concerned in general with biodiversity conservation in the world, we ought to be concerned with what is happening in the tropics.

If most of the tropical world is a patchwork of fragments in a matrix of agriculture, and most of the world's biodiversity is located in tropical areas, should we not be concerned with understanding the reality of that patchwork? Yet the vast majority of conservation work concentrates on the fragments of natural vegetation that remain, and ignores the matrix in which they exist. Even at the most superficial level, it would seem that ignoring a key component of what is obviously a highly interconnected system is unwise.

The bias that favours concentrating efforts on the fragments while ignoring the matrix is far more damaging to conservation efforts than first meets the eye. As we argue in detail in Chapter 2, the matrix matters. It matters in a variety of social and political ways, but more importantly, it matters in a strictly biological/ecological sense. Recent research in ecology has made it impossible to ignore this fact. The fundamental idea has to do with the notion of a metapopulation.[2] Many, perhaps most, natural biological populations do

Figure 1.3 A fragmented landscape in Chiapas, Mexico.

Source: John Vandermeer.

Note
Fragments of natural vegetation are embedded in an agricultural matrix.

not exist as randomly located individuals in a landscape. Rather, there are smaller clusters of individuals that may occur on islands, or habitat islands, or simply form their own clusters by various means. These clusters are 'subpopulations', and the collection of all the clusters is a 'metapopulation'. The critical feature of this structure is that each subpopulation faces a certain likelihood of extinction. Accumulated evidence is overwhelming that extinctions, at this local subpopulation level, are ubiquitous. What stops the metapopulation from going extinct is the movement of individuals from one cluster to another. So, for example, a subpopulation living on a small island may disappear in a given year, but if everything is working right, that island will be repopulated from individuals migrating from other islands at some time in the near future. There is thus a balance between local extinction and regional migration that protects the metapopulation from going extinct at the regional level.

There is now little doubt that isolating fragments of natural vegetation in a landscape of low-quality matrix, like a pesticide-drenched banana plantation, is a recipe for disaster from the point of view of preserving biodiversity. It is effectively reducing the migratory potential that is needed if the metapopulations of concern are to be conserved in the long run. Whatever arguments exist in favour of constructing a high-quality matrix, and there are many, the quality of the matrix is perhaps the critical issue from the point of view of biodiversity conservation. The concept of 'the quality of the matrix' obviously must be related to the natural habitat that is being conserved, but most importantly, it involves, at its core, the management of agricultural ecosystems.

The agroecological argument

When considering agriculture, the research establishment has become mesmerized by a single question: How can we maximize production? Rarely is it acknowledged that such a goal is really quite new, probably with a history of less than a couple of hundred years, and little more than 60 years old in its most modern form. In recent years there have been major challenges to this conventional agricultural system, all of which, in one form or another, involve questioning this basic assumption. Why should maximization of production be a goal at all? We return to this question in Chapter 4. But for now, suffice it to say that many analysts have concluded that environmental sustainability, social cohesion, cultural survival, and other similar goals are at least equally as important as maximizing production, creating two attitudinal poles in what has become as much an ideological struggle as a scientific debate.[3]

This fundamental contradiction in attitudes is reflected most strongly in the alternative agricultural movement, especially among those who study agriculture from an ecological perspective. There is a distinct feature of more ecologically based agriculture that must be recognized here. Ecological scientists, in general, tend to have a certain mind-set, largely derived from the complexity of their subject matter. This mind-set is most easily appreciated by comparing the mind-set of a typical agronomist with a typical ecologist, especially an agroecologist. Both seek to understand the ecosystem they are concerned with, the farm or the agroecosystem. And both have the improvement of farmers' lives as a general practical goal. But the ways they approach that goal tend to be quite different.

Consider, for example, the classic work of Helda Morales.[4] In seeking to understand and study traditional methods of pest control among the highland Maya of Guatemala, she began by asking the question, 'What are your pest problems?' Surprisingly, she found almost unanimity in the attitude of the Mayan farmers she interviewed – 'We have no pest problems.' Taken aback, she reformulated her question, and asked, 'What kind of insects do you have?', to which she received a large number of answers, including all the main characteristic insect pests of maize and beans in the region. She then asked the farmers why these insects were not a problem for them, and again received all sorts of answers, always connected to how the agroecosystem was managed. The farmers were certainly aware that these insects could be problematic, but they also had ways of managing the agroecosystem so that the insects remained below levels that would categorize them as pests. Morales' initial approach was probably influenced by her original training in agronomy and classical entomology, but her interactions with the Mayan farmers caused her to change that approach. Rather than study how Mayan farmers solve their pest problems, she focused on why the Mayan farmers do not have pest problems, in the first place.

Here we see a characterization of what might be called the two cultures of agricultural science: that of the agronomist (and other classical agricultural disciplines such as horticulture, entomology, etc.) and that of the ecologist (or agroecologist). The agronomist asks, 'What are the problems the farmer faces and how can I help solve them?' The agroecologist asks, 'How can we manage the agroecosystem to prevent problems from arising in the first place?' This is not a subtle difference in perspective but rather a fundamental difference in philosophy. The admirable goal of helping farmers out of their problems certainly cannot be faulted on either philosophical or practical grounds. Yet with this focus we only see the sick farm, the farm with problems, and never fully appreciate the farm running well, in 'balance' with the various ecological factors and forces that, in the end, cannot be avoided. It is a difference reflected in other similar human endeavours – preventive medicine versus curative medicine, regular automotive upkeep versus emergency repair, etc. The ecological focus of asking how the farm works is akin to the physiologist's focus of asking how the body works. The agronomist only intervenes when problems arise, the agroecologist seeks to understand how the farm works and thus how it is maintained free of problems. Agronomists seek to solve problems, agroecologists seek to prevent them.

Why is this debate important for the conservation of biodiversity? If we accept the fact that most tropical areas are highly fragmented and that for biodiversity conservation the matrix matters, and we recognize that 'the matrix' consists of managed ecosystems, mostly agriculture, then the way we manage those agricultural systems becomes crucial for biodiversity conservation. If, as we noted above, all populations are metapopulations, migrations among natural habitat fragments is key to their conservation, and those migrations do not occur in a low-quality matrix, which is to say a biodiversity-unfriendly agricultural ecosystem. The rest of this book makes the case for the intricate connection between biodiversity and that human activity we call agriculture. We argue for a new paradigm for biodiversity conservation. That paradigm acknowledges what we now know about ecology: that the agroecosystem in which natural habitats are embedded is important. Recognizing this implies a connection with the effective planning that has occurred and still does occur in the agricultural sector. And this brings us to the question of rural social politics.

The rural grass-roots argument

A traditional way of designing an agroecosystem is by reference to ecological structures that exist in surrounding vegetation, a vision sometimes referred to as 'natural systems' agriculture.[5] Thus, for example, agriculture in grassland areas (which is where the most important cereal production in the world occurs) has recently been less environmentally sound than we might wish because it is designed to be an artificial annual monoculture whereas the native vegetation is a perennial polyculture.[6] The fact that we produce annual monocultures in these regions is simply due to the lack of available domesticated perennial grasses and the knowledge of how they might be productively managed. Other examples could be cited, but the principle is obvious – we should take cues from the natural world as to how to construct agroecosystems.

However, if this sort of ecologically complex 'natural systems' design is to be carried out, it is necessary to look to the actual practices that are currently in place. And when we do that, we are struck with the similarity between 'natural systems' agriculture and the way the vast majority of small-scale farmers in the tropics actually do plan their farms. Naturally the issue is complex, as all socio-political issues are, but the underlying idea behind the complex realities is that small-scale farming in the tropics has been and to a great extent continues to be based on something similar to 'natural systems' agriculture – what many farmers, especially in the Global South, refer to as 'agroecology' (Figure 1.4).[7]

Figure 1.4 Small-scale indigenous farmers in the highlands of Peru participating in the *faena* (collective work). The farmers' traditional agricultural system is based on cultivating a variety of crops that are native to the region and highly adapted to local conditions. Their farming methods incorporate many agroecological techniques.

Source: Asociación Paqalqu.

Overriding this fundamental ecological and economic framework is a political one. The very farmers who practise agroecological methods have frequently been driven off their lands, legally or not, and those who have preserved their farms are today faced with enormous economic, ecological, and political pressures. As a consequence, a large number of rural political movements have sprung up to challenge the international system that, to their minds, has created most of their problems. These movements are large, and tremendously diverse. However, within this diversity there seems to be emerging a framework that is remarkably all-embracing: the idea of food sovereignty. That is, couple the venerable idea of food security with the fundamental right of farmers to land, water, seeds, and other means of production, as well as the rights of rural communities to decide what and how to produce, and the basic plan of food sovereignty emerges.[8] It is a movement for farmers' rights as much as the general right to adequate food, and recounts the old cliché, 'give a woman a fish and she eats for a day, teach her how to fish and she eats forever'. It could be rephrased, 'give a local community big bags of grain and they will eat for a week. Construct socio-political, economic systems that allow the farmers in the community to produce the necessary food, and the community will eat forever.'

While there are many obvious reasons for wanting food sovereignty, for the specific purposes of our argument the crucial aspect of this new social movement is that it normally contains within it the assumption that the production system will preserve biodiversity and be based, to whatever extent possible, on the functioning of the local natural system. This is precisely the recipe that we claim is needed. We examine these broader socio-political aspects, both in theory and practice in Chapters 6 and 7. In Chapters 8–10, using the examples of coffee, cacao, rice, and the MesoAmerican *milpas*, we argue that we already have models for the way in which agriculture ought to be practised if it is to be biodiversity-friendly.

Critical perspective on interim experience since first edition

In recent years, many of the issues we explored in the first edition of *Nature's Matrix* have been investigated further in scientific and conservation literature. Events on the ground have made the necessity of confronting the issues we raised even more urgent. The need for biodiversity conservation has only become clearer and, with the passing years, it is evident that national and international organizations are often failing to meet the challenge. Worse yet, new approaches being made to the conservation of species are in many cases misguided and in others simply serve as masks to disguise the further exploitation of nature in ways that also deepen social inequalities.[9]

Major conservation organizations such as the World Wide Fund for Nature and The Nature Conservancy, whose aims are seemingly consistent with our recommendations, entered into agreements with national and local organizations over recent decades to enlist indigenous and small-farmer and peasant organizations in the tropical world to protect biodiversity. These agreements seemed to be based on a recognition that rural people in biodiverse-rich areas practised livelihoods that had successfully contributed to the preservation of high levels of biodiversity. These agreements also embodied the idea that, with increased financial and technical support, such groups might be even more successful in maintaining or enhancing the species diversity of the areas in which they worked while simultaneously improving human welfare. Some of these programmes have succeeded along the lines of the examples we cited in our first edition. However, scholars have revealed that many of these efforts seem to have fallen far below expectations, demoralizing both conservationists and local folk and discrediting the prospects for future undertakings.[10]

The analysis of the disappointments of such programmes is strongly consistent in diagnosing the problems that have undermined them. In nearly all the cases analysed, the outside organizations initiating the programmes promised that they would not only consult with local groups, but they would obey structures meant to recognize the primacy of the authority of the local people in both the design and implementation of the programmes. Where such assurances were not given, not surprisingly, the programmes hardly got off the ground for lack of local cooperation. Where assurances had been accepted by locals, outside conservation organizations and allied governmental authorities violated the terms of the agreement, usually in both subtle and gross terms.

Typically, in the initial stages, a good deal was said officially and unofficially about the deep wisdom and practicality of local knowledge and practice; but as time went on, when conflicts arose between local ways and the perceptions of the outsiders based on what they saw as 'scientific' knowledge, the ways of locals and the legitimacy of their knowledge were ignored or suppressed. Often, for example, locals believed that hunting (and/or fishing) restrained by local cultural norms played a positive role in the preservation of overall species richness, but in the slightly longer run, conservationists sometimes rebelled against such ideas and supported strong measures to prohibit or severely restrict hunting. Similarly, while officially conservationists had come to recognize that swidden, or in their terms, slash and burn agriculture, could play a positive role in maintenance of species-rich matrices, over time they fell back on their deep intuitive rejection of the idea that cutting and burning of even small areas of forest could really be compatible with species diversity. Accordingly, they took what measures they could to limit forest clearing, contrary to the initial understandings.

In conflicts over such matters as hunting and clearing, conservationists often resorted to support from government authorities in restricting what they saw as damaging activities. In such cases, local folks came to the conclusion that nothing had changed in the existing power structures, and that their knowledge and ways were not being taken seriously or were actively being repressed. Local elites were prone to reassert themselves in such instances, once again appropriating both authority and resources, while reminding conservationist organizations that they had, from the beginning, contended that local poor people could not be trusted. Years of effort to convince people that their ideas and their role in decision-making would be taken seriously vanished.

Reinforcing the sense of betrayal among local people, sometimes the failures of conservationist alliances fell prey to the simple inability to deliver on promises made. Financial subsidies and production assistance failed to materialize or were not maintained over time. Outside technical experts sometimes proved incompetent and/or remained privately unconvinced that they should respect local practice and knowledge.

In some cases, supposed conservation programmes based on alliance between conservation organizations and local peasants and indigenous people were far more stark in their deception. In particular, schemes to limit carbon dioxide emissions from tropical forest regions promised rewards to local folk in return for their cooperation, including respect for traditional practices. Governed by international agreements on climate change and referred to as REDD, REDD+, or REDD++, these controversial measures are meant to be key parts of the way nations reduce greenhouse gas emissions through increasing the amount of atmospheric carbon dioxide through forest protection and regeneration. However, these schemes sometimes masked the privatization of land worked and/or titled communally, leading to the expropriation of such lands by private logging firms. Forest previously protected through traditional practices of peasant producers were converted into private profit-making activities veiled by thin claims to 'sustainable forestry'.[11]

Such cases came to be seen by many scholarly analysts and people in the tropics as part of a larger trend towards 'land grabs'. In much of the tropical and sub-tropical world, corporations and wealthy investors began in the early 21st century until the present to carve out vast new landholdings. These were sometimes meant to take advantage of the collapse of resource and land prices over the last decade or so, and in others to simply speculate that such land would acquire value with time. Much of this land has been destined to be put into intensive agricultural cultivation in the ecologically and socially disastrous forms we discussed in our first edition.[12]

In this sense, the tendency toward land grabbing was simply a replay of the economic imperialism that has plagued the colonial and post-colonial world, as we discussed in our first edition. It also corresponded to the success of the 'neo-liberal' trends we analysed. However, some of what many saw as land grabbing has increasingly been dressed up as protection of land for conservation purposes. Dispossession of local folk was, and is, sometimes disguised with claims to respecting the role of precisely these dispossessed local people in conservation of biodiverse landscapes, sowing further distrust of the cause of biodiversity conservation as well as destroying rural livelihoods. In other cases, dispossession is simply justified by the crude rules of 'market logic' and the legal structures that enshrine it.

Taken as a whole, the disappointments and failings of much of conservationist practice, especially in the tropics and sub-tropics, has now been characterized by many scholars and social justice activists as the result of the formation of a new 'nature industry' bent on further dispossession of the world's poor in the name of conservation. 'Nature, Inc.' is seen as raw capitalist exploitation, justified by a new conservation rhetoric, allied with misguided or narcissistic efforts by the privileged upper and middle classes of the world to preserve a new form of consumer goods – natural landscapes and the species that inhabit them. 'Conserved nature', in the view of these critical scholars and activists, becomes a kind of global theme park to be consumed in the form of beautiful films, artwork, and nature tourism. Ecological science itself becomes suspect as a form of middle-class self-indulgence. While at least nominally recognizing the importance of conservation of species and landscapes, the critique of 'the nature industry' promotes a deep cynicism that could make it more difficult to harmonize the need for conservation with concerns for social justice. On the other hand, it could lead to the greater clarity needed to truly confront the challenges at hand.[13]

The new critique of 'the nature industry' and the disappointments on which it is based give new emphasis to the political and economic analysis we laid out in the first edition of the book. The profound historical and contemporary inequalities of global society, expressed in both social inequities of all kinds and in the distressed landscapes and vanishing species of the planet, challenge all efforts at improvement in the situation. They cannot be ignored. That is why we have insisted from the beginning not simply that conservation practice must recognize the legitimate contributions that largely poor rural peoples have made and continue to make towards biodiversity conservation, but that we also must recognize the legitimacy of social organizations and movements that promote the economic and political claims of the rural poor. Far too often, words such as 'collaboration' and 'consultation' in conservation practice have become shams for the maintenance of pre-existing social structures – as such, they account for much of the failure observed in conservation programmes. Clearly, individual projects and programmes of biodiversity conservation can be expected to perform poorly when prevailing social, economic, and political conditions maintain the continued subservience of the rural poor.

The argument that conservation would be best served by intensification of agricultural production, on the one hand, while greatly expanding areas under protected conservation status, on the other, has become better known in recent years as the 'land sparing' strategy. In one of its strongest statements, it is exemplified by E. O. Wilson's book, *Half-Earth: Our Planet's Fight for Life*[14] in which Wilson argues that half of the planet's land surface should be set aside in 'inviolable reserves'. In conservation debates it is contrasted to the 'land sharing' approach identified with the strategy we advocate. The 'land sparing/land sharing' debate has sharpened and has been the subject of considerable scholarly research in the last decade. A recent meticulous review of this new research by Kremen concludes that the debate,

> limits the realm of future possibilities to two, largely undesirable options for conservation. Both large, protected regions and favorable surrounding matrices are needed ...; they work synergistically and are not mutually exclusive. A 'both-and' framing of large protected areas surrounded by a wildlife-friendly matrix suggests different research priorities from the 'either-or' framing of sparing versus sharing. Furthermore, wildlife-friendly farming methods such as agroecology may be best adapted to provide food for the world's hungry people.[15]

Kremen's conclusion seems sensible and entirely compatible with our original argument.

It would be disingenuous, however, to ignore the strongly political nature of the debate. The argument for a strong emphasis on agricultural intensification is being made on many grounds, as much, or more, within the debate about world food needs as within biodiversity conservation debate, as Kremen alludes to in the last sentence of her abstract, cited here. The intensification argument is heavily supported by agri-business firms interested in selling more chemical fertilizers and pesticides, more genetically engineered seeds, and more farm machinery. The intensification argument also supports what has been seen as an aggressive strategy of concentration of land into fewer hands, including 'land grabbing' in Asia, Africa, and Latin America. Organizations of small farmers, peasants, and indigenous peoples have seen this strategy as a direct threat to their livelihoods, control over land, and, indeed, survival. The stakes are enormous in the immediate world of economics and politics. It is impossible to cleanly separate the debates about conservation strategies from this set of controversies.

Towards a new paradigm

In the rest of this book we develop our argument in great detail. But its outline, we feel, is clear from this introductory chapter. It is a three-part story, with the parts interrelated in complicated ways, each with its own complex historical roots, but fundamentally telling a simple story. First, in fragmented landscapes, seemingly the predominant landscapes in most tropical areas of the world, the balance between extinction and migration is what determines whether a species will survive over a large area and in the long term – there is no question that it will periodically go extinct in particular fragments, but the key issue is whether that extinction event will eventually be countered by a migration event or will eventually become part of a regional extinction. Second, the matrix in which the fragments occur is mainly devoted to agriculture of various kinds, and the particular form of agriculture may or may not be biodiversity-friendly, either in its ability to preserve directly some forms of life or in its ability to act as a passageway for migrating organisms. Third, this matrix is constructed primarily by small farmers, many of whom struggle to survive in the current political and economic climate, and have consequently developed a worldwide movement to transform agriculture from its post-World War II form (so heavily dependent on biocides and chemical fertilizers) to a more ecologically sound form.

Our new paradigm incorporates these three issues into a prescriptive whole. Support for the worldwide food sovereignty movement results in promotion of biodiversity-friendly agricultural methods that encourage high-quality matrices, thus increasing migrations to balance the inevitable extinctions that exist in natural habitat fragments. It is a paradigm that suggests a reorientation of conservation activities, away from a focus on protected areas and towards the sustainability of the larger managed landscape; away from large landowners and towards small farmers; away from the romanticism of the pristine and towards the material quality of the agricultural matrix, nature's matrix. At the centre of this new paradigm is the urgent need for social and environmental justice, without which the conservation of biodiversity ultimately becomes an empty shibboleth.

Notes

1 The term 'technification' comes from the Spanish 'tecnificación' which is used to signify the intensification of coffee farms by eliminating or reducing shade trees, planting new high-yielding varieties of coffee at higher densities, and applying more agrochemicals.
2 Levins (1969); Hanski (1999).
3 van der Ploeg (2009); Rosset and Altieri (2017).
4 Morales and Perfecto (2000).
5 Altieri (1987); Ewel (1999).
6 Jackson (1980).
7 Wezel *et al.* (2009); Ferguson and Morales (2010); Rosset *et al.* (2011); Rosset and Martínez-Torres (2012); Gliessman (2013).
8 'Food Sovereignty: A Right for All' (2002), Political Statement of the NGO/CSO Forum for Food Sovereignty, Rome, 8–13 June, https://nyeleni.org/spip.php?article125, accessed 10 September 2018; Patel (2009); Wittman *et al.* (2010); Altieri and Toledo (2011); Holt-Giménez and Altieri (2013); Agarwal (2014).
9 For a richer discussion of this issues, examined in specific contexts as well as in theoretical depth, see, Hecht *et al.* (2014).
10 This summary discussion is taken from the following case studies: Sauer and Almeida (2011); Doane (2012); Latta and Wittman (2012); Martinez-Reyes (2016); Zanotti (2016); Velázquez Runk (2017).
11 Cobera and Schroeder (2017).
12 There is a large literature and a notable abundance of internet resources and organizations on this issue. A useful start is Oberlack *et al.* (2016).
13 Büscher *et al.* (2014).
14 Wilson (2016).
15 Kremen (2015: 52). On this topic also see Vandermeer and Perfecto (2005b); Perfecto and Vandermeer (2010); Tscharntke *et al.* (2012).

2 The ecological background

The fundamental patterns of biodiversity

When the popular media approach the issue of biodiversity, the subject matter is almost always about charismatic megafauna – tigers, elephants, pandas, and the like. We too lament the probable extinction faced by these evocative creatures. The world will surely be diminished as the last wild gorilla is shot by a local warlord beholden to one or another political ideology, or even as a rare but beautiful bird species has its habitat removed to make way for yet another desperately needed strip mall. The irony is gut-wrenching to be sure. However, such concerns are the very small tip of the very large iceberg. If we simply take mammal diversity as an estimate of the number of creatures that are likely to be thought of as charismatic by the general public, we are talking about approximately 4,500 known species. By comparison, there are currently about 900,000 known species of insects, and that is almost certainly a gross underestimate of how many actually exist. We have no idea how many species there are, since estimates for terrestrial arthropods (most of which are insects) range from about two million to as high as 30 million.[1] While the most recent estimate narrows the range to between 2.5 and 7.8 million,[2] even if there are only two million species of insect, we see that a focus on the 4,500 species that happen to look more or less like us, is limited to a rather small fraction of the earth's biodiversity. And to make the point even more dramatic, consider the biodiversity of bacteria. Microbiologists define two bacterial cells to be in the same species if their DNA overlaps by 70 per cent or higher, which, if applied to mammals, would put all primates (if not all mammals) in the same species. Simply from the point of view of numbers, the world of biodiversity is mainly in the small things, from bacteria to insects, leaving the charismatic megafauna as a rather trivial subplot to the main theme of diversity on earth.

Apart from these dramatic patterns associated with classifying life, biodiversity manifests itself in a variety of other patterns. For convenience of presentation, we identify four:

1   Biodiversity tends to increase and then undergo massive and rapid extinctions at periodic intervals, at least for the past 600 million years – an evolutionary pattern.

2   Species diversity tends to increase with decreasing latitude and altitude –
     a geographic pattern.
3   Species diversity tends to decrease on islands when the island is smaller
     and/or more distant from the mainland – an insular pattern.
4   Species diversity tends to decrease as the intensity of management of the
     ecosystem increases – an intensification pattern.

Each of these patterns is relevant to the general topic of how to conserve biodiversity. We summarize each of them in turn.

Evolutionary pattern: diversity through time

When the earth began, almost five billion years ago, it was void of any life, as
far as we know. For more than a billion years it was bombarded by meteors,
and there is some speculation that life could have originated many times, only
to be destroyed each time by a meteor. Nevertheless, the earliest solid evidence we have for life on earth is in some fossils, structurally similar to bacteria, from about 3.5 billion years ago. The actual process of photosynthesis
evolved at some point around three billion years ago, and can be thought of
as the first major biological revolution in the history of the earth. Before the
evolution of photosynthesis, the atmosphere was virtually devoid of oxygen.
Yet now, and for the past two billion years, not only does the earth's atmosphere have an abundant supply of oxygen, but the majority of life forms
actually depend on oxygen for generating energy.

   During the enormous period in which oxygen was accumulating in the
environment, it is likely that there was some sort of evolutionary pressure for
increasing size, although exactly which environmental force may have produced that pressure is at best a matter of speculation. Nevertheless, in order to
get larger, it was necessary to have more organization in the cell, which gave
rise to the partitioning of the hereditary material within a membrane, what
we today call the nucleus of the cell. Then, in one of Nature's greatest
revolutions, a parasitic bacterium which invaded a cell in the same way bacterial parasites do today, gradually evolved a lower toxicity and the attacked
cells themselves slowly evolved the ability to take advantage of the biochemical apparatus that the parasite used for energy generation, leading to one of
the most important symbiotic relations in the history of life. This was, by
most accounts, the origin of the mitochondria, the small structures inside the
cell that are responsible for producing the energy the cell needs to carry on its
regular activities. A similar story can be told for chloroplasts, the small structures inside plant cells that contain the biochemical machinery for
photosynthesis.

   The result of all this new organization was a new type of cell, with an
obvious high level of organization within the cell itself: a nucleus, mitochondria, chloroplasts (for plants), and many other evident structural details. This
was the eukaryotic cell, which first appeared about 1.75 billion years ago. All

previous life forms were very small cells, the size of today's bacteria, a type of cell organization referred to as prokaryotic.[3]

The next major event in the evolution of biodiversity was the ability of cells to associate with one another, which, in addition to the need for a way of sticking together, also required some sort of communication system among cells. This cell-to-cell signalling is a major area of biological research today, but for purposes of the present discussion, we only need note that the ability to engage in communication between cells that were able to physically stick together was another major evolutionary event: the first multi-cellular organisms. This happened about a billion years ago.

Then the most amazing, and even perplexing, event took place. Around 600 million years ago, there was an explosion of life forms. Almost all the major forms of animals we know today suddenly appeared. Within a very short window of time, perhaps as short as 30 million years, the earth went from almost no multi-cellular life to a rich diversity of such life. This event has been termed the Cambrian explosion, based on the classical name for Wales, location of the site where some of the first fossils of this period were originally discovered and studied. And not only was it a rich diversity of life forms, it was the origin of all the basic body plans we see today, according to some palaeontologists.[4] That is, we went from almost no multi-cellular biodiversity to 'all' the basic diversity of forms the world has since known, in a period of only 30 to 40 million years – a blink of the eye in geological time.

Naturally, when we say 'all' the biodiversity, this implies something about the level of diversity under consideration. We actually had 'all' the world's biodiversity about 1.75 billion years ago when the evolution of eukaryotic cells added a third basic body plan to the mix of only bacteria and a similar group of prokaryotic organisms, the Archaea. At this high level of organization, we have not seen any new innovations in the past 1.75 billion years. But moving to a lower taxonomic level, what we see in the Cambrian explosion is the sudden appearance of all the basic body plans we know today, among the animals, plus a variety of other basic plans that do not seem to have survived into modern times.

An example of the basic body plans that did not survive to the present day is that of the trilobites. These were animals whose segmented body was arranged in three longitudinal lobes. In some sense these were the most successful large multi-cellular organisms that have ever lived. They were incredibly diverse – big ones, small ones, ones with plain bodies, ones with spines and other ornaments. And they were always a numerically significant component of all oceans. Of course, they shared their environment with many other forms, memorialized in the description of the famous Burgess Shale in Stephen J. Gould's delightful book *Wonderful Life*.

The Cambrian explosion was followed by several major diversification patterns, each interrupted by general extinctions. Indeed, the formal geological names given to various segments of life's history in the post-Cambrian explosion world correspond to periods between these general extinction events.

Diversification followed by extinction was the rule of evolution for all post-Cambrian time. But there would be a truly catastrophic event about 330 million years after the beginning of the Cambrian period. For reasons that remain obscure, almost all animal species and many plant species disappeared! This event is referred to as the Permian extinction, and marks the final demise of that basic body plan so common in all oceans for the prior 330 million years – the trilobites.

The ensuing 180 million years saw the rise to dominance of the dinosaurs. If we can categorize the first 330 million years after the Cambrian explosion as the age of the trilobites, the next 180 million can be called the age of the dinosaurs. Dominating the terrestrial environments of the world, the dinosaurs diversified three distinct times, each of which was interrupted by a mass extinction, giving rise to the threesome: Jurassic, Triassic, and Cretaceous, names for the interludes between the extinction events. It should be noted that mammals also first evolved at the beginning of this dinosaur age, but they never diversified to any degree and remained a small and minor component of terrestrial biodiversity during this entire period.

About 65 million years ago the earth was subjected to a calamity of such proportions that we can hardly imagine it. A giant asteroid smashed into a point currently located in the northern Yucatan peninsula of Mexico. This event caused dramatic and sudden changes in our planet's environment such that the dinosaurs, to the very last one of them, were eliminated from the earth's surface. While it is not clear exactly why the asteroid caused such a dramatic extinction event (in other words, it could have been the elimination of their food supply, or the contamination of the air, or perhaps some other catastrophic change), the fact of the asteroid being the causative agent, is no longer contested.

The elimination of the dinosaurs from the planet dramatically lowered worldwide biodiversity for a short period of time, as happens after every major extinction event. However, soon thereafter we see another diversification: the rise of the mammals, birds, insects, and flowering plants to a position of dominance in the terrestrial world. Indeed, the basic structure of terrestrial ecosystems as we know them today originated late in the dinosaur age and became dominant after the crash of the asteroid. Flowering plants capture energy through photosynthesis, and the insects that pollinate their flowers facilitate their main form of sexual reproduction. Birds, mammals, and some insects are frequently responsible for dispersion of their seeds, the second aspect of reproduction.

Overall, in geological time we have seen five major extinctions and four major divisions in the qualitative nature of the earth's biodiversity: first, prior to the Cambrian explosion, the pre-Cambrian era, second the age of the trilobites (formally known as the Palaeozoic era – old life), third, the age of the dinosaurs (formally known as the Mesozoic era – middle life), and fourth, what is commonly referred to as the age of the mammals (formally known as the Cenozoic era – new life, and including flowering plants, birds, and insects). Today we are living in the Holocene epoch of the Cenozoic era, which is still very young. Although some argue that the influence of human beings on earth has been so strong that we should indeed recognize a new epoch, the Anthropocene. In 2016 the International Commission on Stratigraphy voted to proceed with a process to define the Anthropocene epoch formally in the Geologic Time Scale.[5]

This brief overview of biodiversity over time makes it clear that the biodiversity we have today on our planet is the result of an evolutionary process that started 3.5 billion years ago and that has undergone periods of diversification and extinctions, even mass extinctions many times, well before humans first appeared on earth, just a geological instant ago. It has been estimated that well over 99 per cent of species that have ever existed on earth have already become extinct. While even this may be an understatement, it is nevertheless adequate to make the point that extinction is a natural and normal process. We know that a species emerges from some ancestor, lives on earth for a short period of time (geologically speaking, from hundreds of thousands to millions of years), and eventually goes extinct. It happens to all of them. Over the long term we talk of a 'normal extinction rate', the number of extinctions per time period. So, for example, the normal extinction rate for birds is one species per 400 years. Yet, since 1500 we have seen 150 bird species go extinct, which is 116 times greater than the normal extinction rate.[6] Occasionally, for diverse reasons, these extinctions occur in a coordinated fashion, massively among many species such that the title 'mass extinction' is warranted. Some authors argue that today we are entering a sixth mass extinction since the current rate of extinction for all species is estimated at 100 times that of the normal rate.[7] The fact that the extinction rate today appears to be on the order of past mass extinctions is worrisome and is the reason we are all concerned about biodiversity conservation (recall that it took about four million years for evolution to recuperate from the last extinction event).

However, there is no denying that extinction is part and parcel of the overall evolutionary dynamics of life on earth, and not just in geological time. As we shall see later in this chapter, extinction is an important contemporary force, although it is a complicated one that must be appreciated for what it is. The Sisyphean struggle to stop all extinctions in all parts of the globe may not only be frustrating for conservationists, but may also in the end be bad conservation policy – a point we discuss in some detail as our argument unfolds.

Geographic pattern: biodiversity changes with latitude and altitude

One of the most obvious patterns in the biological world is the dramatic difference between temperate and tropical regions. In a one-hectare plot of land in Michigan, we have identified 16 different species of trees. In a one-hectare plot of land in Nicaragua, we have identified 210 different species of trees. And such differences exist for almost all groups of species. The bird guide of Colombia lists 1,695 species, while in all of North America (a much larger area) there are approximately 932 species. Butterflies, ants, mammals, and amphibians all show this same pattern. There are definite exceptions (e.g. sea birds), but the general pattern is one of increasing numbers of species as you approach the equator.

This pattern has been the subject of an enormous amount of speculation and debate in ecology. Many hypotheses have been suggested to explain it. One scientist, combing the literature, found a minimum of 21 distinct hypotheses laying claim to explaining the pattern.[8] At present, there is no final agreed-upon theory that satisfies all biodiversity scientists.

A very similar pattern exists with respect to altitude – there are very few species at higher elevations, and more species at lower elevations. Most of the hypotheses as to the underlying cause of latitudinal gradients have also been suggested for the altitudinal gradient, and are likewise not agreed upon by biodiversity scientists.

The practical political implications of this latitudinal pattern are extremely important, and form one of the foundations for our basic argument. It is in the tropics where the bulk of the world's biodiversity lies, and it is also in the tropics where most of the poor people in the world reside. It is also in the tropics where there is still a large rural population whose livelihood depends on agriculture. The connection between rural people in the tropics and biodiversity conservation is at the heart of the new conservation paradigm presented in this book. We revisit this point several times in the development of our thesis.

Insular pattern: biodiversity on islands

One of the most noted patterns of biodiversity is seen on islands. Generally, there are more species on larger islands and more species on islands nearer to the mainland. Nevertheless, exactly how many more species will be found on an island twice the size of another island, remains highly variable. Despite this variability, it is a general rule of thumb that if you plot the logarithm of the number of species against the logarithm of the size of the island, you get a straight line.

Explanations of this pattern generally fall into three categories. First, it may be simply a sampling problem. This is certainly part of the explanation and is extremely important when it comes to the difficult issue of sampling biodiversity. Even today the technical literature is filled with the elementary error of determining species richness in a series of samples and then taking the average of those determinations as the estimate of species richness in that environment. Because species accumulate non-linearly with increasing sample size, this procedure is fundamentally wrong and can give highly misleading results.

The second category has to do with environmental heterogeneity. It seems obvious that a larger area, be it a sampling unit or an island, will probably contain more microhabitats than a smaller area and thus provide more 'niches'

for species to fill.[9] In the end, this explanation falls within the general category of differing extinction rates (if a new species arrives at an island but fails
to locate its habitat, it becomes extinct), and is part of the theory of island
biogeography, as discussed presently.

The third category is an equilibrium balance between immigration and emi-
gration rates, first fully elaborated by Robert MacArthur and Edward Wilson in
their classic work *The Theory of Island Biogeography*.[10] The basic idea is that populations of organisms are always dispersing in one way or another. Any area is
thus likely to receive migrant individuals at regular intervals, whether in the
form of a dispersed seed of a plant or a migrant individual bird or bat, or the
propagule of a micro-organism. On an island one could, theoretically, sample
all those incoming organisms and determine which ones were new to the island
and which ones had been living on the island already. The rate at which *new*
species arrive on the island is the immigration rate. Also, if you had enough
helpers and technology, you could keep track of all the organisms on the island,
and thus know when any particular species completely disappeared, which is to
say when it becomes extinct, on that particular island. Then, if the rate of
extinction is greater than the rate of immigration, the total number of species
on the island would decline. Similarly, if the rate of immigration of new species
is greater than the rate of extinction, the number of species would increase.
The perfect balance, known as 'the equilibrium number of species', is when the
extinction rates and immigration rates are the same (Figure 2.1).

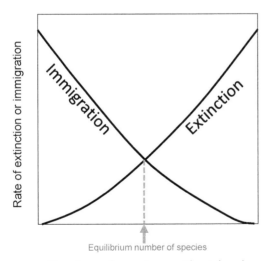

Figure 2.1 The theory of island biogeography developed by Robert MacArthur
and Edward O. Wilson. The number of species found on an island
can be determined by the balance between immigration and extinction rates.

Source: The authors.

The equilibrium theory of island biogeography, has been employed in many different guises in ecology and, especially today, is implicated in all sorts of practical problems ranging from the ideal design of a nature reserve to the underlying reason why tropical rainforests have so many species of trees. In conservation biology, the theory is applied to habitat islands (e.g. fragments of forests), and it is often assumed that the agricultural matrix that surrounds these 'islands' is like the ocean in real oceanic islands and therefore devoid of terrestrial biodiversity. One of our arguments, and one that we elaborate further in subsequent chapters, is precisely that the agricultural matrix is nothing like the ocean, devoid of terrestrial biodiversity. On the contrary, the agricultural matrix can act as a refuge for biodiversity as well as a facilitator of movement of organisms between the 'islands' of forest. When combined with the related subject of metapopulation theory (discussed later in this chapter), it is also a major feature of the most recent refocusing of the conservation agenda, the new conservation paradigm, and the subject matter of this book.

Intensification pattern: changes in biodiversity with management

Over 50 per cent of the terrestrial surface of the earth is covered with managed ecosystems.[11] In the popular and romantic conceptualization of Nature as a Garden of Eden, many conservationists think of agriculture as the defining feature of biodiversity loss. The world gets divided into those areas untouched or minimally touched by *Homo sapiens* as contrasted to those areas 'despoiled' by human activity. One of the main observations that caused a re-evaluation of this prejudice was the correlation between the decline of populations of songbirds in the eastern US and the transformation of the coffee agroecosystem of Mesoamerica. As detailed in the previous chapter, the traditional method of coffee production includes a diverse assemblage of shade trees with the coffee bushes growing as if they were part of a forest understorey. These coffee farms appeared to be forested land if viewed from above and it has now been convincingly demonstrated that they are important habitats for those very bird species from North America when they migrate south for the winter.[12] This key observation has been significant in demonstrating that agroecosystems can be critical repositories of biodiversity, but even more importantly, that the particular type of agricultural practice can be a determinant of the biodiversity the ecosystem contains. Not all coffee farms harbour high levels of biodiversity, and the characterization of what types of agroecosystems generally harbour greater or lesser amounts of biodiversity has only recently emerged as a serious scientific question.

When dealing with managed ecosystems it is first necessary to distinguish between two concepts of biodiversity. First, the collection of plants and animals that the manager has decided are part of the managed system – rice in the paddies of Asia; maize, beans, and squash in the traditional fields of Native American Mayans; carp in the fish ponds of China, etc. This is referred to as the 'planned' biodiversity, or sometimes the 'agribiodiversity'. Yet in each of these ecosystems there is almost always a great amount of biodiversity that spontaneously arrives – the aquatic insects and frogs in the Asian rice paddies, the birds and bugs that eat the Mayan's *milpas*, the crayfish that burrow their way into the sides of the Chinese fish ponds. This is referred to as the 'associated' biodiversity or 'wild biodiversity'. Frequently the managers themselves are determinedly concerned about the planned biodiversity, especially when dealing with genetic crop diversity. However, it is almost certainly the case that the associated biodiversity is the most abundant component of biodiversity in almost all managed ecosystems, and as such, it has received a great deal of attention in recent years.

Although the term 'agricultural intensification' has a very specific and complex definition in anthropology, in the biodiversity literature the term 'management intensification' is taken to be the transition from ecosystems with high planned biodiversity and a more traditional management style, to low planned biodiversity and an industrial management style, such as the use of agrochemicals. So, for example, in the case of coffee just mentioned, the intensification of coffee refers to the reduction of shade trees ending up with an unshaded coffee monoculture. The ecology of the agroecosystem is such that the final stages of intensification usually involve the application of agrochemicals to substitute the functions or ecosystem services of some of the biodiversity that is eliminated. The coffee example will be discussed in more detail in Chapter 8.

The main question of concern is how the pattern of associated biodiversity changes as a function of the intensification of agriculture. Although this question received some attention over the last 20 years, it remains largely unanswered for almost all agroecosystems and almost all taxa. The studies that actually ask this question come up with results that depend on the taxon involved and even the definition of what constitutes greater or lesser intensity. But that said, it is possible to make one central generalization: biodiversity declines with agricultural intensification. Most of the patterns described in the literature can be summarized into two basic prototypes of associated biodiversity change as a function of intensification (Figure 2.2). First, as tacitly assumed by many conservation practitioners, as soon as a natural habitat is altered by a management system, associated biodiversity tends to fall dramatically. Second, as has been shown in several cases, associated biodiversity declines by only small amounts with low levels of intensification and only after much higher levels are reached do we see dramatic declines. Which of these two patterns (or what combination of the two) exists in particular systems is largely unknown because this question has not been a popular one amongst those concerned with biodiversity – a major intellectual lacuna in conservation biology.

Indeed, although many conservation organizations in the past ten to 15 years have come to realize that the agricultural component of a landscape is important to conservation, it remains frequently the case that individual conservation practitioners pay little attention to agroecosystems or forestry

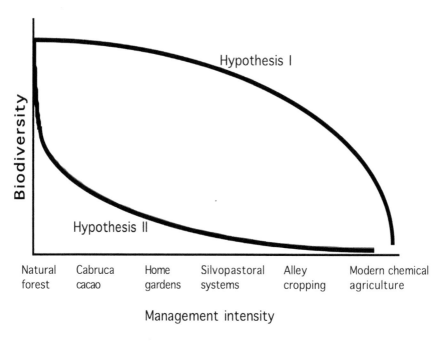

Figure 2.2 Two hypotheses about the relationship between management intensity and biodiversity.

Source: The authors.

systems or aquatic managed systems. Their assumption seems to be that once management activities are initiated in an area, the question of biodiversity becomes irrelevant. From a strictly intellectual point of view this position is simply one of ignorance. But from a practical point of view it is a position that could have devastating consequences when it comes to biodiversity conservation itself. If a minimum of 50 per cent of the world's surface is covered in managed ecosystems, and if managed ecosystems contain even a small fraction of the biodiversity contained in unmanaged ecosystems, ignoring them will be counterproductive, to say the least.

While many agroecosystems are tremendous reservoirs of biodiversity and deserve attention for that reason alone, there is another reason why the intensification gradient is important to biodiversity conservation – agroecosystems form the matrix through which organisms must travel as they move among natural habitat fragments. The pesticide-drenched rice field is hardly convivial to biodiversity generally, but is also not a habitat through which some organisms are likely to travel. An anecdote perhaps helps make our point. In conversation with a local bird-watching enthusiast in the coffee-growing areas of southern Mexico, we mentioned the highland guan, a forest bird species in the area. We asked if the species is ever seen in the abundant shade-coffee farms in the area. After some thought, the answer was 'hardly ever'. We immediately focused on the word 'hardly'. The highland guan will not live in a coffee plantation, but it will cross over one in pursuit of a new patch of forest. Will it do the same in a non-shaded plantation, or a maize monoculture? The issue is not whether species are able to form successful populations in these matrix habitats, it is whether they will ever venture into them at all. If not, the matrix in fact is that impenetrable ocean, devoid not only of permanent biodiversity, but also of any migratory potential.

In addition to the two alternative hypotheses of Figure 2.2, there are two slightly more complicated hypotheses that may more accurately represent what actually happens in the real world. In Figure 2.3a we show how both hypotheses can be simultaneously correct, except they are diverted to the left and right, with a zone of rapid transition between them. We refer to this as hypothesis III. Frequently there is a tacit assumption that changing consequences of some environmental driving variable (in this case management intensity) will have a 'monotonous' relationship, that is, will always be decreasing (or possibly increasing) no matter where the driving variable is situated. But, as is evident in Figure 2.3a, such need not be the case. As we move from low management intensity to higher management intensity, yes we have a sort of smooth decrease in biodiversity, but at a particular point of management intensity, biodiversity declines precipitously. That point is technically referred to as a 'tipping point' and the transition from relatively high biodiversity to relatively low biodiversity is technically a 'critical transition'. The application of this hypothesis in the real world also implies that those who are planning the management intensity, either directly or indirectly, are unlikely to know where precisely that tipping point occurs.

We expect that a related, more complicated, pattern will become more visible as biodiversity relationships to agriculture are more intensively studied. It is the pattern shown in Figure 2.3b, what we might refer to as hypothesis IV. In this particular framework there are two tipping points embracing an important zone, known technically as 'hysteresis'. To the left of this zone, biodiversity remains high; to the right of this zone, biodiversity remains low. However, if one imagines a slow progression of management intensity from low to high (or, alternatively, from high to low), unlike under hypothesis I or II, we do not see a general slow progression of biodiversity from high to low (or, alternatively, from low to high). Rather, we see a sudden change, identical to what we expect under hypothesis III, but here it depends on what biodiversity we start with, at the time management intensity begins to change. Within this hysteretic zone a particular level of management intensity (a particular point on the abscissa) could result in either a relatively high or relatively low level of biodiversity, depending on where biodiversity was located before setting the management intensity point. The consequence of this particular situation is that biodiversity may seem insensitive to management changes over some range of changes. Yet, as management continues changing, eventually a tipping point will be approached, and the system will suddenly change. Yet after the change,

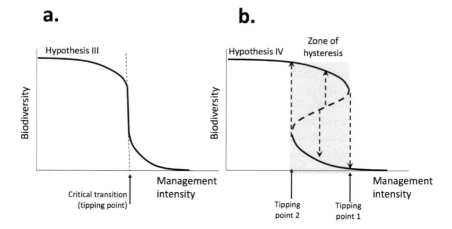

Figure 2.3 Expanded and more complicated expectations regarding agricultural intensification.

Source: The authors.

if the management conditions are reversed, there is no guarantee that the biodiversity will revert to its previous state. In the ecological literature these changes are frequently referred to as 'regime shifts', one regime being relatively high biodiversity, the other relatively low biodiversity. So, for example, if one begins with very high biodiversity in very diverse agroforestry systems intermixed with forests in Southeast Asia, and begins the process of intensifying the agricultural system by planting more and more oil palm, the biodiversity of the region may not be reduced that much at low levels of intensification, but at a certain higher level of intensification there could be a tipping point where there is a dramatic reduction in biodiversity. Ironically, if one begins the process of diversifying the agricultural system again (de-intensification), the biodiversity is not likely to revert to its original state since many species will have gone locally extinct during the time of oil palm domination. It will take a much lower level of intensification to go back to levels of biodiversity comparable to the levels just before the collapse. This time contingency is what is called hysteresis.

The relationship between biodiversity and the intensification of agriculture is a key component of the argument developed in this book, and is further elaborated in future chapters.

Why the biodiversity patterns matter

To some extent these four biodiversity patterns have been examined independently by biodiversity scientists. Here we consider them together because of their importance to the general theme of this book. Each pattern touches on an important aspect of what we are calling a new paradigm of biodiversity conservation.

The pattern of evolution and extinction through time has been the domain of palaeontologists and evolutionary biologists who seek to understand the reasons those patterns exist. Its importance to our general theme is the remarkable fact that probably over 99 per cent of all species that have ever existed are already extinct. Extinction is a major fact of biology. While the massive extinctions associated with the end of each geological period are truly spectacular, a more important fact is that local extinctions occur all the time. These local extinctions are an important part of the ecological side of our story.

The tropical world has attracted the attention of ecologists for years and has led to the posing of some of the most important and still enigmatic questions concerning biodiversity, particularly about how it is maintained. We explain these problems later in the chapter. However, it is worth noting that the vast majority of the world's biodiversity is located in the tropics, precisely where the vast majority of the world's poverty is also located. This conjunction of biological fact with political reality creates important imperatives for the conservation agenda.

The fact that more species live on larger islands than on smaller ones creates a backdrop for understanding how species fit together in ecosystems. Beginning with, as we noted, the basic idea of a balance between extinction and immigration of species, we move into a very similar idea of extinction and immigration of individual populations, and from the patterns that exist on oceanic islands to the patterns that exist on habitat islands, and then to the patterns that exist in fragmented landscapes dotted with patches of natural habitats. This general pattern is also key to the ecological side of our argument.

That biodiversity tends to decline with agricultural intensification is perhaps the most important pattern of all. At a trivial level, since agriculture is invariably implicated in the process of habitat fragmentation, if one is concerned about what happens to biodiversity as that fragmentation proceeds, it would be somewhat naïve to simply ignore the habitats that arguably occupy most of the landscape. But beyond this obvious issue, the matrix within which the fragments will be located is for the most part composed of agriculture. Therefore, how that agriculture either promotes or endangers biodiversity is a key issue when assessing the quality of the matrix.

The ecological background to biodiversity studies

From the paradox of the plankton to metapopulations

There has always been a bit of an enigma concerning highly biodiverse ecosystems, even though its significance has largely remained buried in the back alleys of esoteric ecological theory. According to fundamental ecological principles, there really should not be as much biodiversity as we see in some places. The more than 400 species of trees you can find in one hectare of an Amazonian forest, for example, live in a way that is impossible, according to classical theory. The reason for this is neither elementary nor, in the end, correct, but for years there has been a fundamental contradiction between what most ecologists believed about biodiversity maintenance and the reality of the hyper-diversity of the tropics.[13]

According to classical theory, biodiversity can be thought of as meta-phorically equivalent to a game of musical chairs. If you have ten chairs and 11 people, once the music stops, one person is eliminated and a tem-porary adjustment of number of people to number of chairs is achieved. In the classical theory the chairs represent distinct resources or ecological niches, and the people represent species that need those resources/niches in order to survive. The theory says, simply, there can be no more people than there are chairs, no more species than there are niches (resources needed by particular species), and it is known as the 'competitive exclusion principle'.

The great ecologist G. Evelyn Hutchinson, assuming the principle of com-petitive exclusion, noted that many communities contain species that would be difficult to characterize as having distinct niches in the first place. Con-sider, for example, phytoplankton, the small photosynthetic organisms that passively float in the open water of lakes and oceans. A given lake, or a given section of ocean, will contain many species, all of which more or less pas-sively float, absorb light and use the same nutrients. It is difficult to imagine how so many species in such a uniform environment could be partitioning some aspect of the environment into separate resources or different niches. Here, then, was a paradox. Hutchinson called it 'the paradox of the plankton' and it has become a touchstone for biodiversity studies ever since.[14] Different species of plankton seem to all do the same thing, which, according to the classical theory, means that they cannot coexist. Yet so many of them do. There is only a single chair in the musical chairs game, yet somehow all 11 people seem to keep playing the game.

Contemporary ecology is converging on a body of evidence that chal-lenges the classical theory. If we stick with the musical chairs metaphor, the new theory sees ecosystems as organized according to a collection of musical chair games, with groups of chairs arranged in small clusters around the room, and those people excluded from one cluster managing to run quickly to another cluster to find a seat. That is, the classical equations that described the way species interact with one another have been expanded in several ways, and we now have a more complicated and realistic view of how biodiversity is organized. The spatial distribution of individuals within a species, the non-linear ways they interact with one another, along with their random movements, all fit into a new paradigm of biodiversity main-tenance. It is still the case that the most fundamental ecological principles tell us that there can be no more species than there are resource types, but modern principles add spatial distribution, non-linearities and chance to the mix, resulting in a far better fit between the real world and the theoretical world of the ecologist.

The basic structure of this new formulation has some points in common with the now well-accepted equilibrium theory of island biogeography, the theory that effectively explained the basic pattern of biodiversity on islands, as described earlier in this chapter. What MacArthur and Wilson, co-inventors of the theory, basically did was change the focus of biodiversity studies from an emphasis on the competitive exclusion principle, to an emphasis on extinction and migration processes. Continuing with the musical chairs metaphor, they asked not how individuals were competitively excluded due to the unavailability of chairs, but rather how fast they were excluded from one cluster of chairs, and whether or not they could get from one cluster to another. In other words, they changed the focus from asking about competition among species to asking about the balance between migration and local extinction.

This new focus was morphed by ecologist–epidemiologist Richard Levins to examine the balance between extinction and migration,[15] not of species, but of subpopulations, groups of individuals belonging to the same species, but grouped on islands, or habitat patches, or somehow separated from other such groups of individuals of the same species.[16] The collection of these subpopulations became known as a metapopulation.[17]

Understanding the principles of metapopulations is probably easiest with reference to the classical framework employed by epidemiologists when studying diseases. Normally we consider an individual person as either susceptible to a disease or already infected with it. We can also become immune to many diseases, but we shall ignore this fact for now. Being infected means that the organism that causes the disease has invaded the body of the individual in question. The familiar trajectory for non-lethal diseases is that an individual becomes infected but then his or her body fights off the infection and becomes cured, and, unfortunately, liable to infection all over again, which is to say, they go back again to the susceptible category. So, epidemiologists usually ask, what proportion of the human population is currently sick, how fast do they get cured, and how fast do they become infected again? Becoming infected depends on the transmission rate of the disease organism, which is effectively the same as the migration rate. Becoming cured means that the disease organism became extinct in that particular person's body. The course of the disease is a consequence of the balance between the rate at which new individuals become sick (the migration rate of the disease organism) and the rate at which sick individuals get better (the local extinction rate of the disease organism).

Applying this epidemiological formulation to the problem of how species are maintained, the idea is that sick or well people represent metaphorical habitats, and the disease organism either inhabits them or not. So, rather than think of human bodies, think of islands or patches of particular kinds of habitat that may or may not contain individuals of a particular species. In metapopulation analysis, we ask what fraction or percentage of all the available habitats actually contains a subpopulation of the species of concern. That is the variable of analysis, usually designated with the symbol p. Then we ask what is the rate at which subpopulations become extinct (equivalent to the rate at which individuals become cured of the disease), and what is the rate at which habitats that did not contain individuals of this species actually receive migrants (the migration rate). The most important result of this theory is that over the long run p is equal to 1.0 (100 per cent) minus the ratio of extinction (e) to migration (m), ($p = 1 - e/m$). So, for example, if the migration rate is very high relative to the extinction rate, the ratio e/m will be close to zero and the equilibrium occupation of the habitats will be close to 100 per cent. If the migration is reduced to the point that it is almost equal to the extinction rate, the ratio e/m will be almost 1.0, which means that the occupation of the habitats will be close to zero, and the population, the entire metapopulation, will become extinct (Figure 2.4).

With this theoretical framework we see the importance of distinguishing between local extinctions (which are natural and unavoidable, the e in the basic equation) and 'global' or 'regional' extinction which is the condition where all subpopulations have disappeared. The goal of conservation, at its most elementary level, is to try to make sure that p, the proportion of patches occupied by the species, stays greater than zero. Metapopulation theory tells us that to do so we need to concentrate on m, as well as e.

Although the theory of metapopulations is central to conservation to be sure, the pattern of biodiversity more generally on islands has been in some ways the extension of metapopulation theory to the assemblage of species that exist in an area. That is, the biodiversity in a habitat fragment can be analysed using this basic ecological theory.

As noted earlier in this chapter, the way in which animals and plants are distributed on islands has been a classical topic in biology. The almost universal pattern is that smaller islands have a small number of species and larger

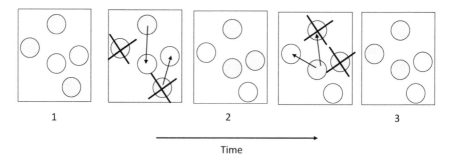

Figure 2.4 Diagram of metapopulation dynamics.

Source: The authors.

islands have a larger number of species, one of the most well known and universal patterns in ecology. MacArthur and Wilson's theory remains one of the few theories that has stood the test of time in ecology. The theory proposes a balance between two rates, the rate of immigration of new species (species that were not already there) and the rate of extinction (that is 'local' extinction, the disappearance of a species on the island). It is of more than historical artefact that the very first example discussed in this pioneering work was of habitat fragmentation, noting that habitat fragments are effectively islands (Figure 2.5).

Indeed, in their introduction they note:

> Insularity is moreover a universal feature of biogeography. Many of the principles graphically displayed in the Galapagos Islands and other remote archipelagos apply in lesser or greater degree to all natural habitats…. The same principles apply, and will apply to an accelerating extent in the future, to formerly continuous natural habitats now being broken up by the encroachment of civilization….[18]

In an early attempt to use this theory in the design of nature reserves, Diamond and May noted that isolated reserves of natural vegetation were

Figure 2.5 The pattern of fragmentation of woodland in Cadiz county, Wisconsin (after Curtis, 1956) as presented in MacArthur and Wilson (2001).

Source: John T. Curtis.

indeed very much like islands.[19] Relative to the thesis of this book they noted:

> If one must settle for several smaller parks, a way to raise the equilibrium number of species in any one such park is to raise the immigration rate into it.[20]

With a slight change in terminology we can resurrect the arguments of MacArthur and Wilson (and Diamond and May) in a form that directly speaks to the relationship between matrix and fragments of natural habitat. In Figure 2.6 we plot the two rates, extinction and immigration, as a function of the species number already in the fragment. However, we note that both the extinction rate curve and the immigration rate curve are presented in two

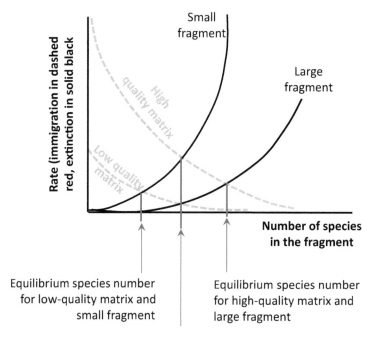

Figure 2.6 Details of the basic equilibrium theory of island biogeography as applied to the situation of fragments in a matrix, considering the fragments as islands of particular sizes in a matrix of particular qualities (and thus determinants of immigration rates).

Source: The authors.

different forms, for high- and low-quality matrix and for a small and large fragment. As we explained earlier in this chapter, the crossing point in those two curves represents the number of species that we expect in the system (where there is a balance between the rate at which new species arrive and the rate at which they go extinct). We show three such points in Figure 2.6.

The importance of extinctions

That local extinctions are a normal and inevitable part of ecosystem dynamics is not an item of debate in serious scientific circles, a fact not always fully appreciated by conservation practitioners. Indeed, there is now an impressive amount of literature on extinctions from fragments. For example, an important result from long-term experiments on forest fragmentation in Amazonia shows that while smaller patches of forest have higher extinction rates of birds than larger ones, the actual extinction rates of even the largest patches are surprisingly high.[21] Indeed, some are so high as to suggest that the only acceptable size for a biological preserve is one that is far beyond reasonable expectation. A similar example comes from the long history of the National Park Service in the US, arguably the country with the best-developed national park system in the world. Tabulating observed mammalian extinction rates for biological preserves in the US, a dramatic increase in extinction is found as the size of the park decreases. But more importantly, the largest national parks were not immune to extinction rates, even though, as expected, the smaller parks had higher rates.[22] Many other examples could be cited.

A particularly interesting example comes from work done on reptile species on islands in the Aegean Sea.[23] As is always the case, the number of species is correlated with the size of the island. However, in this particular case the geological history of the islands is well known and the authors were able to discriminate between the part of the pattern due to local extinction events and that due to the particular position of the island with respect to mainland source areas. Their analysis convincingly showed that the pattern of species diversity on the islands was completely due to extinctions – extinctions were larger on smaller islands. Here, as elsewhere, local extinctions are a normal and inevitable part of ecosystem dynamics.

As we come to understand the way the diversity of trees is maintained in forests, and extrapolate that understanding to other organisms, it is perhaps not all that surprising that local extinction rates are so high. As an example, ecologist Steve Hubbell postulates that one key factor in maintaining tropical tree diversity is recruitment limitation (or failure to disperse to suitable habitat patches).[24] Long-distance dispersal events, under Hubbell's formulation, are important in maintaining species diversity. Therefore, as observed in the examples above, fragmenting the forest and consequently limiting the rare dispersal event from point *x* to point *y* in the original spatially extended forest is likely to cause local extinctions and a concomitant reduction in regional biodiversity.[25] Unfortunately, such expected extinctions are likely to occur far into the future, making the political case for conservation here and now particularly difficult.

There is a certain intellectual cacophony as ecologists have been trying to understand what makes patterns of biodiversity from Darwin to the present day. The theory of island biogeography was a kind of exception in that it reflected one of the truly consistent and universal patterns of ecology, the relationship between area and biodiversity – small areas housed less biodiversity than larger ones, whether sampling quadrats in a forest or islands in the ocean or fragments of one type of vegetation in a matrix of another type. Furthermore, the relationship was quite regular. Almost every time a census was done, if you make a plot of the log of species numbers against the log of the area sampled, a straight line results. This was true of subsamples taken from large areas and, most evident to early ecologists, also true of islands, and is generally referred to as the 'species/area' relationship. This regularity on islands was one of the core pieces of data that propelled the MacArthur/Wilson and Levins/Heatwole teams to come up with their theory in the first place. Yet it is not restricted to islands. Sampling particular areas within a larger context reveals the same phenomenon, a straight line on a log/log plot. We can gain some insight from a look at the comparison of 'mainland' species/area relationships. A square kilometre of land within a mainland area can be presumed to exist in the highest quality matrix possible, which is to say, precisely the same fauna and flora within the sampled square kilometre. But an island of one square kilometre, exists in, by definition, a very low-quality matrix, the ocean. Comparing species/area curves on mainland areas versus islands thus gives an idea of the importance of matrix quality in the conservation of species.

In Figure 2.7 we illustrate the species/area curves from a recent compilation of land plants on both islands and mainland areas.[26] If we take the mainland as representing the highest quality matrix and the islands the lowest quality (the intervening matrix in this case is the ocean, in which no land plants can survive), the dramatic effect can be seen. So, for example, with ten reserves of ten square kilometres each (a total of $100\,km^2$), a total of about 900 species will be conserved. The same ten reserves in the lowest quality matrix would conserve only 180 species.[26] Ignoring the matrix quality issue thus would result in about an 80 per cent extinction rate. Naturally this is an extreme example, but some of the most extreme fortress conservationists do indeed suggest such a strategy.

Admittedly these sorts of extrapolations are meant to shock. In the real world both fragments and matrices are not likely to behave like averages (which is what the analysis in Figure 2.7 would suggest), and even an intensely unfriendly (to biodiversity) matrix of sugar cane monocultures or Brazilian soybean fields would probably permit some inter-fragment migration as well as sustain some species even within their pesticide-drenched environment. But the point is that real data here and elsewhere, if judged in a dynamic context, illustrate the importance of the matrix. In a conservation

Approximately 900
species with highest
matrix quality

Figure 2.7 Relationship between biodiversity (number of species) and area for islands (what we refer to as the low-quality matrix) and for mainland areas (what we refer to as the high-quality matrix).

Source: The authors.

programme that preserves 100 km^2 but allows the matrix to be of extremely low quality only 20 per cent of the original plant species would have been preserved. Obviously if we establish this conservation area today and assess its plant biodiversity we will find 900 species, and, since local extinctions do not happen overnight, if we come back five or ten years later, we might indeed see the same numbers and be tempted to pat ourselves on the back for having preserved so many plant species, since the extinctions that will eventually happen could hardly have taken place. But conservation is supposed to be over the long term. When the final inevitable extinctions are realized, we would find that 720 of the original 900 species would have gone extinct. A patch or patches of an equivalent size within a forested matrix would likely contain the original 900 species. The difference, of course, is that the fragment or fragments in the conservation area existed in a very low-quality matrix (the ocean), while the matrix of the forest patches is of very high quality (forest).

The landscape mosaic

So far, we have presented the theory of metapopulations in its most elementary fashion. Applying the theory to fragmented landscapes we have 'natural habitats' embedded within an agricultural matrix. Local extinctions occur in the patches of natural habitats but they are counteracted by migration from other patches. The migration occurs through the agricultural matrix, with some types of agriculture promoting more migration than others. The implicit assumption here is that the matrix, although potentially of variable quality, is homogeneous (that is, homogeneously good, intermediate, or bad). With some exceptions, this is almost never the case (Figure 2.8).[28] In this section we explore the more realistic situation where the landscape is composed of a mosaic of habitats of varying quality, including fragments of native habitat.

Suppose that we categorize habitat types in order of their quality, either in terms of biodiversity directly or in terms of their migratory potential (or permeability). A matrix is then composed of a collection of habitats of various qualities, and the quality of the matrix is effectively the sum or average of the qualities of the habitats within it. In Figure 2.9 we abstractly illustrate four categories of matrix.

Thus the qualities of each habitat type somehow combine to make a matrix either conducive to migration or not. From the point of view of landscape planning, in the case of either stepping stones or corridors, the basic question is one of the probability of connecting the two forest fragments. For example, if the probability of an organism migrating out of a low-quality habitat (the lightly shaded habitats in Figure 2.9a or b) is zero, and assuming that the organism can move only to an adjacent square, then the probability that the organism could travel from one forest fragment to the other in Figure 2.9a is zero. On the other hand, that probability (that an organism could travel from one forest fragment to the other) is 1.0 in the case of Figure 2.9b, since all the high-quality habitats are connected. It is an interesting mathematical exercise to find out the probability of a pathway existing from one patch to the other (which is equal to the probability of an organism travelling from one patch to the other), given a random allocation of high-quality patches in the matrix (Figure 2.9a). It turns out there is a threshold effect where the probability goes from virtually zero to virtually 1.0 at a density of high-quality habitats of 41 per cent (below 41 per cent the probability of connecting the fragments is close to zero and above 41 per cent it is close to 100 per cent).[29]

Why the percentage is exactly 41 is somewhat difficult to explain, but the basic idea is evident from the spreading-forest-fire model.[30] Think of the black squares in Figure 2.9d as individual trees in a plantation, the canopies of which touch one another only if the squares themselves touch. If you set a fire in the left patch, the question is how dense the tree plantation has to be for the fire to spread to the other side of the matrix. Assuming that the trees are planted randomly in the plantation, with two or three trees it is obvious that the fire will not spread. With almost all the positions occupied by trees it is obvious that the fire will spread. But what will be the pattern as the density of the trees increases from zero to 100 per cent. It turns out that once the density reaches 41 per cent it is almost certain that there will be a pathway

(a)

(b)

Figure 2.8 Two types of landscape mosaic: (a) A homogeneous landscape dominated by maize monocultures in Moore Township, Northampton County, Pennsylvania, (b) Maize as a component of a landscape mosaic, with surrounding patches of vegetables, fallow lands, and forest, in Chiapas, Mexico.

Sources: (a) John Vandermeer; (b) CC Public Domain.

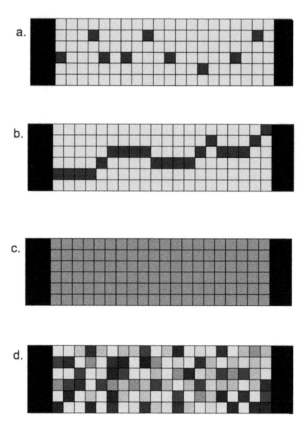

Figure 2.9 The various types of matrix that connect two forest fragments.
Source: The authors.

that allows the fire to percolate from one side of the plantation to the other. The important point is not the precise figure, but rather that there is a threshold, a percolation threshold, at which the probability of percolation jumps to almost one. Using this formula, we see how there is a metaphorical connection between the so-called percolation threshold and the idea of a corridor.

When the sub-habitats in the matrix are neither perfect natural habitats nor completely 'poisonous', but rather of some intermediate quality, the fundamental question to be asked is different. Rather than the probability of percolation (probability of an effective corridor, which is the same as the probability that an organism *could* travel from fragment to fragment), we ask what is the probability that a dispersing organism *does* travel from fragment to fragment, which is a function of every possible pathway in the matrix. Thus, for example, in Figure 2.9c there are 20 columns of habitat that need to be traversed in order to migrate from one fragment to the other. If each of the squares has a probability of 0.95 that an individual entering will successfully migrate through the habitat, the probability of getting from one fragment to the other, for any individual moving in one direction, is $0.95^{20} = 0.36$. In other words, for every 100 individuals trying to migrate from one fragment to the other, 36 will be successful.

The situation becomes a bit more complicated when the matrix is a landscape mosaic.[31] Each possible pathway has a distinct probability, as illustrated for an extremely simple example in Figure 2.10. For individuals that randomly disperse, the average of those distinct probabilities (average of all possible pathways, not just two as illustrated in Figure 2.9) will be the probability of dispersing from fragment to fragment.

It is thus evident that the landscape mosaic is conceptually distinct from the stepping stone or corridor idea. In the case of stepping stones or corridors, the interesting question is 'What is the probability that a pathway exists between the fragments?' In the case of a landscape mosaic, the basic question is 'What is the probability of migration from fragment to fragment?' The answer for stepping stones or corridors will be either close to zero, or close to one. The answer for a landscape mosaic will be some figure between zero and one, but not restricted to one extreme or the other.

It is evident with this particular formulation that the habitat with the lowest quality is likely to have an especially strong effect on the overall quality of the matrix. For example, a habitat type that receives a large dose of biocides at regular intervals, such that any insect entering that habitat is almost sure to die, will act as a major impediment to the migration of insects from fragment to fragment. Thus, even though it is the landscape as a whole that

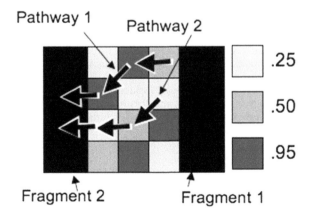

Figure 2.10 Extremely simplified landscape mosaic in which the matrix between two fragments is composed of three types of habitats.

Source: The authors.

determines the overall migration rate,[32] it is still the case that the qualities of individual habitat types are important and worthy of much attention in whatever planning programme is involved, whether the anarchic mode of the neo-liberal economic model or some more organized planning. As we move to integrate more design into landscape ecological research,[33] formulations that better reflect the complexities of human-altered landscapes could be helpful.

For most agricultural landscapes, especially in the tropics, there is ample evidence to suggest that the mode of production can determine various degrees of habitat quality with respect to biodiversity.[34] For example, in tropical rice fields on the island of Java, an experimental study confirmed the expectation that so-called autonomous biological control in rice exists in a landscape mosaic that included habitat heterogeneity. Of particular importance is the fact that a large number of insect species feed on plankton and detritus within the rice fields themselves.[35] Any procedure that reduces plankton in the rice paddies (e.g. pesticides) will increase the effective extinction rate in that field. Not surprisingly, pesticide use causes declines in biodiversity of a variety of organisms;[36] cereal fields managed organically generally have higher weed species diversity than conventional fields,[37] and fields managed with organic fertilizers have a more diverse microbiota.[38] Some of these examples will be explored in more depth in Chapters 8, 9, and 10.

Notes

1 Erwin (1982); May (1988).
2 Stork *et al.* (2015).
3 Wakeford (2001).
4 Gould (1990).
5 Carrington (2016).
6 Loehle and Eschenback (2012).
7 Ceballos *et al.* (2017).
8 Palmer (1994).
9 The ecological niche of an organism is its role within the ecosystem and all the required resources and characteristics needed for that organism to survive.
10 MacArthur and Wilson (2001).
11 McNeely and Scherr (2003). This percentage includes only agriculture, pastures for livestock production, tree plantations, and urban areas. It does not include fallow areas, secondary forests, or forests that are used and managed by humans.
12 Perfecto *et al.* (1996); Greenberg *et al.* (1997).
13 Hubbell (2001).
14 Hutchinson (1961).
15 The term *migration* here is used to refer to the rate of arrival of new individuals to a patch. Technically it is the immigration. We use both of these terms interchangeably.
16 Levins (1969).
17 Hanski and Gilpin (1991).
18 MacArthur and Wilson (1967), pp. 3–4.
19 Diamond and May (1976).
20 Diamond and May (1976), p. 185.

21 Ferraz *et al.* (2003).
22 Newmark (1995).
23 Foufopoulos and Ives (1999).
24 Hubbell (2001).
25 Ehlers Smith *et al.* (2018); Palmeirim *et al.* (2018).
26 Data from Kreft *et al.* (2008).
27 You can assume that, with no overlap of species, 100,000 fragments of 1 km^2 will harbour the same biodiversity as a reserve of 100,000 km^2, or half the species that are found in New Guinea. However, the presumption of no overlap is highly unrealistic.
28 A close-to-homogeneous matrix can be encountered in areas dominated by a particular type of crop, for example, the banana-growing region of the Atlantic coast of Costa Rica, where a single banana plantation can extend for tens of thousands of hectares (Vandermeer and Perfecto, 2005a).
29 Vandermeer *et al.* (2010b).
30 Turcotte and Malamud (2004).
31 Vandermeer *et al.* (2010b).
32 González González *et al.* (2016).
33 Nassauer and Opdam (2008).
34 Swift *et al.* (2004).
35 Settle *et al.* (1996).
36 Brittain *et al.* (2010); Geiger *et al.* (2010); Krauss *et al.* (2011); Beketov *et al.* (2013).
37 Hyvönen *et al.* (2003).
38 Hartmann *et al.* (2015).

3 Ecological theory and political realities

A fragmented landscape, extinctions and migrations

Conventional wisdom has it that the natural habitats of the tropics have been devastated. Who can be against conserving the ones that remain? Yet, as we argue in great detail later, the simple loss of habitat is only a very small part of the problem. It is true that much of the original forest and savannah cover of the terrestrial parts of the tropics has been dramatically altered since European invasion and colonization. But in the vast majority of cases that alteration has not been absolutely complete. The impressive scenes of devastation so common on conservation websites are actually not as widespread as one might think. In fact, as agricultural frontiers open, forests and savannahs are cleared as needed, and as technology and topography allow. The consequence is that much, perhaps most, of the tropics can be characterized not as 'deforested' but rather as 'fragmented' (Figure 3.1).

Anticipating a more detailed analysis and case studies (Chapters 4–10) we here note that this pattern of fragmentation has created a landscape that, viewed through foggy lenses, looks a lot like an archipelago of islands in the ocean. But here, the islands are fragments of forest, savannahs or other natural ecosystems in the ocean of agricultural activities. A real conservation programme cannot ignore this reality! Even as the world struggles to protect the few remaining large areas of tropical habitats from further exploitation, we must acknowledge that a large fraction has already been exploited and that perhaps most of the world's biodiversity is located not in those few remaining protected natural areas, but in the far more extensive landscapes in which thousands of islands of natural habitat exist in a matrix of myriad agricultural activities. Our purpose in this book is to examine exactly what must be done in such a situation, if biodiversity conservation is to be successful in the long run.

If habitats are generally like islands in an archipelago, we are naturally led to ask whether it would be possible to treat the ecology of those islands as we would treat real islands. And here the body of ecological theory described in the previous chapter can be brought to bear on the problem. As a normal part of the theory of island biogeography, the biodiversity on any given island is

Figure 3.1 A fragmented landscape near Campo Grande, Brazil. Darker areas are remnants of natural vegetation embedded in a heterogeneous agricultural matrix.

Source: Google Earth, Digital Globe.

proportional to the rate of immigration of new species to that island divided by the rate of extinction on that island. We emphasize that extinction at a local level is a perfectly normal process, one that can be seen, albeit using subtle clues, on many islands and in many fragments.[1] But even though species become extinct regularly on individual islands, they are rescued from becoming extinct over their entire range through migration among the islands.[2]

This picture, a picture accepted as prevailing wisdom amongst ecologists today, has extremely important consequences for biodiversity conservation. In a landscape that has already been fragmented, we face an island-like situation. For biodiversity conservation we must be concerned with the local extinction rate and the immigration rate. The prevailing wisdom amongst many conservationists is that the political agenda is uniquely determined by the need to protect remaining natural habitats, which is to focus on the islands in the landscape. For the most part, the extinction rate on an island is determined by the size of the island, smaller islands having higher rates of extinction than larger ones. And the critical fact we wish to drive home is that there is not much we can do about local extinction rates, given a fragmented landscape. Thus, protecting the remaining patches of natural habitat is certainly an important component of a conservation programme, but it is not nearly enough! Fixing the landscape at a particular level of fragmentation means fixing the extinction rates at a particular level. Nothing beyond preserving the patches of natural habitat can help in lowering extinction rates, if indeed local extinction is strictly a function of patch size.[3] Extinction is a normal biological process and will go on, no matter what we do.

We need remain mindful of the fact that extinction is only one side of the biodiversity equation. The other side is immigration. And immigration rates are indeed something we can do something about. Immigration rates are set by what happens in the 'ocean', which is to say in the matrix within which those habitat fragments occur. And for the most part, at least in the areas of the globe that are currently tropical, the 'ocean' is an agricultural matrix. This means that the type of agroecosystem we allow to exist in the matrix will determine the immigration rate to the islands of natural habitat. This elementary fact about the ecology of biodiversity has enormous socio-political consequences. It means that a serious conservation programme should focus on the type of agriculture practised within the matrix, rather than on what happens solely in the fragments of natural habitat.

Completing the general picture, we note that promoting a high immigration rate means focusing on the agroecosystems, which in turn means examining what is happening socio-politically in the agricultural matrix. This apparently unavoidable conclusion brings us face to face with the political turmoil that occurs today in the agricultural sector of the tropical world. Ignoring that turmoil is ignoring the only part of the biodiversity question that matters in a practical sense: the immigration rate.

Early conceptual models: SLOSS, Forest Transition Theory, and Integrated Conservation Units

As we noted in Chapter 2, in one of the first attempts at applying ecological theory to the problem of conservation at a landscape level, Diamond and May employed the then new theory of island biogeography to the problem of wildlife conservation.[4] Most importantly this early attempt recognized the issue of island (fragment, or reserve) size, number, and surrounding matrix on the problem as a whole, effectively acknowledging that conservation programmes could benefit from a consideration of the characteristics of both the fragments and the surrounding matrix. This book, at the technical ecological level, is clearly a descendent of these initial attempts at applying ecological theory to this important practical issue.

Yet the early history of these attempts is less than enlightening. One of the issues that became a centrepiece of debate was the argument as to whether a single large or several small reserves (SLOSS) would be most conducive to the preservation of biological diversity.[5] The argument from island biogeography theory seems to have emphasized a very small part of what the theory actually said, perhaps due to the underlying and perhaps unconscious bias of some conservationists who sought refuge in their artificially constructed idea of 'wilderness'. The argument, in brief, was that since the theory and empirical evidence says that larger areas (islands or fragments) will contain more species at equilibrium, it is reasonable to expect that the larger the individual reserve, the more species it will contain. As pointed out quite early on by Dan Simberloff and Lawrence Abele,[6] the theory of island biogeography simply did not speak to the question of a single large or several small reserves. It simply said that the total area of 'NOT ocean', would be proportional to the equilibrium number of species preserved. Thus, a biological reserve of 1,000 km², would preserve (according to the theory) precisely the same number of species as two reserves of 500 km². Debates on this issue were intense in the late 1970s and early 1980s. They were conceptually quite simple but complicated in their details, as summarized succinctly by Simberloff and Abele. After intense debate, the whole issue of SLOSS seems to have died a quiet death. It is worth noting, however, that an implicit assumption shared by all sides in this debate to one degree or another, is that the space between the islands (fragments) was largely desert, that is, area which was in principle uninhabitable (and perhaps even untransversable). Our argument begins with a challenge to that assumption.

In a slightly more historical vein a parallel set of arguments emerged in the 1990s, mainly concentrated on forested ecosystems. Geographer Alexander Mather, in viewing the historical record of forest cover in France and Hungary, proposed first an empirical generalization and then a mechanism for that generalization, the forest transition model.[7] Most importantly he sought to use the lessons of the developed world to understand and possibly intervene in the ongoing deforestation of the tropical world. In brief, the argument is that when a forested region is first occupied by people, the needs for food production will promote the expansion of the agricultural frontier and therefore the cutting of forests. Some of the soils under the forest will be good for agriculture, others poor, but as agriculture expands, all these areas are cut to open up space for agriculture. However, as 'development' proceeds, urbanization grows and cities acts as a magnet for farmers, especially those who accidentally settled on the poorer soils, thus resulting in abandonment of many farms, after which the forest returns. We thus generally expect a U-shaped curve if we plot the amount of forest over time, first declining but then increasing as urbanization and, in many cases, industrialization, continues its rise. It is basically the pattern observed in New England and most of Western Europe. Assuming this to be the case generally for the Global North, can it be expected to occur also in the Global South? Furthermore, for conservation purposes, can we use a knowledge of its causes to intervene to accelerate the rate of increase in forest cover? Some say yes.[8]

Inherent in the argument is the 'death by inefficiency' of a large sector of the farming community. The soil heterogeneity assumption of Mather's original model became a metaphor for whatever caused 'inefficiency' in the farming sector. Originally, those farmers who just happened to settle on bad soils morphed into those farmers who were unable to take advantage of more efficient technologies or conform to new rural realities. And the expected growth through urban industrialization would eventually form that magnet that would pull the inefficient farmers into the cities, leaving the land to regenerate through natural forest succession. The temptation to incorporate such reasoning into intentional adaptive programmes was certainly seductive, leading to questions of how much food is produced per unit of labour as surrogate for the efficiency of a farming operation, broadly defined.[9]

Although the apparent regrowth of forest in many tropical countries, such as Puerto Rico and Vietnam, seems to conform to these ideas,[10] at least qualitatively, most tropical areas remain in the process of further deforestation or fragmentation.[11] Furthermore, much of the data on which forest cover is based, ignores the basic question of 'what is a forest', a much more difficult question to answer for the tropics than for forests within higher latitudes.[12] Much of the enthusiasm of the forest transition model seems to be based on a narrow focus on those things that can be easily quantified, ignoring many issues that are evidently determinant of deforestation and reforestation.[13] Such complexities suggest a need to consider more complicated social issues. One of the more useful analyses to emerge in the past decade is the rediscovery of the work of Soviet agrarian economist and rural sociologist, Alexander Chayanov, and, especially, the new paradigm of the 'new rurality'[14] or the 'new peasantry,'[15] both of which will be discussed in Chapter 4.

While much of the original forest transition model was based on experience in higher latitudes, attempts to apply a similar analysis to the tropics have been mixed at best, as mentioned above. One exceptionally documented case is that of El Salvador, perhaps the most deforested country in Central America. Using detailed Landsat data and basic economic indicators, Hecht and Saatchi[16] demonstrated that the expected rural to urban migration and depopulation of the countryside did not correlate with increased forest growth. Rather, analysing department by department, these authors demonstrated that the best predictor of reforestation was familial remittances. When a member the family lives in the US, for example, and sends a small part or his or her 'developed world salary' back to the family, that monthly or weekly bonus amounts to a major piece of the family budget and means the family in El Salvador does not have to eke out a living on the last corner of the farm where the soils are too rocky or wet or acidic. As a consequence, this all-important piece of many national economies in much of the Global South,[17] may operate in an indirect way to result in lower deforestation and more afforestation. It is an alternative take on the Forest Transition Hypothesis.

Regardless of whether or not tropical regions were experiencing forest regrowth, the main conservation strategies until the 1990s were based on the 'fortress conservation' model inspired by neo-colonial attitude and the dichotomy between nature and people.[18] However, the failure of this model to achieve true conservation, not to mention the implied racism and the social injustice intrinsic in such programmes, led to a transition to a seemingly new paradigm, Integrated Conservation and Development (ICD) programmes. This idea effectively takes the pure conservation goals of debates like the SLOSS and combines them with the inherent socio/political ideas of the forest transition model into programmes of biodiversity conservation that integrate rural development components. The overall idea of the ICD programmes would seem much in line with the arguments we make in this book. Yet there is a key difference. Ross Hughes and Fiona Flintan identified several key features of these programmes that depart significantly from the approach that we are proposing in this book. First, ICD programmes are usually associated to a protected areas (usually a national park) and do not seek to address issues that are peripheral to the park; second, they do not strive to devolve control or ownership of the resources to local communities; and third, the majority of the ICD programmes are funded and motivated by external organizations (bilateral or multilateral funders, international conservation organizations, wildlife departments within the national government).[19] Given these features, it is not surprising that so few of them have had any positive impact on conservation.[20] As a general rule. it is safe to say that, overall, programmes that are imposed from above have very little long-term success. Two of us worked for 17 years on a project within the MesoAmerican Biological Corridor (MBC), a much ballyhooed project theoretically connecting the many remaining forest fragments on the Caribbean coast of Central America. Very much like the ICD programmes, the MBC was proposed as a conservation programme to facilitate animal movement and conservation across Mesoamerica while promoting sustainable development. The MBC consisted of four main types of zones: core zones, buffer zones, corridor zones, and multiple use zones. It was initially funded by the World Bank and the United States Agency for International Development (USAID), with technical support provided by the University of Florida, United Nations Environment Program (UNEP) and the Central American Commission for Environment and Development (CCAD). The buffer zone had several categories of permitted human use, including 'extractive reserves'. Extractive reserves were such that human beings could engage in activities such as selective logging, sustainable hunting, perhaps even some small-scale agricultural activities, but no large industrial-scale operations. The whole idea seemed, at the time it was initiated, excellent. However, down on the ground, in our 17 years working in what was formally designated as 'extractive reserve' in Nicaragua, we never met anyone living within that reserve that was aware of the fact that they were living in an extractive reserve. Indeed, most people we encountered in the region did not even know what an extractive reserve actually was. And this is one of the main criticisms of ICD programmes. They are usually conceived and initiated from 'above'. 'Desk men', as many of our Nicaraguan friends called them, made maps and drew lines on those maps affecting people's lives – that was their job. Little attention was paid to the hopes and aspirations of the people who lived in the area, their cultural traditions, their insecurities, even their concern, or not, with conservation.

Although the evidence of the impacts of ICD programmes on conservation is weak and remains hotly contested, these programmes continue soaking up large amounts of conservation dollars. Now under the mantra of Reduction Emissions from Deforestation and Forest Degradation (REDD) and REDD+,[21] these programmes have turned to a more direct payment for ecosystem services (PES), which are even more controversial than the indirect approach of the old-style ICD programmes. Rather than using conservation dependent economic activities (such as ecotourism) to stimulate development, PES involve direct payment for so-called ecosystem services, mostly carbon stocks. Although REDD and REDD+ focus on reducing emissions of greenhouse gases, it has been argued that these programmes, by reducing deforestation, also have biodiversity benefits.[22] However, like any market-based approach to conservation, the economic viability of these programmes depends on the profitability of alternative land uses. Some consider these programmes to be the latest conservation fad and predict that they will eventually disappear from the conservation discourse and practice.[23] Meanwhile, entire communities in the Global South have their livelihood in the balance.

Land sparing *versus* land sharing: the imperial mind confronts biodiversity conservation

The most recent manifestation of this sort of argument has emerged as the debate between 'land sparing' versus 'land sharing'. Ultimately it is perhaps as transient as its previous cousins, but as of this writing it seems the most evident iteration of these kinds of arguments. Furthermore, it is more directly related to the theme of this book since it effectively minimizes the need for attention to the matrix. Although the debate itself is simplistic and short sighted, it nevertheless has received a lot of attention in the conservation literature, and is thus worthy of a degree of attention. As in previous versions of the topic (e.g. SLOSS, Forest Transition Model, and ICD programmes), there are clear socio/political issues that are involved, regardless of many conservationists' desire to ignore them, as we discuss below. However, our position is that at the most technical level there are obvious difficulties with the argument that have been tacitly ignored. We first elaborate on those technical difficulties before pursuing in depth the probably more important socio/political issues.

Initially formulated by Green and colleagues, it contrasts a strategy of agricultural intensification and sparing land for conservation (thus land sparing) with a mixed tragedy of producing food and conserving biodiversity on the same land (thus land sharing).[24] In one of the strongest statements of the land-sparing ideology, biologist E. O. Wilson's book, *Half-Earth: Our Planet's Fight for Life*, argues that in order to rescue the world's biodiversity, half of the earth needs to be put aside into conservation preserves.[25] In order to do that, he argues, the other half has to be extremely efficient at producing food, and if that means conventional agriculture, with its monocultures, agrochemicals, genetically modified seeds etc., so be it. Perhaps Wilson is extreme, but his position is emblematic of the general theme of land sparing.

As an exercise, we might look at some real-world data and ask how much area, in theory, based on biological reality alone, we would need to preserve ALL species. If we presume that what we see today represents some long-term average, the answer is usually a surprisingly small area. For example, in Figure 3.2 we show the species accumulation curves for tree species in two tropical rain forests.[26] As is evident, the number of species goes up as the area sampled goes up, but there is a clear asymptote. By the time you sample 30 hectares, you are approximately at 99 per cent of the tree species that are there. A naïf might suggest that since we seek to preserve the species that are actually in a particular area, and since we can do that by setting aside 30 hectares, we need not concern ourselves with something larger. This is effectively what much of the empirical work on land sparing actually does, regardless of what they may claim. Examine the current species diversity in two natural preserves, one with a low quality (but presumably highly productive) matrix and one with a high-quality matrix. Whatever biodiversity is observed is assumed to be the biodiversity that will be there in perpetuity. So, if the current biodiversity data tell us that we in fact accumulate all the species by the time we have sampled 30 hectares, with the assumption that today's

Figure 3.2 Species area curves for trees species in Barro Colorado in Panama Island (filled symbols) and Pasoh Forest Reserve in Malaysia (open circles).

Source: The authors; data from Condit *et al.* 1996.

biodiversity represents biodiversity in perpetuity, it is indeed correct that 30 hectares is sufficient.

To be sure, the argument is naïve, and not likely to be actually made by anyone who understands ecology. Yet it is an error frequently made among proponents of the land-sparing idea. The data used are static. Indeed, it is true that 30 hectares would preserve 99 per cent of all the species, but over a very short time frame. Let us take an argument directly from Dr Wilson's own work, the theory of island biogeography, as we summarized in Chapter 2. He pointed out that the number of species on islands (and a biological preserve must be thought of as a kind of island) is a dynamic variable, meaning that any static picture is of necessity only temporary. There are always migrations of new species to the island, and there are always extinctions of species from those islands. For reasons that are beyond the scope of this text, those extinctions are dependent on the size of the island. But the migrations are dependent on what is between the islands, and what is between the island and its mainland, which is to say the matrix within which the islands are located. As with any dynamic process, an observation at a point in time does not necessarily reveal anything about the long-term state of the system. For instance, in the context of tropical forests, censuses from a 50-hectare forest plot in Panama from 1982 to 2010 show a very non-static picture. For example, the palm species *Bactris coloradonis* had 38 individuals in 1982, 17 in 1985, 6 in 1990, 2 in 1995, and then disappeared completely (zero individuals encountered) in 2000, 2005, and 2010. Therefore, an isolated 50-hectare reserve would not have maintained this species. On the other hand, the species *Cecropia longipes* was absent in the censuses of 1982, 1985, and 1990, but then a single individual was encountered in 1995, 12 in 2000, 14 in 2005, and 13 in 2010. So, the species diversity census from a single year, let's say, 1982, would have shown the presence of *B. coloradonis* but not *C. longipes*, while the census in 2010, 18 years later, would have shown the reverse. In fact, what these data show is that some species went locally (within the 50-hectare plot) extinct, while other species 'migrated' into the plot (their seeds dispersed from surrounding forest), over the 30-year period.[27]

We emphasize that this is what normally happens. It happens within very large 'natural' protected areas as much as in fragmented landscapes. Species always go locally extinct. Sometimes the extinction may be of a group of trees living on a hillside in the middle of the Amazon rainforest, other times on a tropical island. Extinctions are normal at a local level. The only thing stopping them from turning into extinctions at a broader regional scale is subsequent migrations from other subpopulations. This is why a 50-hectare plot of forest in Panama contains about 200 species of trees, while the island of Ocoquili, off the coast of Panama, with its area little more than half the size (30 hectares – the point at which the mainland plot asymptotes – see Figure 3.2), contains a mere 34 species.[28] Local extinctions are real, and it does not matter how big the reserve is, if it is surrounded by low-quality matrix, whether an ocean or intensive industrial agriculture. With no migration, the natural local extinctions will accumulate and could turn into regional extinctions.

Beyond these narrow technical issues, the land-sparing *versus* land-sharing strategies are inherently difficult to evaluate in practice. As we have argued here, continuing localized extinctions are to be taken for granted while survival of most species will typically take place across vast landscapes, provided the potential for inter-fragment migration is substantial. Any evaluation of policies or programmes to protect species thus must be taken over long periods of time and over large areas. As such, confounding factors beyond the simple ideas of sparing versus sharing are bound to be numerous and complex.[29]

The problem of defining time and spatial boundaries for studies in a way that would guarantee convincing results is itself a profoundly difficult task. These problems are reflected in current research, where advocates of one strategy or the other routinely find comparative evaluations to be inadequate and unconvincing. Authoritative studies capable of resolving the debate are not likely to settle the debate any time soon. The best that can be hoped for are careful studies that help clarify particular issues in specified regions and times. Meanwhile, current efforts to protect species can be seen to have failed in a great variety of circumstances. As a planetary matter, species extinction continues apace.

A historical view of conservation efforts

Beyond the scientific difficulties of evaluating these strategies, there is a larger problem. Looked at as a human problem, in the sense that successful protection strategies require societal change, it seems clear that the sharing versus sparing debate reflects profoundly different historical traditions, philosophical positions, and political loyalties. These have been partly shaped by ongoing scientific research, but they represent societal questions that are not likely to be solved simply by better research.

Begin with the fact that the idea of pursuing human activities with a primary goal of preserving biodiversity is, for the most part, a very new phenomenon.[30] Kings and nobility protected lands in European and other civilizations going back thousands of years, but they did not do so for the protection of the broadest possible array of species; rather, they sought to protect game species for their own enjoyment and prestige, and forest resources for future use. This sometimes resulted in protection for a much wider range of species, but this was almost never, if ever, the goal.[31]

Awe versus biodiversity in US-protected areas

Beginning in the 19th and early 20th centuries, national governments began to set aside protected areas that often had as a consequence a good deal of species protection. But while certain iconic species earned the attention of protection efforts, broad species protection was not on the agenda. For example, the National Parks system of the United States, which until recently provided the highest available level of protection for land and wildlife, was created for several reasons that only incidentally protected anything but a handful of storeyed species. Railroad companies lobbied for National Parks that would promote mass tourism and thus sell train tickets and room reservations in large hotels often owned by the railroads themselves. Sublime peaks and waterfalls became symbols of national character associated with the task of what would become known as 'nation building'. At the same time, state and federal policies encouraged the slaughter of bison, antelope, prairie dogs, bears, cougars, coyotes, and wolves – even offering bounties to hunters who worked steadily towards species extinction. Further, in many if not most cases, the creation of 'natural parks' required the usually violent removal of indigenous peoples.[32]

Citizens inspired by the very influential 19th-century ideas of romanticism and transcendentalism lobbied for the protection of places that would inspire awe and pose a counterpoint to an otherwise shallow commercial and narrowly pragmatic culture – such lobbying usually focused on magnificent mountain, coastal, or waterfall features thought to inspire reflection and provide solace to the soul. The animals living there were at best appreciated as scenery and not primarily as ecological phenomena, and often the most magnificent places favoured as parks were markedly poor in species in contrast to surrounding landscapes. Nationalists, railroad owners, and romantics were interested in salvaging the magnificent, but had relatively little interest in saving a wide variety of species, most of which tended to avoid the awe-inspiring but sparse landscape features of National Parks.

For example, the protection of the sublime landscapes of the high Sierras, and the later more extensive and more partial protection of California's mountain forests almost completely ignored the idea of preserving the vastly more species-rich landscapes of the foothills and valleys, which were totally transformed to make them suitable for farming and cities. What became Yosemite National Park was a partial exception, but an exception that tends to prove the rule. The valley floor of Yosemite was species rich. However, the valley floor's value to the Park was appreciated largely as a place from which the visitor could gaze upward at mountain grandeur. The indigenous people who had treasured the valley for its species richness were removed violently or with the threat of violence. Then, species of the valley were poorly protected because of the intense tourist presence there.[33]

Elsewhere in California, valley rivers originating in mountain snowfall were resources to be dammed at the expense of huge populations of salmon while exotic shad and striped bass were imported at the expense of at least dozens of native species. Hundreds of low mountain or foothill valleys resembling Yosemite Valley became species-poor reservoirs, stocked with often non-native species, and managed in ways that discouraged the survival of many native species, including salmon.

Only in the 1980s did conservation officials in California begin to think about creating spaces that might protect valley and foothill species. When they did, they mostly did so under the prodding of the Endangered Species Act, not because they suddenly realized that their predecessors had destroyed a great array of California species as they had successfully protected sublime, but inanimate and mostly barren granite peaks. Some of the very recently protected valley areas are an improvement, but so small in area and so compromised in a variety of other ways as to be mere pathetic reminders of what was lost.

The powerful image of the alpine fly fisherman casting his line over a high mountain lake only began to seem a little foolish when in the late twentieth century it became recognized that such lakes surrounded by bare granite were woefully lacking in nutrients and therefore in naturally occurring trout. Trout were present mostly as trout varieties not found naturally in the high Sierras, placed in mountain lakes and streams by a state agency anxious to sell fishing licenses. The business of enjoying the idea of nature was not the same as the protection of nature.[34]

The overwhelmingly largest areas of protected areas in the United States are not under the jurisdiction of the relatively protective National Parks Service, but, rather, are administered by the National Forest Service and Bureau of Land Management. Both were created in a sense of well-justified panic in the late 19th century as commercial firms had stripped the nation of most of its forests and wrought havoc on non-forested lands. The purpose of the Forest Service and Bureau of Land Management was distinctly not species protection, but rather resource conservation in the sense of promoting rational resource use for recovery and long-term economic use of the land and its resources.[35]

Such agencies, backed by commercial enterprises, were aggressive in asserting that they were in the business of managing selected natural resources rather than in 'merely' protecting them – simple protection of land and wildlife being seen as naïve and, in any case, lacking a viable political constituency. A young wildlife management specialist and Forest Service employee named Aldo Leopold realized that these agencies were actively hostile to anything remotely resembling an ecological perspective promoting broad species protection; his disillusionment inspired the essays that are often seen as the foundation of the modern environmental movement.[36]

Beginning in 1965, significant areas were made wilderness areas which have the effect of affording high levels of species protection, but the areas are not chosen primarily on the basis of overall species richness nor on particular species habitat. Most wilderness has been chosen for its availability as such and for the remoteness and beauty of its landscape, not primarily for biological reasons. In 2008, a more propitious time for environmental initiatives than ten years later, at this writing, a Forest Service analyst wrote that the long history of 'resource management' had only slowly and very partially given way to an ecosystem management perspective. His well-supported view was that public lands agencies, such as his own, remain largely very resistant to habitat and species protection as primary goals, if for no other reason than the lack of a large and well-organized constituency necessary to force more significant change.[37]

The small fraction of wildlands that has been in recent decades put into reserves intended primarily for broad biodiversity protection are usually severely compromised by their limited size and by competition with surrounding land uses – think, for instance, of the newly ambitious purposes of the Florida Everglades constantly undermined by a handful of powerful sugar growers and their allies among the more short-sighted residential and commercial developers.[38]

Contemporary sharing and sparing in the tropics

If we move to the relatively species-rich areas of the tropics and sub-tropics, the history of setting aside 'protected areas', is at least equally as problematic. In Africa, as we have noted elsewhere, the grand wildlife parks were mostly created as part of an imperialist agenda imposed on the species-rich landscapes of indigenous people. The most active and influential people promoting these parks were those interested in 'big game hunting' and/or the control of diseases affecting cattle and people. As such, the areas selected were biased toward large, charismatic species. The creation of the parks was often severely harmful to the interests of local peoples who had little choice in the matter. While in many cases we can celebrate the fact that, given 20th-century human population growth rates and the destruction of native economies, these parks have provided very significant species protection that otherwise would not have occurred, such protected areas, as we have discussed, are immersed in controversy. The creation of large new areas devoted to strictly protection strategies is extremely unlikely.

In Brazil and some other nations, extremely large protected areas have been created, partly because of the sheer immensity of virtual wilderness areas, but also for a variety of other reasons ranging from the commercial to the militarily strategic. Some of the largest of these reserves are governed by legislation that explicitly protects indigenous extractive and agricultural activity as well as such activities practised by more recent settlers. Such reserves recognize the 'nature's matrix' idea in their very conception, being examples of land-sharing strategies.

In contrast, in October 2018, the newly elected President of Brazil, announced his intentions to reduce protection for land in the Amazon and the *cerrado* in favour of development of rapid extractive and agricultural activities. He has also announced his intentions to declare the MST,[39] the principal social movement advancing agroecological land-sharing strategies, to be a terrorist organization whose members he promises to jail or exile. These measures clearly represent potential reverses for land-sharing strategies. But is this a victory for land-sparing strategies because it means rapid expansion of intensive agriculture, or a defeat because it reduces the areas and levels of species protection? This dilemma reveals a fundamental problem for land sparing, because it demonstrates that the single largest measure by which intensive agriculture in the world can be practically expanded, by developing agricultural frontiers in Brazil, represents the single largest threat to the most species-rich regions of the planet. Further, it demonstrates that in practice those forces most active in promoting intensive agriculture are also those most determined to compromise protected areas.[40]

What can be said without fear of contradiction is that the creation of new and significantly large protected areas will almost everywhere face strong opposition that in a post-imperial world will not be overcome. An apparent exception might be the vast new areas recently put into protected status in Chile and Argentina as a result of the initiative of American philanthropists. In fact, this expansion of protection across much of the Southern cone is not an exception. These sparsely populated areas were given national protected status only after decades of negotiation and compromise, with the explicit intention of both the Tompkins, the American philanthropists, and of the national governments that most of the ranching and farming activity in the region would be protected as such, just as the forests and plains of the region will be protected from excessive logging and destructive mining. The result will again be a matrix of activities arising out of historical development as well as elements of conservation planning, precisely as we advocate here in the spirit of land sharing.[41]

Land-sparing advocates assume a degree of consensus and/or exercise of sheer political power that is unrealistic in the extreme. Their proposals for putting as much as half of the earth's surface in protected areas is simply unhinged from human history in general, but also from the history of conservation. It glosses over the fact that most of the approximately 15 per cent of the land surface of earth that is in 'protected' areas is everywhere protected in various very limited ways – the idea of just extending the 15 per cent to 30 per cent, or even half of the earth's surface can only seem possible if one not only ignores political reality but also fails to deal with the restricted sense in which protected areas have been protected.

Just as important is the failure to recognize the effects of intensive agriculture outside field boundaries and across regions, as we have argued. Genetically modified corn and soy grown in the American Midwest reduce the monarch butterfly populations that migrate to Mexico. Pesticides and fertilizers create massive dead zones in coastal waters. Sharp global declines in the global populations of insects, amphibians, migratory birds, and coral reefs all have significant connections to intensive agricultural activities of the sort that land sparing intends to further intensify. Human-induced climate change is caused to some degree (perhaps 20–25 per cent) by intensive agriculture though it could be reduced by a variety of agroecological strategies of the sort we propose for the matrix. The chemical and fossil fuel intensity of modern agriculture must be seen as part of the problem, not part of the solution.[42]

Taken as a whole, the land-sparing strategy arises from a false sense of power and control. Protected areas, to work as imagined, must be utterly isolated from the real complexity of human society, much as they are imagined to be isolated from the actual dynamics of other species' populations moving across landscapes. Agriculture is imagined as an activity that may be practised without important effects beyond field boundaries. Only a radically simplified view of both the requirements of species protection and the effects of intensive agriculture can support such a view.

The land-sparing vision of vast new protected areas complemented by a sacrifice zone of chemically intensive agriculture is an essentially imperialist vision both in its historical roots and in its present context. It seeks to lay across the earth an extremely simplified bi-polar scheme of complete protection on one hand and intense destruction on the other. While supposedly based on ecological expertise, it in fact flies in the face of the much more complex view of a sophisticated ecological perspective. While supposedly based on hard-headed political realism, the land-sparing strategy is in fact based on a political fantasy which has no agent capable of carrying it out consistently and no significant human constituency to support it.

In contrast, the land-sharing vision does not rely on a totalizing dichotomy that obliterates ecological and cultural complexity. It is not a grandiose plan but an approach to problems and opportunities. It seeks to work within the framework of complex landscapes and complicated politics. Rather than seeking to eradicate complexity, it seeks to work within continually evolving local, regional, and global situations.

Within the contrasting views are contained, not simply scientific disagreement, but a range of philosophical and political value choices. While we can continue to try to assess the degree to which these contrasting views work to protect biodiversity in practice, the land-sparing vision's realization can only really be assessed on its own terms once we have carried out a kind of one-time experiment on a planetary level. The land-sharing approach can never be expected to demonstrate total success, but neither will it carry such a high risk of irreversible failure.

The model forest example

In some ways what we are proposing in this book is the antithesis of Wilson's half Earth, but effectively more extreme in its analysis, based on ecological, sociological, economic, and political analyses. Rather than 'half earth', our extreme would be 'whole earth'. Terrestrial biodiversity conservation is a large landscape project, involving not only whatever fragments of 'natural' vegetation exist, but also the matrix in which those fragments occur. Our concern is thus with the entire landscape and the long term. Our biodiversity unit is the landscape. Thus, arguments of the size of the reserve (SLOSS), or transitions from forest, to agriculture, to forest (the FTM), or ambitions of integrating development goals into conservation programmes (ICD programmes), are really inadequate in and of themselves. What we strive for incorporates elements of all of those narratives toward the goal of insuring that local extinctions are balanced by migrations and that people have digni-fied livelihoods. In other words, we seek to guarantee that conservation of biodiversity occurs in the long run. In the rest of this book we argue that the only way to do this is by maintaining a biodiversity-friendly agricultural matrix, managed by small-scale farmers practising agroecology.

One existing framework, albeit not especially well known, is the Model Forest framework.[43] The principles of Model Forests include a participa-tory governance model, management that promotes community empower-ment, as well as local economic development, food security, initiatives for sustainable development, and initiatives that address the challenges of climate change. The main idea is that of 'a forest with people', managed by local communities for their own benefit and that of biodiversity. We tentatively suggest it represents an interesting and possibly revolutionary discourse that could eventually guide conservation planning. The concept of a Model Forest is a territorial initiative for the sustainable development of regions where forest ecosystems play an important role. The Canadian model, as summarized by John Hall, stated that a model forest is characterized by,

> The wise and sustainable management of our forest resources to conserve their biodiversity, productive capacity, and ecological integrity while, at the same time, maintaining the social, cultural, and economic benefits we derive from them, is a concept that has brought many to work together who might not have otherwise done so. This also means that we must look well beyond our present needs and think about future generations.[44]

Based on what were perceived as early successes in Canada, the concept emerged internationally at the United Nations Conference on Environment and Development in Rio de Janeiro in 1992, where Canada's prime minister announced its formation. Since that time the seven model forests that exist in Canada have been joined by almost 60 other forests in more than 31 coun-tries. It is a project in motion, with clearly uneven development and, due to its grass-roots nature, is unpredictable. The goals of the idea encompass most of what seems to have been in existence in a fragmented form in the various other conservation and/or development programmes proposed in the past. Thus, directly from its web page, a model forest:

1   Is both a geographic area and an approach to the stainable manage-ment of landscapes and natural resources.
2   Must encompass a land base large enough to represent all of the for-est's uses and values.
3   Is a fully working landscape of forests, farms, protected areas, rivers and towns.[45]

The web page goes on to state:

> The approach is based on flexible landscape and ecosystem management that combines the social, environmental and economic needs of local communities with the long-term sustainability of large landscapes. The approach links a comprehensive mix of stakeholders, natural resource users and land use sectors within a given landscape and helps create a common vision of sustainability and constructive dialogue involving all stake holders. Those involved define what sustainability means in their own context, identify a common vision and set of goals, devise a governance structure and strategic plan, then work collaboratively to achieve the goals set out in that plan. In turn, the process nurtures a depth of trust and transparency and increases the willingness to implement innovative solutions.[46]

Clearly the programme differs significantly from the traditional conservation approaches of dividing nature into 'with people' and 'without people'. The human being is put back into the picture as an integral part of the ecosystem, and all the socio-political structures that inevitably emerge from our nature as a social species are acknowledged as participating in the development of the conservation structure. Indeed, the conservation programme is not distinguished from the livelihood programme. The call for participatory governance models, management that promotes community empowerment, as well as local economic development, food security, initiatives for sustainable development, and initiatives that address the challenges of climate change, are all included in the basic idea of a model forest.

Notes

1 Brooks *et al.* (1999); Sax and Gaines (2008); Wyatt *et al.* (2008); Gibbons and Harcourt (2009); Laurance *et al.* (2011); Gibson *et al.* (2013); MacPhee and Sues (2013); Benchimol and Peres (2014).
2 Holyoak and Lawler (1996); Kuussaari *et al.* (1996); Ingvarsson (2002); Hufbauer *et al.* (2015); Phillips and Swanson (2018); Terui *et al.* (2018); Wheeler *et al.* (2018).
3 Here, there is a legitimate question as to how the matrix may actually affect extinction rates. Some authors have suggested that larger migration rates can actually reduce the extinction rates within islands or habitat fragments (Brown and Kodric-Brown, 1977), what has been referred to as the 'rescue effect'. We more or less accept the more classical idea that local extinction rates are only a function of the size of patches, although our general argument is even stronger if the rescue effect is taken into account.
4 Diamond and May (1976).
5 Diamond (1975).
6 Simberloff and Abele (1982).
7 Mather (1992); Mather and Needle (1998).
8 Wright and Muller-Landau (2006); Grau *et al.* (2008); Redo *et al.* (2012); Melo *et al.* (2013).

  9 van der Ploeg (2009).
 10 Rudel *et al.* (2000, 2005); Parés-Ramos *et al.* (2008); Meyfroidt and Lambin (2009); Sloan (2015).
 11 Hansen *et al.* (2013); Brinck *et al.* (2017).
 12 Chazdon *et al.* (2016).
 13 Matteucci *et al.* (2016).
 14 Hecht (2010).
 15 Van der Ploeg (2013).
 16 Hecht and Saatchi (2007).
 17 As noted by Hecht (2010), the economics of familial remittances in the age of globalization are enormous. As she notes, if migrants were a country their GDP 'would be placed among all but the very top European economies'.
 18 Adams (2013).
 19 Hughes and Flintan (2001).
 20 Kremen *et al.* (1994); Barrett and Arcese (1995); Hughes and Flintan (2001); Brooks *et al.* (2013).
 21 REDD+ (or REDD plus) are programmes that extend REDD to include sustainable forest management, conservation of forests, and enhancement of carbon sinks.
 22 Harvey *et al.* (2010).
 23 Redford *et al.* (2013).
 24 Green *et al.* (2005).
 25 Wilson (2016).
 26 These data are from Condit *et al.* (1996).
 27 These data are from Condit *et al.* (2012), wherein is an excellent technical discussion of the importance of regional migration in this particular ecosystem.
 28 D'Arcy and Hammel (1985).
 29 Perfecto and Vandermeer (2008a, 2010); Fischer *et al.* (2011); Tscharntke *et al.* (2012); Kremen (2015).
 30 A possible exception to this is the Sacred Groves of India. Associated with the Hindu tradition some of the Sacred Grooves are considered to be sanctuaries for flora and fauna and no human use is allowed; Prime (2002).
 31 Williams (2003).
 32 For an excellent discussion of the evolution of exploitation and protection in the National Parks, see Worster (2016). For an excellent discussion of the removal of indigenous people to open lands for national parks see Spence (1999).
 33 Spence (1999).
 34 Knapp (1996).
 35 MacCleery (2008).
 36 Leopold (1949).
 37 MacCleery (2008).
 38 National Academy of Sciences (2018).
 39 MST, Movimento do Trabalhadores Sem Terra, or the Landless People Movement, is regarded as the biggest social movement in Brazil and one of the largest in Latin America; Wright and Wolford (2003).
 40 *Guardian* (2018); BBC News (2018).
 41 *New York Times* (2018).
 42 Vermeulen *et al.* (2012).
 43 Website for the International Model Forest Network: www.imfn.net/international-model-forest-network, accessed 12 September 2018.
 44 Hall (1996).
 45 See the website of the International Model Forest Network on www.imfn.net/international-model-forest-network-global-forum/, accessed 13 March 2019.
 46 International Model Forest Network, www.imfn.net/international-model-forest-network-global-forum/, accessed 13 March 2019.

4　Historical patterns in agriculture

The general argument of this book involves agriculture. Since almost all terrestrial ecosystems are now fragmented, resulting in the likelihood that most species are forced into a metapopulation structure, the areas between fragments will determine the degree to which organisms migrate from fragment to fragment, and thus their survival probability. Those areas between the fragments are composed, to a great extent, of agroecosystems. Thus, any serious conservation effort must have at least part of its focus on agroecosystems. Within that focus we ultimately conclude that the industrial agricultural system is not nearly as likely to provide the high-quality matrix that is necessary for effective conservation as small, farmer-based sustainable systems. Because our argument is so dependent on this agricultural component, we present an in-depth elaboration of precisely what that industrial system is, where it came from, and why it is in crisis.

The development of agriculture

Pre-industrial agriculture

Worldwide, many non-industrial agriculturalists still practise an agriculture that may be similar to that practised by early agriculturalists, variously referred to as 'slash and burn' agriculture, swidden agriculture, or shifting agriculture. This form begins with the felling, or slashing, of natural vegetation. The slashed vegetation dries and is subsequently burned, eliminating much of the vegetation which otherwise could have formed a weed community. Into the nutrient-rich ash-covered ground the desired crops are planted and usually grow vigorously, at least in the first year. The next year, or sometimes sooner, the accumulated post-harvest vegetation is burned, and crops are planted again. In this annual cycle, crop production declines each year, due to a reduction in soil fertility, accumulation of pests, and a variety of other factors. Crop production eventually declines to unacceptable levels, and the field is left fallow. The consecutive number of years in production varies widely, as does the time necessary to leave the area fallow, depending on local ecological, cultural and socio-economic conditions.

During the fallow period, enough time is needed to replenish the soil nutrients and organic matter, to eliminate whatever disease agents have built up in the soil, and to develop the vegetation that once again will permit fire to eliminate weeds – generally, fire-resistant plants are eliminated by higher, more woody vegetation through the process of ecological succession. Another relevant factor is the size of the local human population and its food requirements, since that will dictate how much land must be put into production at any given point in time, which indirectly establishes a limit on the length of the fallow period. For example, if a total of 100 hectares is in the land pool, farming one new hectare per year permits the potential of a 100-year fallow period, while farming ten hectares per year permits the potential of only a ten-year fallow period. Contrarily, the intensity of production itself determines, to some extent, the number of people needed to maintain that level of productive activity. That is, there are two relevant population densities that need to be considered – the number of people necessary to make the production–fallow cycle function and the number of people that production–fallow cycle will support.[1]

It is worth noting that in the contemporary world there is a form of 'slash and burn' agriculture that differs substantially from more traditional forms.[2] The more traditional form is usually adapted to local conditions with fallow periods that are long enough to permit the recovery of the ecosystem before entering a new crop cycle, thus making it sustainable over the long term. The traditional form is also frequently associated with land tenure arrangements that provide some level of security over land use. Contrarily, recent migrants into a new area sometimes practise what is also called 'slash and burn' agriculture, but with less than perfect knowledge of local ecological forces, a shorter fallow period, techniques that are not conducive to sustainability, and little or no land tenure security. It is useful to distinguish between this migratory agriculture that uses 'slash and burn' techniques and traditional 'slash and burn' agriculture, which is arguably more sustainable. Traditional 'slash and burn' has a particular territory in which the farming family or farming community operates; the fallow lands are treated as part of the agricultural system; and only rarely is old-growth vegetation cut. In contrast, migratory agriculture involves the cutting and burning of whatever vegetation is on the land, frequently old-growth forest at the edge of the agricultural frontier, farming the land as long as it's productive, and then moving on to new land, causing fragmentation and deforestation.

A characteristic of traditional 'slash and burn' that is often not fully appreciated is the care and attention given to each phase of the agricultural cycle. A striking example of high-level management is the tradition of *mal monte* and *buen monte* (bad weeds and good weeds) developed by the Mayans of southern Mexico.[3] Farmers actively observe their fallow land, and have a detailed knowledge of the sorts of wild plants that invade, recognizing some as harmful to the overall agricultural system, but others as beneficial. Sometimes farmers will actually weed out 'bad' plants from the fallow land and plant good ones. Exactly what makes some plants good and others bad is not always obvious. When asked why particular plants are good, farmers frequently answer 'they make the soil healthy', or 'it is easy to cut'. Legumes are usually considered good plants and one can speculate that the underlying ecological mechanism is nitrogen fixation. Spiny vines are usually regarded as bad, probably because they are annoying when cutting the vegetation during land preparation. The important point is that the fallow land itself is not just land that is left idle for nature to take its course, but is thought of as a regenerative part of the agricultural cycle.[4]

Managing the fallow cycle is just one example of the sophisticated technology, based on underlying ecological knowledge that is frequently employed in these more traditional systems. Furthermore, 'slash and burn' agriculturalists and other more traditional farmers are excellent experimentalists, always trying out new techniques, new varieties, new crop combinations, and so forth. Even though their farming methods are steeped in tradition, that tradition is flexible, and actually includes innovation. Such an attitude is of obvious importance when searching for alternative ways of doing things. Local farmers may not understand the terms used by the academic ecologist or, for that matter, the rural sociologist, but their understanding of the natural world and the society around them is deep.[5] Above and beyond the obvious ethical imperative of incorporating stakeholders into any development plan, ignoring the knowledge base of local farmers is simply bad science, as well as bad politics. Fortunately, the importance of local knowledge is now being recognized by the international community as exemplified by the recommendations of the International Assessment of Agricultural Knowledge, Science and Technology for Development (IAASTD)[6] and the FAO in their document about agroecology and the Sustainable Development Goals.[7]

The evolution and domination of European agriculture

We can think of the year 1492 as a watershed in the story of how agriculture has transformed the face of the earth. When Columbus stumbled on the Americas he unleashed a process of profound change that was as important for the biology of the earth as it was for the history of human society. We can also use the date as a rough marker of the start of another related process of change that was to be just as profound for the fate of the biological diversity of plants and animals. Europe was not only poised to reshape the Americas and Africa and much of Asia through imperial expansion and trade, but Europe was itself beginning to change its own ways of using the land and its resources. These changes, in turn, would be exported to the rest of the world where European colonies were established. Taken together, the European encounter with the Americas and the new ways Europe began to use its land would have enormous effects on the relationship between agriculture and biodiversity.

At the end of the 15th century, many regions of Europe were already beginning to see changes that would become steadily more dominant over the following five centuries. As seen in the landscape, fields and pastures were becoming larger. More of the land was being used to produce goods to be sold in cities and transported relatively long distances, rather than being consumed by those who produced, sold, or bartered them locally. The conversion of forests to pasture and cropland accelerated. Concomitantly, those who worked the land began to be paid wages rather than produce as part of a complicated system of serfdom that granted serfs access to land in return for a series of traditional obligations to their lords. Nobles had traditionally regarded their land as a source of power and prestige, treating their forests and marshes as reserves for hunting pleasure, hunting itself being thought of as a peculiarly noble activity that both signified power and granted prestige. As the European economy changed, the nobility began to wonder if they could make better use of wild places as sources of cash in the new economy of trade and manufacturing. In particular, the manufacture of textiles had become the leading edge of a technological frontier of industrialization, and those who controlled land began to see more value in it as a way of producing saleable wool in the newly forming economy of money. The nobility that failed to sense the power of the changes underway were often pushed aside by urban entrepreneurs who had a better understanding of the new role the land was coming to play. In short, agriculture was becoming capitalist agriculture.[8]

The process was self-reinforcing. The production of wool required less labour per unit of land area than the production of wheat and other food crops. In England, a set of legal changes, known as the Enclosures, were designed to push peasants off cropland. The Enclosure movement and similar events elsewhere in continental Europe generated conflict and violence over roughly two centuries as peasants were forced to relinquish land used for their sustenance so that it could be used to provide wool for manufacturing. Textile manufacture required workers – and at this stage of primitive industrialization, it required a lot of workers to produce any quantity of cloth. Increasing numbers of people worked in manufacture, trade, and transportation of goods, while at the same time cropland, forests, and marshland were converted to sheep pasture. Consequently, demand grew to produce more food on less land and with less labour. Those who could grow more food were rewarded with increased income. Furthermore, as trade grew, there was more demand on forests for wood to make boxes, barrels, wagons, and ships. As cities grew, more wood was also needed for houses and other buildings, and great quarries and mines were cut into the earth to provide a vast range of materials.

For a long time, the changes in the use of agricultural land use and the more intense use of forest products was largely localized and haphazard. But by the 18th century, people began to speak of 'scientific agriculture', seeking a systematic approach to the problem of feeding more people with less available cropland and fewer farmers and farm workers. By the 19th century, the term 'scientific forestry' came to describe attempts to increase the amount of forest products the land could produce. Similar self-conscious efforts to use the land more intensively grew up on the European continent. Inventors, entrepreneurs, scientists, and farmers began applying new scientific and technological advances to the problems of maintaining or enhancing soil fertility, making tools more durable and efficient, providing better drainage and irrigation, selecting and breeding crops and animals that would yield more, and improving transportation and motive power.

While all this was occurring, the European settlements in the Americas provided both a stimulus and an outlet for making agriculture both more capitalist in nature and more intensive in its use of land and labour, in many cases slave labour. Several decades before Columbus's voyages, the Portuguese were already experimenting with growing sugar on islands off the west coast of Africa. They found it profitable to capture or buy slaves in Africa to plant large fields with the single crop of sugarcane, a crop which, at that point, brought fabulous prices as a luxury good sold in Europe. The model of the large plantation specialized in a single crop and cultivated with forced labour would come to transform regions in the Caribbean, in much of North America, in Central America and much of South America. Sugar, cotton, tobacco, and later, coffee, cacao, and banana plantations were great engines of social and biological change. In most places, monocrop plantations replaced tropical and sub-tropical forests of enormous biodiversity. For example, all along the northern and central coast of Brazil, planters destroyed a great swath of the Atlantic Coast rainforest, or *Mata Atlântica*, to plant sugar cane. Later, in the 19th century, the coffee boom in Brazil would nearly finish the job of destroying Brazil's Atlantic Coast rainforest, perhaps the most diverse in the world. Plantation owners did the same in much of the Caribbean and parts of Mexico and Central America and in the south of the US. Most of the forest would simply go up in smoke, but a significant amount of wood went to further fuel European expansion by providing dye-woods, ship timbers, and building lumber for new cities in the Americas and for the old, rapidly growing cities in Europe. Although some crops have been displaced by others, tropical monocrop plantations continue to this day to advance further deforestation and/or prevent regrowth of forests throughout the tropics.[9]

As they devastated the forests and other natural habitats of the Americas, monocultural plantations also had powerful indirect effects on European agriculture. Some of the money earned from American plantations was invested in hurrying forth the changes in industry and agriculture that were reshaping the European landscape. Plantation products from European colonies were among the most important items in the growing world commerce and provided essential food and materials for European economic growth. In addition, it is notable that the plantation system formed a model for agriculture as an economic enterprise that assembled masses of labour and large expanses of land for profitable exploitation. Such plantations were not new to Europe – the Romans had used a comparable model – but the American plantation revived the form in the specific context of the rapidly expanding and increasingly technological capitalist economy of Europe. Greater attention would be paid in Europe to the possibilities of replacing small-scale, diverse farming operations with large agricultural estates that would specialize in one or a few crops. In turn, inventions such as the steam engine and railroad would bring about greater and more ambitious changes in the Americas and across the globe.

Confronting declining soil fertility

A central challenge for agriculture has always been the maintenance of soil fertility. In the medieval world, those who worked the land maintained fertility through a number of techniques: crop production was physically close to and integrated with raising livestock and draught animals, so the manure from animals was relatively easy to return to the land. Crop rotation and rotation between grazing and crops was routine. Nutrients were acquired from the seas in many areas by harvesting seaweeds and other marine organisms and incorporating them into cropland. Leaf litter from the forest floor was composted with other plant waste materials. Wood ash, bones, shells, lime, and other minerals were used as soil amendments. Human waste was sometimes used in the fields. The relatively small scale of farms and fields, and the integration of animals, crops, forests, and fisheries, facilitated all of these techniques. The work of the whole family was applied to the multitude of tasks and. in many areas, cooperative work among families and villages was common and essential. Demands on the fertility of the soil were limited by supplementary food from hunting, fishing, and gathering of wild plants. The whole system only had to provide for the self-sustenance of farming families, plus some additional food for a relatively small number of people who were not directly engaged in agriculture. It was a system strongly based on the mutual interdependence of agriculture, pasture, and forest. These elements were usually further integrated with streams, lakes, and the sea.

The so-called 'Norfolk four-course rotation' is an example of this system. In the first year, a root crop was sown. The second year would be oats or barley, followed in the third year by a legume (usually clover), and finally, in the fourth year, wheat. Sheep were usually pastured within the system, consuming crop stubble and cover crop (e.g. clover), and, as a matter of course, recycling the nutrients by directly depositing manure in the fields. Most of Central Europe used some sort of three- or four-field system, usually with two cereals (e.g. oats and wheat) alternating with a legume, and sometimes pasture.

To this day, everywhere, even in Europe and North America, some elements of these rotational systems survive. Many of the techniques are now being rediscovered and adapted for modern organic and other alternative types of agriculture, a point we emphasize later as part of the overall thrust of our argument about conservation.

In addition to the development of complex rotational systems and the application of various soil amendments, one of the most important approaches to the problem of soil fertility was the opening up of new agricultural lands. By seizing vast new territories, Europeans were effectively able to acquire soil fertility to provide the agricultural needs for the new world economy that they were constructing. In North America, Argentina, Brazil, and Australia, for example, Europeans brought into production vast areas of very rich soils that had never before been cultivated. The accumulated fertility of virgin soils, due to the biological interaction of perennial grasses and trees with the mineral substrate, was perhaps as important to the emergence of Europe in dominating the world economy as was the gold and silver looted and mined from Peru, Mexico, California, and Brazil. It provided wealth that ramified throughout the entire world system, including the waves of European immigrant farmers. In addition, Europeans were able to enlist the land and labour of established farming economies of colonial possessions to enrich the European economy. And all of this was accomplished through the subjugation or genocide of the people who had lived in these landscapes until the European conquest.

As the appropriation of the fertility of the world's soils for European purposes proceeded, Europe itself began to experience serious problems of declining soil fertility at least as early as the mid-18th century. This was mainly due to the breakdown of the farming systems that had integrated all the elements of the landscape: cropland, pasture, forest, waterways, and the sea. Specialization of most of the agricultural economy in those crops that were most profitable for provisioning growing urban populations, and specialization of individual farms according to local comparative advantages tended to displace the more integrated systems of the past. The land had to be forced to produce far more than the subsistence needs of the farming family and a small marketable surplus. Collecting and spreading animal manure and composts involved longer distances and much greater energy expenditure. Crop rotation and fallow times were neglected or abandoned in order to maximize immediate profits. The important array of wild products from the forest declined with the advance of forest clearing. Perhaps most importantly, on the specialized farms, the old knowledge of how to adapt and how to integrate the various elements of nature in a system capable of maintaining fertility, was often lost. In many areas, as it was lost, it was also scorned as quaint, old-fashioned and useless – a mere relic of the illiterate peasantry who had to be displaced by more 'advanced' and 'scientific' farmers and estate managers.

The new farmers and estate managers to some extent became agricultural experimenters in the new kind of more specialized and intensive crop systems that were taking over. Declining fertility was both a private economic problem for agriculturalists, and a public concern of society interested in maintaining food supplies for growing and ever-more urbanized populations. In the 18th century, in search of private wealth and public welfare, various kinds of non-farming scientists, academics, inventors, and entrepreneurs began to investigate the problem of soil fertility and how it might be solved. A good deal of work was done to attempt to make composting cheaper and more manageable for the large and specialized farm. People began to experiment with new combinations of soil amendments of various kinds, including lime and mineral supplements. By the mid-19th century, experimenters began applying the impressive scientific advances in chemistry to the problem of plant nutrition.

A key conceptual breakthrough occurred in 1840 when Justus von Liebig articulated what has come to be called Liebig's Law of the Minimum, in which he recognized that nutrients were required by plants in specific proportions.[10] Thus a single nutrient in short supply would prevent plant growth. The nutrient that was limiting was the one that was at its 'minimum' with respect to plant growth.

The Law of the Minimum led to the idea that still dominates soil management today, that soils need to be provided with specific nutrients, the limiting ones, so as to increase plant growth. While contemporary understanding of nutrient cycling makes this idea essentially anachronistic, as we discuss below, the idea nevertheless formed the underlying logic for the development of an industrial approach to soil fertility. Thus, in the mid-19th century, an incipient fertilizer industry arose producing superphosphate through a simple chemical procedure, and importing phosphate rock, guano, sodium nitrate, and potassium directly from natural sources. Mainly through the process of producing superphosphate, the fertilizer industry had become the British chemical industry's largest customer by the dawn of the 20th century.[11] Fuelled by the need for nitrate explosives during the world wars, the industry grew ever larger and became one of the major components of the chemical-industrial complex we see today.

Although early capitalization was based on the production of superphosphate, phosphate was not the major limiting nutrient in agriculture in most temperate areas of the world. As Liebig's Law of the Minimum emphasized, all nutrients would eventually become limiting if agriculture continued its development along the lines required by the developing capitalist system. He was particularly concerned, as were many analysts of his day, with the separation of town and countryside. He wrote emotionally about the ecological irrationality of a system in which the nutrients of the countryside (agriculture) were transported to the city (in the form of food) and then dumped in the local waterways (in the form of excrement), only to create the foul conditions of 19th-century European cities, at the same time as it created an ecological crisis in agriculture – something we might consider in today's supposedly more sophisticated world, as we continue pumping agricultural runoff into our waterways the world over.

The use of mineral fertilizers was at least a temporary and partial solution to the problem of declining soil fertility that had troubled Europe since the 18th century. Furthermore, based on imported minerals, sulphuric acid (for the production of superphosphate), guano, and sulphate of ammonia (a by-product of the gas works), the fertilizer industry in Britain was already big business by the turn of the century. By the late 19th century, farmers noticed that soil fertility was declining rapidly in the exceptionally rich soils of the central plains of the US and in other nations where Europeans had settled. At the same time, global demand for food was increasing, and competition among producing regions around the world became more intense. As in Europe, the application of mineral and then synthetic fertilizers became a central practice of agriculture in these regions.

An important breakthrough for the industry was discovering how to synthesize ammonium directly, the most important technique being the Haber-Bosch process.[12] With the new process of ammonia production, the fertilizer industry expanded rapidly. An especially important element of this expansion came, not from the needs of agriculture, but from the need for nitrogen-based explosives in World War I. In fact, the world wars provided temporary solutions for companies like BASF and IG Farben as overproduction had begun to reduce their profits. The need for large quantities of explosives on all sides of the conflict was a godsend for the chemical industries of all countries involved. During the interwar years, improvements in technology again created a crisis of excess capacity which was largely solved by World War II. As might be expected, the end of World War II created another under-consumption crisis. This time, the wartime production levels were sustained by the dramatic expansion of industrial agriculture throughout Western Europe and North America in the post-war years, interrupted only intermittently until the present time.

The gradual emergence of the elemental approach to soil fertility, while enabling dramatic increases in production with little appreciation of the complex nature of nutrient cycling in ecosystems, had two negative consequences – disrupting the overall functioning of the soil nutrient cycling system, and pollution of groundwater, waterways, and water bodies in general, including the ubiquitous 'dead zones' responsible for killing millions of fish and other sea life throughout the world.[13] Furthermore, there is nothing in the physics or chemistry of soils to suggest, even approximately, that the direct application of low ionic forms is somehow better than more complicated forms, such as are found in compost or other organic fertilizers. Indeed, much analysis associated with the newer forms of more ecologically based agriculture begins with a challenge to this discourse, noting that from the plant's 'point of view' it matters little if its source of nitrogen is directly applied ammonium or ammonium derived from natural decomposition processes.

Plant breeding emerges to serve ideology

Parallel to the increasing use of fertilizers, farmers and the new breed of specialists known as agricultural scientists began to be much more aggressive in selecting and breeding crop plants for higher yields. Plant breeding was often tied in principle as well as in historical time with the use of mineral and synthetic fertilizers, because newly developed crop varieties often depended on and could make better use of the high levels of nutrients available through the new fertilizer technology. Plant varieties that could produce more wheat or corn per acre usually required more nitrogen, phosphorus, and potassium, and plant breeders began to breed for the ability of plants to use these nutrients more efficiently. In the first half of the 20th century, plant breeders were remarkably successful in creating varieties adapted to the availability of higher levels of these nutrients to produce more grain. Plant breeding and crop variety selection became intimately tied up with the story of soil fertility.

Plant breeders began to reach a limit, however, in their ability to convert inorganic fertilizers into greater yields. Specifically, while the new varieties could convert much more nitrogen into grain, there was a point at which increasing applications of nitrogen did not result in higher yield and could even become toxic to the crop. Working out how to raise the ceiling for fertilizer application to gain yet higher yields became a major challenge to plant breeders in the 1930s.

It was more than slightly strange that in the very years when agricultural scientists focused strongly on the need to raise yields of major grain crops, the world was in the grips of the Great Depression, an event caused at least in part by overproduction.[14] These were the years when the US government, and others around the world, began to pay farmers to retire land from production. They were the years when pigs and other livestock were bought up by the government and then were killed and buried in great trenches to keep them off the market. Farm families abandoned the land in great numbers even though there was mass unemployment in the cities – there was no point in remaining on the farm if there was no market for the crops. In spite of this glaring contradiction, agricultural scientists continued to work hard on developing crops with higher yield potential that could make use of high levels of fertilizer application.

The argument for continuing to invest money and time in higher-yielding crops continued through the 1940s and 1950s, but in the 1960s it was reshaped to attempt to deal with the contradiction between high yields and low demand. Instead of emphasizing the advantages to richer nations of higher-yielding crops, the newly developing discourse focused on the poorer countries. These nations and colonies or former colonies, what we call today the Global South, were largely in the tropical and semi-tropical regions of the world, and were under the direct or indirect domination of the richer nations of the north. Here, it was argued, widespread hunger and famine were due to the lack of agricultural productivity. In addition, as the earliest stages of sanitation and public health measures began to bring down death rates in poor countries, there would be a need for ever-increasing food supplies for what would soon become rapidly growing populations. Many of these arguments carried substantial elements of guilt-mongering about the past and emphasized the need for agricultural scientists of richer nations to carry 'the white man's burden' of resolving the problems of societies that had been profoundly disrupted by colonialism and economic intervention by rich nations. If agricultural scientists could banish hunger through greater productivity then surely all the sins of the past could be forgotten, if not forgiven.

The counter-argument was that hunger and famines were primarily due to injustice and not to the low productivity of agriculture in those regions. Land and wealth were concentrated in a few hands. The best agricultural land was controlled by local elites and/or foreigners who used it to produce food and fibre for the markets of rich nations rather than for local populations who had little purchasing power. The majority of rural people were not only excluded from the best land but were subjected to a combination of low wages, high land rent (often in the form of share-cropping arrangements), confiscatory taxes, lack of credit, and lack of access to technical information and markets. Because the real money to be made in agriculture came from distant and usually foreign markets, the low purchasing power and illiteracy of rural agricultural producers and workers were not a problem to land owners but rather constituted a way of ensuring continued low production costs.[15] Attempts to change these conditions met determined political opposition by those in control of resources and land. Often, keeping control of land and labour involved every manner of violence, from individual beatings and torture to mass slaughter.

The argument for social justice as the main approach to resolving hunger was not merely academic – it was being put forward by powerful nationalist, socialist, and communist movements throughout the Global South. These movements, some with dramatic success, were challenging local power structures and the international world system based on colonialism and neo-colonialism in nation after nation. In China, India, and the countries of Southeast Asia, Africa, and Latin America, the specialized European-style monocrop plantation was a major symbol of historic and continuing injustice, and the monopolization of the best land for production for distant markets was always a central grievance.

Control of pests

By the turn of the 20th century there were a variety of methods of pest control that had become well established, ranging from strategic rotations to resistant varieties of crops, release of predators and other natural enemies, and others. It is important to note that there was no particular emphasis on any one technique, the discourse of pest management being one of using all methods available, frequently in combination with one another. Farmers tended to view their farms as either healthy or not, almost as if the farm were the patient and the farmer the doctor, with pests and diseases as metaphorical germs that made the farm sick. Maintaining the farm (the patient) in a healthy state was the principal goal of the farmer. The emergence of petroleum-based biocides, and especially the socio-political and economic realities of World War II, would change all of that.

With the outbreak of war in 1939, governments generally became interested in pesticides as possible biological warfare agents, and a great deal of government-sponsored research into chemical poisons followed. The US was particularly active in biological and chemical warfare research, and as a result came up with a product that would see its application not in war but in the war against weeds, 2,4-D (although later, as a component of the infamous 'agent orange', it would see war service in Vietnam). In 1945, in the US, 917,000 pounds of 2,4-D were produced, and by 1950 the total had risen to 14 million pounds.[16]

But the real revolution was the synthesis of the chlorinated hydrocarbon, dichloro-diphenyl-trichloroethane (DDT) in the late 1800s, and the discovery of its insecticidal properties in 1939. It was quickly adopted by British and US armed forces. Tropical diseases, many of which were vectored by arthropods, were more important killers than the enemy in the practice of warfare. Finding effective solutions was a priority of war research, and DDT seemed to provide the solution.

By 1945 other chlorinated hydrocarbons had been developed, and the arsenal of insect-fighting weapons was well established as part and parcel of war preparedness. Germany had made the same preparations, emphasizing organophosphates (e.g. parathion) rather than chlorinated hydrocarbons. World War II had thus created a high capacity for the production of biocides, with the general class of carbamates joining the chlorinated hydrocarbons and organophosphates to make the three major classes of insecticides we know today. Combine that with the herbicides that were originally developed as a by-product of biological warfare research, along with further research into nitrate explosives that easily translated into more efficient methods of producing nitrogen-based fertilizers, and the charge could legitimately be made that World War II was the seed that germinated the agrochemical revolution.

The war was a watershed for the chemical industry and a tremendous productive capacity was the result. However, it was apparent that peace would present a problem in that the capacity that had grown so explosively during the war would suddenly become an excess capacity. The industry thus saw that it was due to face an under-consumption (or overproduction) crisis and began searching for a solution. Agriculture was the obvious target. The industry developed some ingenious marketing strategies in post-war US and Europe. With war fever having reached a pitch, the public was especially susceptible to wartime rhetoric. What had originally been an argument that we needed the chemicals to defeat the enemy in war, was easily translated into the need for these chemicals to defeat the new enemy in agriculture.[17] The importance of this advertising blitz cannot be overemphasized. With wartime and especially post-war propaganda, agricultural pests came to be seen as enemies to be vanquished. A 'war metaphor' replaced the previous 'health metaphor' (the idea of maintaining a healthy farm).

The weapons in this new war were the new chemical pesticides produced by the same corporations that produced them for the war effort. The wartime-induced productive capacity of the chemical industry was thus saved by changing the attitude towards pests, seemingly subtly, but in the end in such a way as to transform agriculture dramatically. The new metaphor meant that farmers changed from stewards who maintained the health of their farms, to warriors who vanquished their enemies on the battlefield of the farm. The consequences were the massive spraying of biocides in the years following World War II.[18]

The Green Revolution

In one nation, Mexico, these three tendencies – chemical fertilizers, improved varieties, and pesticides – all came together forcefully in the creation of the research programme that would come to be known as 'The Green Revolution'. In the three years before the US entered World War II, the US and its future allies became acutely concerned about a Mexican government that had become serious about promoting land reform. The reform threatened specific US and British investors and presented a then widely admired example to other peoples of Latin America and the world. In addition, the Mexican government in 1938 expropriated and nationalized the oil industry, one of the world's most productive, which had been owned by US and British corporations. Serious attempts to convince the US government to invade Mexico were rebuffed, but the US did become convinced that it needed to do what it could to blunt the force of the Mexican movement for social change. In the 1940 Mexican Presidential race, the US supported the conservative candidate who was running against Lázaro Cárdenas, of the Party of the Mexican Revolution. With the help of the US, Manuel Avila Camacho won the elections and undertook various commitments to help Mexico follow a path of economic growth consistent with US economic and strategic interests on the eve of World War II. One of these commitments was a programme of agricultural research.[19]

The agricultural research programme, with a heavy involvement of the Ford Foundation and the Rockefeller Foundation, explicitly promised to resolve hunger and poverty in Mexico, and the world, by increasing agricultural production. The early researchers focused on the problem that they had been trying to solve previously: how to develop grain crops that could raise the ceiling on nitrogen absorption. If varieties could be developed that would make use of greater quantities of nitrogen fertilizer, yields could increase dramatically. The story of how this was done is intriguing and has been often told. Indeed, US researcher Norman Borlaug, who won the 1970 Nobel Prize for his work, spent decades telling and retelling the story of his work in Mexico in the development of high-yielding wheat. The success in wheat would soon be more or less duplicated in maize, rice, and other crops.

The high-yield crops, however, brought with them a variety of problems. They were designed to make greater use of nitrogen fertilizers, which meant that they were also dependent on these fertilizers. Fertilizers would be toxic even to the new varieties if they were not combined with large amounts of reliably delivered water. The densely planted, nitrogen-rich, irrigated plants made an ideal field environment for insect pests and plant diseases. Pest and disease vulnerability were to be solved by heavy reliance on the newly available synthetic pesticides. In most situations, combining all of these elements meant the use of more and more elaborate and expensive machinery and, in many regions, it meant the building of vast new networks of dams and irrigation canals. So, what would become the 'package' offered by Green Revolution agricultural researchers and the companies who quickly commercialized their product was made up of newly developed high-yielding varieties (HYVs), synthetic fertilizers, synthetic pesticides, and, wherever rainfall was unreliable, irrigation. The nature and requirements of this technological package would transform agricultural landscapes around the world in the last half of the 20th century and create massive growth in industries manufacturing the seeds, chemicals, irrigation equipment, and machines designed to make the package economically viable and technically reliable. It also represented massive profits for what is now a well-established agroindustry.

What made this Revolution a 'green' one? After all, fields were green before this revolution in agricultural technology. The point was that the revolution was *not* 'red'. A publicist working with researchers realized that there needed to be an attractive way of summarizing the meaning of the whole package. It was about green plants and crops, but it was also very crucially about a particular approach to hunger. It posited that it was the lack of productive capability, and *not* social injustice that caused hunger, and it was increased productivity, and *not* social and political change that would end it. In nearly every poor nation of the Global South where it was adopted, it was posed quite specifically as a technological solution to the social economic and political problem of hunger. It was said to be better than a political solution that would be based on redistribution of land and other productive agricultural assets, mass education, and political enfranchisement of the poor. In most poor countries, the aggressive promotion of the Green Revolution began during years when the US State Department and local governments began to worry about the prospects for success of political movements demanding land reform and the redistribution of wealth.

In Europe and the US, what amounted to the Green Revolution package was seen and adopted largely as a continuation of existing trends in agricultural technology. In the US, and later in Europe, the package gave new force to the persistent problem of overproduction that plagued commercial farmers throughout the second half of the 20th century. Every farmer felt compelled to out-compete his and her neighbours in yields, while painfully aware that as each farmer did so, agricultural prices would continue to plummet. Each drop in price created a new incentive to try to produce more to earn sufficient income to survive. While some countries adopted actual controls on how much a farmer could market, supply control (or supply management) in the US focused on restricting the area of land a farmer could cultivate in major surplus crops. This gave rise to increased intensification on the land the farmer was allowed to use, increasing the amounts of fertilizer, pesticides, and water used, and intensifying environmental damage on cultivated land. As a result, surpluses remained an ever-greater problem in most years. The US government aggressively promoted food exports, including 'food aid'. Such food aid sometimes did genuinely feed the hungry, but its more important effect was in destroying local markets for many small-scale farmers, most of whom had not benefited from the Green Revolution package, usually because they could not afford to do so. Sometimes they resisted adoption because of the destructive effects the package had on local environments, cultures, and communities. As the majority of farm families in the world were excluded from participation in the Green Revolution because they did not have the credit to buy the necessary inputs or access to the high-quality land and water necessary to make it viable, they had to abandon the countryside by the tens of millions. They crowded into cities poorly able to absorb their presence and with insufficient employment opportunities.[20] In many countries, most notably India, they often had to rely on food handouts produced by the richer farmers in their own country or in the US or Europe, whose domination of the market had driven the poor families off the land. In other cases, they went hungry or became chronically malnourished. Around the world, the countryside sent a tidal wave of migrants to cities. It is no exaggeration to say that the Green Revolution was one of the principal forces for drastic social change in the 20th century, perhaps the most important of all. In this sense, it was indeed revolutionary.

As Green Revolution agriculture increasingly dominated agricultural landscapes all over the world, the older style plantations, with their roots in the colonial era, continued to monopolize much of the best land in the tropics and sub-tropics. Many plantation managers began to adopt chemical-dependent techniques similar to those developed for Green Revolution grain crops. Almost everywhere, they continued to depend on low-wage agricultural workers. In many places, such plantations went through cycles of relatively high productivity followed by soil exhaustion and the onset of extreme poverty. For example, the cotton and tobacco plantations of 'The Old South' in the US – Virginia, the Carolinas, and Georgia – had experienced both phases before the US Civil War, pushing slave plantation agriculture further west to Louisiana and Texas. Some of Brazil's richest sugar plantation areas experienced steady decline due to soil exhaustion in the 19th century from which the regions have not yet fully recovered. Coffee plantations in Brazil left a broad swath of forest destruction to be followed only decades later by soil exhaustion during the late 19th and the 20th centuries. Beginning in the early 20th century and continuing to the present, a combination of plantation agriculture and 'modern' chemical-dependent agriculture has left similar destruction across extraordinarily species-rich areas of humid rainforest and savannah in Brazil's Amazon region. As environmentally destructive chemical-dependent Green Revolution agriculture combines with the older cycles of plantation production, much of the tropical and sub-tropical world (the planet's great reservoir of species) has been impoverished in biological richness, and often in soil fertility and human welfare.[21]

Elsewhere, we have written about the enormous, and mainly negative, implications of taking the Green Revolution approach to resolving the problem of hunger.[22] Even so, many observers believe that intensified chemical-based agriculture has had a substantial positive effect with regard to biodiversity. They argue that intensification of production on the best land allowed regrowth of forests and other wild vegetation on more marginal land.[23] As we argued in Chapter 3, this 'land-sparing' argument is at best ill-informed, both from an ecological and socio-political point of view.

The industrial model

From peanuts to peanut butter

For years, political economists asked why it is that in agriculture we don't see the same process of capital penetration that we see in other sectors of the economy. In other words, why small farms have not been completely displaced and turned into industrialized corporate mega-farms. A partial answer to this question lies in the distinction between farming and agriculture. Farming is the process of turning seeds into harvestable crops with the use of labour, energy, and other resources, like land and water. Agriculture is not just farming, but also the production of agricultural inputs and the processing, packaging, transportation, and marketing of the outputs. As evolutionary biologist Richard Lewontin said, 'Farming is growing peanuts on the land. Agriculture is making peanut butter from petroleum.'[24] When examined in this manner, there is no doubt that capital has indeed penetrated the agricultural sector. However, until recently, farming, the most vulnerable, risky, and unpredictable component of agriculture, was left to the farmers.[25]

The post-World War II capitalization of agriculture was accomplished primarily through the substitution of inputs that were generated from within the farm itself, with inputs that were manufactured outside the farm and needed to be purchased. Starting with the early mechanization of agriculture that substituted traction power for animal power, to the substitution of synthetic fertilizer for compost and manure, to the substitution of pesticides for cultural and biological control, the history of agricultural technological development has been a process of capitalization that has resulted in the reduction of the value added within the farm itself.[26] In today's farms, the labour comes from Caterpillar or John Deere, the energy from Exxon/Mobile, the fertilizer from DuPont, and the pest management from Dow or Beyer. Seeds, literally the 'germ' that makes agriculture possible, have been patented and need to be bought mostly from a few large corporations. This stifles, and in some cases even ends, the millenary practice of seed saving and crop improvement by farmers.

The output side of agriculture is not very different. Once the 'product' leaves the farms, it is increasingly processed, packaged, transported, marketed, and sold by large corporations. As an example, Figure 4.1 shows the UK wheat bread bottleneck that illustrates how both farmers and consumers are dependent on a few traders, millers, bakers, and retailers. Similar bottleneck graphs can be drawn for many other agricultural products.[27]

The last 30 years have seen a process of consolidation and vertical and horizontal integration of agrobusiness that has resulted in only a handful of corporations supplying farm inputs and buying farm outputs. Just over the last few years, there have been three mega-mergers in the agribusiness sector: China-Chem (China National Chemical Corporation) merged with Syngenta, Dow Chemical with DuPont, and Bayer with Monsanto.[28] It remains the case that the direct producers of food, the farmers, are squeezed between powerful oligopolies (like Monsanto [which was acquired by Beyer in 2018], Exxon/Mobile, and Syngenta) that control the prices of farm inputs, and powerful oligopsonies (like Nestle, Kraft-Heinz, and Tyson Foods) that control the prices of farm outputs, and increasingly firms such as Cargill that operate extensively as both input suppliers and monopsonist buyers. When analysed in this fashion, it becomes evident that the billions of dollars in commodity subsidies in the US actually represent an indirect subsidy to corporate agrobusiness.

This general structure can be seen in almost any part of the world where agriculture is heavily capitalized. In recent years retailers, such as Walmart,

The UK bread wheat 'bottleneck' (flour milling and baking)

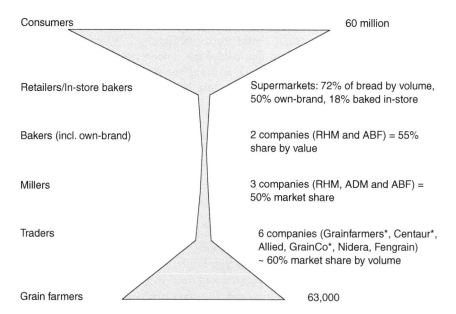

Figure 4.1 The bottleneck for wheat bread in the UK.

Source: Bill Vorley and the United Kingdom Food Group.

have also consolidated as mega-stores, and are playing a crucial role in the commodity chain. For example, in Latin America, by 1993 large supermarket chains already controlled 10 to 20 per cent of food sales, and by 2003 only five corporations were in control of 65 per cent of the sales.[29] It is partially because of this structure that the farmers receive only about 10 per cent of what the consumers pay for the food they buy in the supermarkets.

The industrial system in crisis

The view that industrial agriculture has spawned significant problems is hardly contestable, and has become general knowledge.[30] The modernization of agriculture included two very general goals: more frequent use of the same area of land (e.g. decrease the time of fallow periods) and increased specialization of productive species (loss of plant biodiversity, usually in the pursuit of higher yields and ease of mechanization). The extended fallow periods of many forms of traditional agriculture were abandoned in order to use the same area of land every year or for continuous production. The increased frequency of resource withdrawal was subsidized with increased use of external inputs.[31] The loss of fallow periods and use of agrochemicals, including mineral fertilizers, can lead to loss of soil fertility, in part through disruption of the soil microbial systems, and decreases in macrofaunal and plant-associated biodiversity.[32] The ecosystem functions often provided by such associated biodiversity (e.g. nutrient retention, biological pest control, pollination) are necessarily lost as well.[33] Biodiversity in the ecosystem is decreased further when traditional, diverse, locally adapted crop varieties with resistance to native diseases and pests are abandoned and intercropping is halted in favour of high-yield monocultures amenable to mechanized production methods. Furthermore, higher-yielding crop varieties are often less resistant to pathogens and pests, especially given the highly dense and structurally simple concentration of plants, and are more prone to the spread of insect pests and diseases. In the transition from diverse crops and intercropping to monoculture many natural enemies of the pests are lost, increasing the need for pesticide applications.[34]

Less diversity, more pests; more pesticides, less diversity

The first indication of environmental problems due to this new technology was the publication of Rachel Carson's *Silent Spring* in 1962.[35] Carson suggested that the massive use of pesticides was having a dramatic negative effect on the environment. Previously there had been much popular commentary about the human health effects of pesticides, a concern initially brought up by farm workers in the US and shared even by the pesticide manufacturers. But *Silent Spring* was the first popular account of the environmental consequences of pesticides, contributing not only to concern about environmental poisons, but perhaps providing the main springboard for the emerging environmental movement.

What Carson said is now well known, and, in retrospect, not surprising. Pesticides indeed do what they were designed to do: they kill. However, broad-spectrum pesticides kill not only the targeted pests, but also many species that are not targeted. The poisonous effects of pesticides and their residues may persist for a long time in the environment. Due to their persistence, pesticides also concentrate in the higher trophic levels, thus making non-lethal doses at lower trophic levels quite dangerous at higher levels. Pesticides stop working when pests evolve resistance to them. These were the basic themes of Rachel Carson's book.[36]

Silent Spring was an extremely well-documented book. Despite a Herculean effort at finding errors, mainly by representatives of pesticide manufacturers, only the most trivial errors were eventually found. Nevertheless, an immense and coordinated attack against the book and the author was orchestrated by the pesticide industry, including an attempt by Velsicol, Monsanto, and indeed the entire chemical industry, to pressure Houghton Mifflin, the publishing company, not to publish the book in the first place.[37] Book reviews were generally harsh and, as was later discovered, frequently written by scientists receiving monetary rewards from the chemical industry. As was carefully documented in subsequent works, most independent scientists received *Silent Spring* positively. Indeed, rereading the book today even suggests that its actual problem was understatement.

Silent Spring raised a great deal of awareness about environmental pollution and propelled the environmental movement in the 1960s. However, the chemical industry has been counterattacking ever since. Rachel Carson's predictions have come true in many areas of the world, and further problems have emerged. Industrial accidents in pesticide manufacturing plants, plus worker exposure to toxics in such plants, have added to her concerns. The environment is now filled with chemicals that mimic oestrogen, causing not only feminization in reptiles, but lowered sperm counts in humans, as documented in the remarkable book *Our Stolen Future* by Theo Colborn, Diane Dumanoski, and John Peter Meyers.[38] Unfortunately, the final chapter on pesticides has yet to be written. What began with the need to curb an underconsumption crisis has been transformed into an environmental crisis of unprecedented proportions.

While Carson's book was a watershed, a less acknowledged, but equally important book was *The Pesticide Conspiracy* by Robert Van den Bosch, published in 1978.[39] This book accomplished three things simultaneously. First, it reinforced all of the analysis of Rachel Carson, sometimes updating information and pinpointing problems caused by pesticides that Carson had not foreseen. In this, Van den Bosch was far more critical of pesticides than Carson had been, partly because 16 more years of research had made more and better data available. Second, it made explicit what Carson had only left for the reader to conclude, that pesticides were being used largely for the purpose of making profits for those who manufacture them, and that the 'conspiracy' involved government and university scientists receiving direct and indirect financial rewards from this industry. Third, it elaborated a generalized mechanism that explained the biological reasons that pesticides came to be such a problem, strongly implying that the problem would continue, and could even become worse. Van den Bosch called this mechanism the 'pesticide treadmill', and it has since become part of the general discourse of the environmental movement.

The 'pesticide treadmill' includes three forces operating simultaneously:

1 Pest resurgence – the fact that a pest will usually come back in force after the pesticide kills not only it but also its natural enemies.
2 Pesticide resistance – the unavoidable fact that pests evolve resistance to whatever poison is thrown at them.
3 Secondary pest outbreak – the surprising tendency of the pesticide to kill the natural controls of herbivorous insects that were not previously pests (because of those natural controls that had kept them from becoming pests earlier), causing these insects to reach outbreak levels. Each of these forces had been well known even before *Silent Spring*, but they were first put together as a coherent whole by Van den Bosch.

Putting these three forces together – resurgence, resistance, and secondary pest outbreaks – we have a pattern that is devastating in the long run for farmers. A resurgent pest suggests to the farmer that more pesticide is needed since the problem is worse. Because of resistance, the pesticide is no longer as potent as it once was, suggesting to the farmer that more pesticide or stronger pesticide is needed, and the new pests that mysteriously appear suggest once again that more pesticide is needed. It seems much like a drug addiction story. The more pesticide you use, the more you need – hence the name, the 'pesticide treadmill'.

World consciousness of the environmental problems of industrial agriculture grew rapidly after the publication of *Silent Spring*. By acknowledging Van den Bosch's 'pesticide treadmill' and taking some action to curb the most blatant negative effects of pesticides, the horrible environmental consequences envisioned by Carson have been partly averted. Nevertheless, much of her concerns did come to pass. Agricultural areas, such as the San Joaquin Valley in California, are largely devoid of diversity. A drive through the vegetable producing areas, for example, is ominously similar to the opening sentences of *Silent Spring*, with the absence of birdsong, the absence of small flying creatures, the feeling of desolation (Figure 4.2). The impacts are not manifested only in the most intensive agricultural landscape. For example, scientists from the National Centre for Scientific Research in France reported bird numbers declining by a third over the last 15 years, and in the UK it has been reported that bird declines in agricultural lands have been as high as 56 per cent.[40] German scientists documented an astonishing 75 per cent decline in insects in natural areas surrounded by agriculture.[41] But the casual motorist in the San

(a)

(b)

Figure 4.2 (a) Ansel Adams photograph, 1943, illustrating the Manzanar concentration camp where Japanese Americans were imprisoned during World War II; although Adams took this picture in the mid-1940s, the landscape of most of California's Central Valley looks very similar today. (b) A recent photo of a broccoli farm in the Salinas Valley of California.

Source: (a) Ansel Adams (b) John Vandermeer.

Joaquin Valley or the European farming landscapes, does not really see the whole picture. Between 0.5 per cent and 15 per cent of the pesticides that are applied to the crops end up in aquatic systems. Persistent pesticides, like DDT, linger in the sediments of rivers, lakes, and oceans decades after they have been banned. Pesticide residues are now known to occur in most food we eat.[42] While the levels of their occurrence are small and usually within government safety standards, their potential synergistic effects are not known. While particular incidents may stand out in public memory because of sensational press coverage, far more ominous in the long term is the continual small dosage received each day by the majority of the population. Carcinogenic potential has long been acknowledged, and more recently the potential problems associated with hormone-mimicking pesticides have come to the public's attention.[43]

From fertilizer application to 'dead zones'

There is another problem with industrial agriculture. It probably started in the early 19th century when Justus von Liebig first discovered that you could put inorganic chemicals on the ground to increase plant growth, as described earlier. The problem of soil fertility was thus, in principle, solved. But soil management proved to be far more difficult than had been expected. Once farmers started applying synthetic fertilizers they reduced or abandoned many of the old practices that ensured an ample supply of organic matter in the soil, such as cover crops, rotations with legumes, and others. With time, the more synthetic fertilizer was applied, the less organic matter was left in the soils and the more synthetic fertilizers were needed. While not as well understood as the mechanisms of the 'pesticide treadmill', it is generally understood that provisioning the soil with large amounts of inorganic ions disrupts the normal nutrient cycling processes in the soil, thus making it even more necessary to apply fertilizers. While it does not seem to have been referred to this way before, it seems justifiable to refer to it as the 'fertilizer treadmill'.[44]

As noted earlier, the process of nutrient cycling is complex. For the farmers this complexity is compounded by the variable needs of plants at different times. As a crop grows, its nutrient requirements change. So, for example, when a bean crop is at the point in its life cycle where it is filling its seeds, a large amount of nitrogen is needed. If that nitrogen is not available at that time, the seed-filling process will be slower, and may come to a halt if the local amount of nitrogen available is not sufficient. This is why frequently, a side dressing (an extra amount) of nitrogen is added to a field at the time of seed filling. However, applying exactly the right amount of nitrogen to perfectly satisfy the needs of the crop, could very well be a week by week, or even a day by day, process if the supply is with nitrate or ammonium (two ions that are directly taken up by the crop's roots). Thus, it makes sense simply to apply an overabundance of either of these ions, such that there will always be as much as the crop needs, no matter what its growth phase might be.

Indeed, this was the general philosophical position of the soil fertility management establishment. Supply as much of the nutrient as is necessary, but be sure that the minimum amount is always there, which automatically leads to the application of more than is necessary. The problem is that nutrients that are not absorbed by the plants do not stay in the soil forever, they go somewhere else. Depending on the structure of the soil, the nutrient ions may be leached out of the system extremely rapidly, especially nitrate. The unabsorbed nutrient ions either go deeper into the soil or into the groundwater, and out of the reach of plant roots, or into the waterways as part of the agricultural runoff.[45]

Fertilizer runoff from farm fields in the Mississippi River basin has created a 22,730km² (8,776 square mile) 'dead zone' area of oxygen depletion (hypoxic zone) in the Gulf of Mexico.[46] These so-called 'dead zones' are now popping up at the mouths of many of the world's major rivers, at least those that catch the drainage from farm fields engaged in the industrial model of agriculture. So far, 'dead zones' have been identified for more than 400 aquatic systems covering an area of more than 245,000km² (94,595 square miles) and posing a significant threat to aquatic biodiversity (Figure 4.3).[47]

In conclusion, an unsustainable system

At a most general level, the extensive use of biocides and inorganic fertilizer in the industrial model may in the end prove to be self-defeating, since such use may be undermining the resource base on which agriculture is founded, especially on many tropical soils. The natural enemies that naturally control the populations of herbivorous insects are destroyed by pesticides, and the herbivorous insects evolve resistance. The bacterial and fungal communities of the soil are thrown out of whack by the massive application of nitrates and phosphates, interfering with the natural process of nutrient cycling. And the destruction of other ecosystems and the potential livelihoods they support,

Figure 4.3 Global distribution of more than 400 'dead zones'.

Source: Robert J. Diaz and Rutger Rosenberg (figure from Diaz and Rosenberg, 2008).

from fresh water fisheries to the previously bountiful ocean resources, is monumental. Increasingly, this industrial agricultural system appears similar to an out-of-control adolescent for whom a concern for the future is at best a minor irritant. An additional ominous trend, and directly relevant to this book, is the loss of biodiversity from the expansion of the industrial agricultural system. Industrial agriculture, with its emphasis on large monocultures as opposed to the usually highly diverse traditional systems, on large-scale mechanization and irrigation, and on chemical pesticides and fertilizers, impacts both on planned and associated biodiversity. The consequences of this trend are yet to be fully appreciated, but analysts the world over are in agreement that the loss of biodiversity in general[48] and specifically in agroecosystems,[49] is somewhere between severe and catastrophic.

Notes

1 Boserup (2005).
2 Tinker *et al.* (1996).
3 Chacon and Gliessman (1982).
4 Padoch and Pinedo-Vasquez (2010).
5 Martínez-Torres and Rosset (2014); Reyes-García *et al.* (2014); Altieri and Nicholls (2017).
6 The IAASTD (2009b) was an intergovernmental initiative sponsored by the World Bank, various UN agencies, and various participating countries. It had more than 400 expert authors and was approved in Plenary by delegations from 54 countries.
7 FAO (2018).
8 Polanyi (2001).
9 The interest in bio-fuel has led to a rapid increase in oil palm plantations throughout much of the tropics (Danielsen *et al.*, 2009; Gutiérrez-Vélez *et al.*, 2011; Koh *et al.*, 2011). However, tropical fruits, such as pineapples and bananas, continue to devastate many areas in the tropics (Fagan *et al.*, 2013; Tapia-Armijos *et al.*, 2015).
10 The Law of the Minimum was a principle first developed by German botanist Carl Sprenger in 1828, but further developed and popularized by Justus von Liebig (Brock, 2002).
11 Goodman *et al.* (1987).
12 Smil (2004).
13 Diaz and Rosenberg (2008); Karstensen *et al.* (2015).
14 Himmelberg (2001).
15 De Janvry (1981).
16 Peet (1969).
17 Russell (2001).
18 World Health Organization (1979).
19 Wright (2005).
20 Davis (2006).
21 IAASTD (2009b).
22 Wright (2005).
23 Phalan *et al.* (2011).
24 Lewontin (1982).
25 Recent technological and institutional changes in some sectors have facilitated the takeover of 'farming' by large corporations. This is most evident in factory farming where the 'farmer' contracts with a large corporation that stipulates all conditions

of production (Lewontin, 1998). More recently, this corporate takeover is reflected in the large numbers of 'land grabbings' in the Global South (Rulli *et al.*, 2013; White *et al.*, 2013).

26 Lewontin and Berlan (1986).
27 Vorley (2003).
28 *USA Today* (2017).
29 Reardon *et al.* (2003).
30 National Research Council (1989); Pimentel *et al.* (1995); Pimentel (1996); Matson *et al.* (1997); Tilman (1999); Heller and Keoleian (2003).
31 Buttel (1990).
32 Curry (1994); Gunapala and Scow (1998); Neher (1999); Finegan and Nasi (2004); Zhong *et al.* (2010); Geisseler and Scow (2010); Yuan and Chen (2015).
33 Matson *et al.* (1997); Tscharntke *et al.* (2005); Flynn *et al.* (2009); Culman *et al.* (2010); Power (2010); Laliberte *et al.* (2010); Allan *et al.* (2015).
34 Letourneau *et al.* (2009); Chaplin-Kramer *et al.* (2011); Zhao *et al.* (2015).
35 Carson (1962). We recommend Steingraber (2018).
36 Carson's book focused on persistent pesticides (those that last in the environment for decades). Different problems emerged with non-persistent pesticides because they tend to be more toxic (Perfecto, 1992).
37 'Silent Spring is now noisy summer', *New York Times*, 22 July 1962, www.mind fully.org/Pesticide/Rachel-Carson-Silent-Spring.htm, accessed 29 December 2008.
38 Colborn *et al.* (1997).
39 Van den Bosch (1978).
40 'Shocking decline in birds across Europe die to pesticide use, say scientists', *Independent*, 21 March 2018; www.independent.co.uk/environment/europe-bird-population-countryside-reduced-pesticides-france-wildlife-cnrs-a8267246.html, accessed 13 March 2019,
41 Hallmann *et al.* (2017).
42 Pimentel and Lehman (1993).
43 Bretveld *et al.* (2006).
44 Drinkwater and Snapp (2007).
45 This problem is substantially different in organic production, since the source of all nutrients is from decomposition, effectively a slow-release nutrient-supplying system. However, the technical problem is similar in that the challenge is to provide exactly the right amount of nutrients needed at any point in time, and to avoid excessive nutrients from leaking out of the system.
46 National Oceanic and Atmospheric Administration (2017).
47 Diaz and Rosenberg (2008).
48 Kolbert (2014); Wilson (2016).
49 Black and Okwakol (1997); Giller *et al.* (1997); Matson *et al.* (1997); Donald (2001); Kremen *et al.* (2002); Benton *et al.* (2003); Tscharntke *et al.* (2005, 2012); Perfecto and Vandermeer (2008a); Flynn *et al.* (2009); Allan et al. (2015); Hallman *et al.* (2017).

5   Alternatives to the industrial model

Alternative agriculture movements

Facing the evident problems generated by the industrial agricultural system, especially with its penetration into the Global South as the Green Revolution, not surprisingly, a vigorous alternative movement has emerged. We have written extensively about this movement elsewhere,[1] and here, provide a summary as it relates to the specific topic of this text. First, we note that there is a long history associated with alternative agriculture movements, beginning with practitioners who foresaw the future problems. Second, we summarize the more recent search for models that these movements seek to emulate, either wholesale or in part.

A long tradition

While agriculture was rapidly being 'modernized' along the industrial pathway described in Chapter 4, millions of small-scale farmers all over the world were practising traditional forms of agriculture. These traditional practices serve as suggestive models for an alternative agriculture movement that started in the early part of the 20th century. Indeed, it is not all that surprising that one of the first and most important moves toward the development of an alternative to industrial agriculture occurred in one of Britain's main colonies, India. Sir Albert Howard, along with his botanist wife, Gabrielle Matthaei, travelled to India under the assumption that they would bring all the modern scientific techniques to the Indian farmers. Howard became Director of the Institute of Plant Industry in India early in the 20th century, and was the first to describe the organic system as something distinct from the pathway that world agriculture seemed to be taking.[2] Howard noted, we think perceptively, that von Liebig's approach to the soil was strongly contrasted to that of Darwin, as reflected in the latter's treatise on earthworms.[3] Whereas Liebig insisted on a reductionist approach, reducing the soil to nothing more than its chemical constituents, Darwin correctly saw the soil as a complex biological system. Perhaps Darwin was really the first advocate of an ecological approach to the study of soil.

The Howard/Matthaei team carefully observed local customs. In particular they observed the techniques applied by traditional Indian farmers in all aspects of their production. Howard was especially interested in the composting process commonly practised on traditional farms. He translated those practices into a system of composting (called Indore composting), which effectively remains the underlying idea behind modern composting.[4] He regarded Indian traditional farmers as his main instructors in agriculture, noting, for example: 'by 1910, I had learned how to grow healthy crops, practically free from disease, without the slightest help from [...] all the [...] expensive paraphernalia of the modern experiment station.'[5] He also noted that traditional Chinese farmers were perhaps even more efficient than their Indian counterparts. His approach, which we can presume came from extensive observations of, and conversations with, Indian farmers, was focused on the health of the soil, arguing that a 'healthy' soil, by which he meant one that contained a well-balanced mixture of worms, fungi, bacteria, and other micro-organisms, would produce healthy food; while a soil devoid of those healthy elements would not. Indeed, the connection between ecological health on the farm and the health-promoting qualities of the food produced there was a key element of the early organic movement, and, as we shall see, is emerging as a key element in most alternative agriculture movements.

A variety of other tendencies began to emerge during the early 20th century in Europe, one of the most important of which was the work of Richard St. Barber Baker. Serving the British Empire mainly in Africa, St. Barber Baker, partly due to his upbringing as an evangelist's son, felt a religious tie with forested lands. After obtaining his forestry degree from Cambridge in 1919, he was assigned to Kenya, where he found, unlike the 'healthy' farming practices of India observed by Howard, a scarred landscape, devastated by centuries of wheat farming introduced by the Romans and later by the introduction of goats by the Arabs.[6] He set out to reforest large sections of the country, a task for which he formed the 'Men of the Trees' movement (which remains active today as the International Tree Foundation). Much of interest is to be found in St. Barber Baker's writings about forests, but for the purposes of this text, we note that he saw in the forests lessons that would become extremely important for the emerging organic agriculture movement: the natural cycling of nutrients occurred in natural forests, but not in many of the agricultural systems that displaced them. This is virtually identical to the ideology encompassed in the more modern 'natural systems' agriculture, as discussed below. In 1938, the first Men of the Trees summer school was held, and one of the featured speakers was Sir Albert Howard.

For real institutional representation we must wait until the end of World War II, when the modern industrial system really took off. In the 1940s, Lady Eve Balfour decided to convert her farm in England to a completely organic operation and began a series of long-term observations on the consequences. These became quite famous in organic agriculture circles and are known collectively as The Haughley Experiment, described in her book *The Living Soil*.[7] At about the same time, Swiss biologist Hans Müller, and German doctor Hans Peter Rusch, founded the Organical-Biological Association in Switzerland in 1949. They soon expanded to Germany and Austria and were influential in promoting the idea that the health of the soil was something that needed to be maintained to ensure proper nutrient cycling in the agroecosystem. The Organical-Biological Association eventually gave rise to what have become the largest organic producer associations in the world: Bioland in Germany, VSBLO in Switzerland, and Ernteverband in Austria.

In 1921, Jerome Irving Cohen, son of a Jewish immigrant family settled in New York, changed his name to Jerome Rodale. By the late 1930s he came under the influence of the organic farming movement, and in 1942 began publishing 'Organic Farming and Gardening'. Sir Albert Howard became an editor shortly after the magazine's first appearance. Even today, the magazine remains an important source of information on specific organic techniques. The Rodale Institute carries out cutting edge research on organic production techniques, and remains an important source of research results for organic agriculture.

Throughout this short historical narrative one can see a continual critique, even if it is muted, of the Liebig approach. The soil cannot be seen as a passive sponge through which chemicals pass to plant roots. Modern soil science has come to accept this view on how the soil actually works, even if its origins in Liebigism compel it to continue its support of the chemical-bias in the practice of agriculture. However, this is slowly changing, and the importance of what have been dubbed 'organic' methods is gradually making its way into the mainstream of soil science. The biological side of soils is now recognized as not only a legitimate component of soil science, but as essential. Indeed, in the UK, one of the premier promoters and certifiers of organic produce is the Soil Association.

Throughout the early development of an alternative movement, the problem of pests was dealt with in a consistent fashion. It was a common assumption that a 'healthy' farm would resist pests in the first place – the so-called health metaphor, as described earlier. When pests arose, the proper way of dealing with them was first to eliminate them if possible, but, more importantly, to change production methods so the pest would not reappear the following year. Planned rotations in the developing alternative systems were as much associated with pest control as with soil fertility. As detailed in Chapter 4, the evolving industrial system changed all of that when it began using a war metaphor favouring the application of biocides, but the alternative movement remained on course, thinking of pests as a (preventive) health problem – something that should ordinarily be handled by the complex ecological interactions that have always maintained ecosystems in a quasi-balance. Thus, adopting modes of biological and cultural management of potential pests was part and parcel of the alternative movement.

On the other hand, there is a certain institutional inertia that has more to do with politico-economic assumptions than with either chemistry or biology. The various soil and pest management schemes, normally used by either indigenous or organic farmers, usually consist of a set of rules, sometimes cultish,[8] yet sometimes surprisingly rational once properly understood (e.g. planting with the moon phase so as to coordinate crop phenology over a large area, thus satiating herbivorous pests). Since the dominant social form is the commodity production form of modern capitalism, traditional practices do not fit into the particular set of rules used in that system. Purchasing nitrate to satisfy the nitrogen demands of the crops means a product can be purchased by an entrepreneur and sold to a consumer, thus satisfying the underlying necessity of the system. If nitrate is supplied through biological nitrogen fixation, the entrepreneur is simply not needed. Converting the pest management system to one based on an integration of biological, physical and chemical forces operating in an autonomous fashion is likewise not expected to be popular under capitalism. A brute-force chemical approach is, almost always, more profitable according to capitalist accounting assumptions.

Finally, it should be noted that the history of the emerging 'formal' organic agriculture movement described above is Euro-Anglocentric mainly because the formal challenge to industrial agriculture emerged from those places where industrial agriculture was prevalent: Europe and the US. However, many of the ideas that were formalized in books and academic journal articles came from a myriad of traditional systems practised by indigenous and small-scale farmers all over the world. Indeed, as mentioned earlier, the origins of the movement can be traced back to Indian farmers practising traditional agriculture. In a sense, these ideas were the germ of what eventually became the science of agroecology, which combined general scientific knowledge about the ecology of agroecosystems with specific local and indigenous knowledge about particular agricultural systems.[9]

A wealth of models

In searching for a general plan for the transformation of agriculture from the industrial system to an ecologically sound and sustainable one, we most naturally look for models, both for how the new agriculture will look and for how to transform the present system. Traditional or indigenous systems present us with one model, since by definition they were not subject to the same forces that produced the modern industrial system. Extant organic and agroecological farms, isolated amidst the ocean of industrial agriculture, provide another. And a third model emerges from the idea that the design of agroecosystems should be based on the ecological dynamics of local natural ecosystems: 'natural systems' agriculture. All three of these models are discussed below. But before proceeding to that discussion, it is imperative to note that in all three of these models we are presented with systems in which some of the key social variables have been cancelled out of the equation. Traditional systems, by their very nature, do not exist in the context of industrial agriculture, but rather in pockets where the industrial capitalist system has not yet fully penetrated. Extant organic or ecological farms are mainly run by people whose attitudes have already been transformed, and who are largely unconcerned with promoting a global transformation, but are more interested in the survival and prosperity of their individual farms. And 'natural systems' agriculture is based on the local natural habitats before they were modified by the activities of humans. While it is clear that much is to be learned from these three models, we expect that most of this knowledge will be of a technical sort, leaving us still to ponder the vexing social and political questions about the transformation, a topic we cover extensively in later chapters.

Traditional agroecosystems

Many authors have promoted the idea that traditional or indigenous forms of agriculture, since they have not undergone the alterations of the industrial system, offer models wherein particular practices can be modified to fit into the transformation. Most claim that traditional systems are likely to contain rules of operation that have survived the test of time and may represent signposts along the road to an ecological agriculture, or agroecology. The general attitude is summed up by Miguel Altieri, an agroecologist at the University of California, Berkeley, US:

> ecological principles extractable from the study of traditional agroecosystems can be used to design new, improved, sustainable agroecosystems [...] traditional farming systems have emerged over centuries of cultural and biological evolution and represent accumulated experiences of interaction with the environment by farmers without access to external inputs, capital, or scientific knowledge. Such experience has guided farmers in many areas to develop sustainable agro-ecosystems....[10]

While this is certainly a good working assumption when studying traditional agroecosystems, it needs to be tempered with some scepticism. Although many traditional systems, when kept from external influences that alter them, have persisted for hundreds and perhaps thousands of years being productive and with little negative environmental impacts, others have proven to be less sustainable. Examples exist from the salinization of croplands by the Hohokam in southwestern North America[11] and Mesopotamia,[12] to the massive soil erosion that destroyed Greek ecology during the classical Greek civilization.[13]

Traditional systems share a surprising number of tendencies at a very general level.[14] They tend to have a very highly planned and associated biodiversity, and to employ that biodiversity to utilize all available habitats in the agroecosystem. Nutrient cycling tends to be closed, with kitchen wastes and animal manures as a key part of the cycle. The management of pests tends to be incorporated into the overall management of the system. Inputs tend to be mainly derived from recycling on the farm, with very few inputs from outside the farming system.

Whether such tendencies can or should be applied to the development of modern systems will obviously be conditional and site-specific. Some are relatively obvious or are well sustained by scientific evidence, and are already part of the principles of sustainable agriculture (e.g. the entire philosophy of 'Low Input Sustainable Agriculture' is to recycle so as to reduce external inputs to as low a level as possible). Others, especially those that require more labour, evoke ridicule when suggested to farmers in the developed world (e.g. try suggesting to a European farmer that he or she should use animal traction). But one thing is clear: specific technologies used by specific cultures represent, in a sense, the raw material of ecological development. It is not that we should try to convert all the mountains of southern Mexico to intercrops of beans and maize (the system frequently employed by traditional farmers there), but rather that we should view this method of maize production (intercropping) as a principle that could be adapted for use elsewhere.

We suspect that the general principles outlined above have already been indelibly etched in the minds of all those who seek to transform agriculture. The real value of traditional systems as models is in the specific and particular techniques they use. But, since traditional farming is not static, and farmers are always experimenting and incorporating new techniques into their 'traditional' farming systems, the agroecologist, in collaboration with these traditional farmers, is constantly studying, experimenting, and incorporating these new techniques. The value of traditional knowledge and wisdom has recently emerged as an important component of the critique of the conventional, industrial agricultural system. Indeed, the International Assessment of Agricultural Knowledge, Science and Technology for Development (IAASTD) highlights traditional and local knowledge as a key component of the process of transformation of the current agricultural system.[15] Their report notes: 'Traditional and local knowledge constitutes an extensive realm of accumulated practical knowledge and knowledge-generating capacity that is needed if sustainability and development goals are to be reached'. Farmer–researcher participatory approaches have been found to augment and sometimes even replace conventional technology-generating approaches.[16] In this regard, examples abound where traditional and indigenous farmers and researchers work side by side to develop and implement sustainable agricultural technologies (Figure 5.1).

The experience of Cuba in its transition to a more ecologically sound form of agriculture is also instructive.[17] In the early 1990s, with the collapse of the Soviet Union, Cuba lost its major trading partner, and with it, its main supplier of cheap fertilizers, pesticides, petroleum, and animal feed. This proved to be a disaster for Cuban agriculture, which was among the most intensive of Latin America. Within a year, yields of all crops collapsed and the Cuban population went hungry for the first time since the triumph of the Cuban Revolution 30 years earlier. It became evident to most Cubans that their intensive agriculture, previously a source of pride, had made them vulnerable to external forces beyond their control. For the agricultural research system in

Figure 5.1 Community work of farmers earthing up the farms in Yunguno, Puno, Peru.

Source: Asociación Paqalqu.

Cuba it became imperative to search for alternatives. A major impediment to the development of an alternative model was the lack of locally tuned knowledge about particular ecological circumstances, a consequence of 30 years of almost religious commitment to the industrial model. At this point agricultural researchers sought the help of the few traditional farmers that were left in the country and began a process of learning and integration of traditional and scientific knowledge. To a great extent, it was the intimate relationship between local or traditional knowledge and systematic scientific knowledge that laid the foundations of the new alternative agricultural system in Cuba.

As ecologist Richard Levins has noted, academic knowledge is general but shallow, while local knowledge is specific and deep.[18] Local knowledge is frequently 'flawed' in terms of modern ecological understanding, yet it is set in a world view that provides it with context. The fact that a local farmer in Nicaragua 'knows' that grasses 'burn' the corn, is not ultimately different from the scientific ecologist's 'knowledge' that the corn 'dries out' because of 'competition' from grasses. Whether we say 'burn' or 'competition' really only identifies the discourse under which the actual observation of the corn's performance in association with grass is communicated from one person to another. The fact that the farmer knows the details of what happens to the corn when grasses grow around it, is key to developing local alternatives that work. The fact that the ecologist uses the word 'competition' to describe the phenomenon brings the rest of the experience of the world, to the extent that it is catalogued, as possible knowledge for the planning process. The farmer knows what happens in his or her field. The ecologist can generalize that knowledge, compare it with what happens in other fields, effectively expanding the potential knowledge base.

The deep and local knowledge of the farmer, especially the traditional farmer, is essential to the ultimate development of the alternative model. The ecologist may help generalize and contextualize that knowledge to actually make it richer, but the local knowledge is imperative due to the unpredictability and locality-specific nature of many ecological processes.[19]

However, much traditional knowledge has been lost during the past 50 years due to the hegemony of the industrial model.[20] We are faced with the problem that today's world is filled with urban workers, while what we need are rural ecologists. While this problem is not as universally recognized as the lack of appropriate technology, it may turn out to be a more important one.

We are thus faced with a major contradiction in the move to ecologically based agriculture. On the one hand, we do not have a complete catalogue of techniques that are tried and proven to work under all circumstances – the technical side of the contradiction. On the other hand, the destruction of rural society has taken with it the knowledge base and labour force that will be needed for the transformation – the social side of the contradiction. We confront the task of building a new system based on incomplete technical knowledge and with a society designed to function only under the current system. The truth is that we do not know, technically, how to control, for example, the whitefly in vegetable production in Central America. Yet, even if we did have a technical solution, it is not clear that the local farmers or farm workers would be available and/or willing to undertake that procedure. As we strive to transform the current system, this is a principal contradiction that must be resolved. Local and traditional knowledge and wisdom may aid in resolving it.

Practising organic and agroecological farms

Extant organic and agroecological farms probably offer the best models for the way individual farms will look after the transformation. It is important here to make a clear distinction between organic and agroecological farms. Although there are many small-scale diverse organic farms, the majority of the organic production for commercial purposes, especially in the industrialized countries, comes from large-scale monocultural organic production that replaces toxic and synthetic inputs with organic and biological ones, mostly purchased off farm. Although these organic farms may contribute to a reduction in the use of toxic substances in agriculture, they do it through an input substitution model and do not challenge the industrial approach.[21] Agroecological farms, on the other hand minimized off-farm inputs by harnessing natural processes through functional diversification, and by doing so, they challenge the industrial model.

It is worth noting that there has been an explosive growth of organic agriculture in many parts of the world. Although there are acknowledged problems with further expansion of this sector, it is worth noting that earlier warnings that the growth would soon halt, were clearly premature.[22] Indeed, the area devoted to certified organic agriculture continues to grow in all parts of the world. In 2015, when the last survey on certified organic agriculture was conducted, there were almost 60 million hectares in organic production (including conversion areas) worldwide (Figure 5.2).[23] This is 22.6 million hectares more than ten years earlier.

Although large-scale industrial organic agriculture has grown, so have small-scale agroecological farms.[24] To some extent, agroecology is largely based on what not to do, and arises as a counterpoint to industrial agriculture. Management of soil fertility with compost and natural fertilizers from within the farm, as opposed to chemical fertilizers; management of potential pests through crop diversification and natural as opposed to chemical methods; and increased planned and associated biodiversity, are three main principles. Reversing the tide of chemical usage requires very different management skills than contemporary farmers generally have (especially when local and traditional knowledge is long gone), including crop rotations, cover crops, and careful management of soil nutrients. In Europe, organic farms tend to be mixed farms, with animals providing what is effectively an acceleration of decomposition so that nutrients can be recycled more quickly. Feedstock produced in one field can be recycled through a cow and provide nutrients to another field through the cow manure. In principle, it is precisely the philosophy that was the basis of the rotational systems of Europe at the dawn of modernity.

Much of the benefit of contemporary organic farms is in the favourable position of organic products in the marketplace. On the one hand, we can probably expect an increase in demand for organic products as consumers become more educated about environmental and health issues related to industrial agriculture; but on the other hand, further expansion of organic production will inevitably lead to increased supply and a gradual erosion of the favourable market position. This strongly suggests that further refinements in organic methods, which will require serious ecological research, will be critical to the expansion of the organic sector in the future.

There is a point of view that tends to counteract political support for organic and agroecological farming, namely that they generally produce lower yields than conventional agriculture, and thus could never 'feed the world'. This point of view is wrong on several counts. First, we already produce enough food to feed the world.[25] The problem with world hunger is not the quantity of food that is produced but its accessibility to those that most need it.

Second, in a careful review of the literature, a study from the university of Michigan found that, contrary to the conventional wisdom, organic/agroecological agriculture *could* potentially produce enough calories to feed

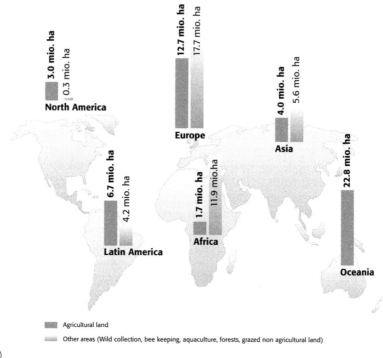

Agricultural land

Other areas (Wild collection, bee keeping, aquaculture, forests, grazed non agricultural land)

(a)

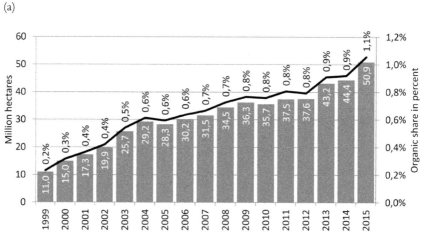

(b)

Figure 5.2 (a) Distribution of global organic land by continent, in 2015 (b) Growth of global organic land from 1999 to 2015.

Source: Willer and Lernoud (2017).

the world.[26] By examining almost 300 studies from all over the world that compared yields in organic or agroecological farms with conventional farms they found that, on average, organic farms produce as much, if not more, than conventional farms. Of course, there is some variability depending on site, crop, and the actual management of the farms, but overall, organic and conventional farms have similar yields. One unexpected result of their study was that the yield ratios between organic and conventional farms were much higher for developing countries than for developed countries. In retrospect, this should not have been surprising since in developing countries the 'conventional' farms are far from efficient. For example, many conventional farmers apply fertilizers when they can get them, not when the crops need them the most. These results simply say that a well-managed agroecological farm can be much more productive than a badly managed conventional farm. However, subsequent meta-analyses have reported yield gaps between organic and conventional agriculture ranging from 5 per cent to 34 per cent.[27] But, it is important to point out that these comparisons are somewhat misleading since they are limited to comparing only the yield of a single crop produced with organic or conventional methods. This may be appropriate for comparisons between monocultures, but falls apart when comparing conventional monocultures with diverse systems where the yield of a single individual crop may be lower than in a monoculture but the total food production of the system is higher.[28]

Proponents of industrial agriculture would argue that yields in developing countries could be increased dramatically with more chemical intensification. And they are right. However, the Michigan study suggests that farmers in developing countries can also increase yields dramatically with agroecological and organic techniques, for less money, using local materials, and without the negative environmental and health impacts of industrial agriculture (and if their crops are oriented toward the market, with the added benefit of a premium price). Again, their results are hardly surprising. Looking at the problem as one of basic biology, a growing crop plant does not care about whether its nitrogen comes from compost or synthetic ammonium, or whether its natural enemies are killed by a parasitic wasp or a synthetic biocide. It will simply grow faster and yield more if it has sufficient nitrogen (and other nutrients) and is protected from its natural enemies. A crop plant has no politics. So, from basic biology we really do expect equal yields from organic versus conventional farming, and the popular notion that organic is less efficient comes from a socio-economic argument and a programme of political propaganda.

The third reason that it is wrong to think that organic and agroecological agriculture could never 'feed the world' stems from the basic assumption that conventional agriculture will be allowed to continue its practice of externalization, while organic agriculture will receive no substantial support from governments. This is a frequently hidden ideological assumption that may be true, but, depending on political arrangements, may very well change.[29] As in

many other industries, those with economic interests behind industrial agriculture are powerful and influential, promoting their own interests through
official and extra-official channels. The environmental consequences – from the
creation of secondary pests to the elimination of biodiversity, to the creation of
'dead zones', to soil erosion and loss of soil fertility, to secondary environmental
contamination, etc. – have been allowed to be written off as externalities: a cost
that is absorbed by society. While no study has tried to quantify all of the costs
actually associated with the industrial system of agriculture, some have quantified particular aspects of it. For example, Pimentel and colleagues estimated that
pesticide use costs the US tax payers approximately eight billion US dollars
every year, most of which is located in the hidden subsidies in the externalities.[30] If popular pressure builds for a more equitable accounting, conventional
agriculture may come under more careful scrutiny and be forced to pay for
some of the real costs of production that are currently absorbed by society. In
US superfund sites, for example, many of which result from dumping the by-
products of pesticide production, clean-up costs are so extreme as to render the
original production process a net drain on the overall economy. Only through
the indirect subsidy provided by a government subjected to intense industry
lobbying are these products able to compete. If sufficient political pressure were
applied, such subsidies could become a thing of the past, and the conventional
agricultural system would probably collapse.

On the other hand, if awareness of the importance of conserving the
environmental base of agriculture grows, political pressure could build for
governmental support for organic and agroecological conversion and production. If and when, and how much, are impossible to predict, but the possibility that such a change could occur cannot be discounted.

Finally, the idea that industrial agriculture is needed to feed the world is
wrong because most of the food that people actually eat, especially in the
Global South, is produced in small-scale diversified farms.[31] It is frequently
claimed that 70 per cent of the food is produced by small-scale farmers.[32]
Although the actual data to support this statement is debated,[33] there is no
doubt that many small-scale farmers that cultivate small plots of land in the
Global South, produce a large percentage of what they consume, and this
figure does not enter into the national statistics. But even using national-level
data, several researchers that have attempted to quantify food production by
the small-scale family-farm sector support the importance of this sector for
food security. For example, a study using national agriculture census data
concluded that globally, 98 per cent of the farms were family farms and that
they worked 53 per cent of the agricultural land globally.[34] Using the case
study of Brazil, they reported that this sector represents 85 per cent of the
land holding, working on only 23 per cent of the total agricultural area, and
produces 62 per cent of the total agricultural output, and, in 2006, was
responsible for 70 per cent of the total domestic food consumption.[35]
Although small-scale family farm is not equivalent with organic or agroecological farms, many if these farms are diversified farms that use low–input

technologies. If they systematically incorporate agroecological techniques, they could be producing even more.

Finally, we emphasize that when discussing organic agriculture as a model, a flag of caution should be raised. Our discussion has emphasized the benefits of organic production for the environment in general, and for the issue of the sustainability of agriculture itself. However, for the purposes of our basic argument, there is something more transcendent about organic production, and that is its extent: the way in which it articulates with neighbouring managed and unmanaged parcels of land. The growing tendency for organic production to morph into large monocultural agribusinesses, due to the increasing demand for organic food, may sometimes be antithetical to the needs of a high-quality matrix. In a capitalist economy, where growth is an imperative of any business, organic agriculture is bound to follow this imperative. Ironically, the success of organic agriculture could become the seed of its own destruction. This is already happening to a great extent in the processing and retailing sector, both of which have seen significant acquisition and merger activity over the last few years.[36] How much of this consolidation is also happening in the organic farming sector is hard to know, but the trend seems to be in the same direction. More frequently, large industrial farms are turned into large organic farms to take advantage of higher prices. The organic mega-farms are characterized by an input substitution model rather than an agroecological model. This means that instead of diversifying the system and harnessing the biological processes generated from within the farm to maintain fertility and regulate pests, the farmer specializes in one or a few crops, applies organic fertilizers and pesticides, and keeps pretty much the same type of management as the conventional farms.[37] If, as we suspect, the agroecological model is the one that will most likely lead to landscapes that are biodiversity-friendly, this tendency of growth and consolidation of the organic sector is not positive. Its overall effects on associated biodiversity are probably negative and, more importantly, the social and economic impacts it generates are as devastating for rural communities as those of conventional agriculture.

A conservation agenda for New England

The challenge of biodiversity conservation is very severe in the context of regions dominated almost exclusively by highly mechanized, chemically intensive agriculture, typically concentrating on production of a small variety of crops, often referred to as the Iowa model, for obvious reasons. The difficulty of this challenge is what has led many conservationists to favour strategies in which agricultural land is regarded as a 'sacrifice zone'. In this perspective, the biological deserts created by agriculture are to be offset by designating large areas of land as protected reserves. As discussed before, this strategy begins by accepting the current style of agricultural production in the American Midwest, the Argentine pampas, and the Brazilian *cerrado*, and in some other areas of the world as a given. This kind of agriculture is assumed to be the dominant model nearly everywhere and is assumed to be the future of virtually all agriculturally important regions. Such a view ignores the existing diversity of agriculture around the world, the fact that at the very minimum, half of the agricultural lands are managed by family farmers,[38] and also demonstrates a lack of imagination about practical alternatives.

The idea of a high-quality matrix that enhances biodiversity, is fundamentally a *regional* perspective that includes but goes far beyond field borders and individual cropping systems. The point is to create not simply crop fields and systems consistent with species-rich environments, but regional environmental effects that incorporate all kinds of land use, as we have emphasized largely within a tropical context where most of the world's biodiversity is concentrated. In the same spirit, others have asked how conservation, including biodiversity conservation, might be far more effective in temperate regions.

In New England, a collaboration between groups of conservation organizations, government planners, academic scientists, and business people has produced complementary visions for food production and wildlands and woodlands that is economically viable and consistent with growing production of both food and forest products. Although one of the most densely settled regions in the United States, New England is also one of its most densely forested. The clearing of the New England landscape for farming reduced forests from more than 90 per cent of the land to around 60 per cent in the last half of the 19th century, but competition from grain production from the Midwest and declining fertility of some farmland in the region led to reforestation, such that the region is now about 80 per cent forested. The human population of the region nearly quadrupled from around four million in 1900 to nearly 16 million today, while the area covered in forest grew from 60 per cent to 80 per cent of the land area. Over roughly the same time period, New England produced a declining portion of its own food, with New England agriculture producing less than 10 per cent of its own food today. Forest area has more recently gone into modest but potentially threatening decline as a result of suburban growth.[39]

The Harvard Forest publication *Wildlands and Woodlands: Farmlands and Communities*,[40] argues it would be possible to permanently maintain 70 per cent of the New England landscape in forest while accommodating likely levels of population growth and significantly increasing the productivity of New England woodlands and agricultural land. They argue that the quality of woodlands could be enhanced in terms of economic productivity and conservation value by more careful management.

> Most land would be actively and sustainably managed for wood, food, and other values while continuing to deliver clean water and air, wildlife habitat, recreation, and support for human lives in a changing environment. About a tenth of the forest (7 percent of all of New England's

land), along with associated wetlands, streams, ponds, and other habitats, would comprise large wildland reserves shaped by natural processes.[41]

Most of New England forests are held by private owners, rather than, as in the West and much of the South, where government holds a large percentage if not most forest land. In order to secure their vision for the future, the Harvard Forest advocates for large-scale expansion of conservation easements in land trusts, with largely willing sellers of such arrangements at a substantial but realistic expenditure of foundation and public funds. More careful, but entirely conventional, land use controls could accommodate projected population growth in the region to minimize forest and agricultural land encroachment.

Prior to the publication of the Harvard Forest's vision, another overlapping and similarly diverse group produced their complementary vision for New England agriculture. Inspired by the increasingly popular idea of localizing food production, this group asked whether New England could produce enough food to satisfy its own regional demand.[42] The answer was a highly qualified 'no', with the qualifications more notable than the negative. New England now produces roughly 7 per cent of its own food. 'A New England Food Vision' argues that using genuinely sustainable production techniques, not now typical of regional production, New England could produce as much as half of its own food while encroaching very modestly on forest land, in a way consistent with the Wildlands and Woodlands report. Putting the two reports together, as their authors intend, we can see a future in which New England remains predominantly forested, sustainable techniques in forestry and food production prevail, and human population is accommodated in communities in which both human health and conservation values are enhanced.

Another kind of vision is being shaped for the American Midwest, the most important food-producing region in the global economy.

Natural systems agriculture

With a focus specifically on biodiversity conservation, it seems a no-brainer to suggest that an agroecosystem should reflect the natural vegetation in the region in which it is practised.[43] Perhaps the most vocal proponent of this idea is Wes Jackson, who reasons that agriculture in grassland areas (which is where the most important cereal production in the world occurs) has gone wrong because the natural ecosystem in such an environment is a perennial polyculture, unlike the artificial annual monoculture that has replaced it.[44] The fact that we produce annual monocultures in those areas is simply due to our inability to domesticate perennial grasses. Jackson's vision is to produce genetic variants of natural perennial grasses that will combine in a 'permaculture' ecosystem that nearly mimics the natural state of the grasslands (Figure 5.3). This strategy is beginning to yield some results with the development of perennial grasses that, while not producing yet as much as their annual

Figure 5.3 Root system of an annual and a perennial wheat.
Source: Land Institute.

counter parts, hold great promise for the eventual combination of perennial polycultures.[45] The particulars of Jackson's vision are somewhat more specialized than the others mentioned in this section, but his logic warrants considerable attention, in particular because 80 per cent of global agricultural land is devoted to annual cereal grains, legumes and oil seed crops.

It must be acknowledged that, in one way or another, the idea of 'natural systems' agriculture has been part and parcel of the organic movement since its popularization in the 19th century. For example, when Ehrenfried Pfeiffer examined the problems of Florida's citrus groves, he looked to the nearby native forests for insights into what was wrong with the agroecosystem, finding, not surprisingly, that monocultural production would likely lead to diseases. He suggested planting native trees among the oranges so as to protect against such problems: a clear example of learning from the local natural system. In fact, according to Conford,

> This story fits well into the organic mythology in the morals that it offers: respect for the natural order as revealed particularly by the wilderness brings economic benefit to those who are not fixated on short-term gains; true science goes out from the laboratory and studies the ecological context, observing rather than trying to dominate; variety is more productive than monoculture; industrial products bring disease and waste.[46]

It is worth noting that some other already extant agroecosystems correspond quite well to Jackson's vision of 'natural systems' agriculture. As noted by various authors, traditional tropical agroecosystems tend to be complex multi-storeyed polycultures that have certain features that resemble the natural forest that they replaced, including the so-called *cabruca* cacao of Brazil (see Chapter 9), the jungle rubber of Southeast Asia and almost all traditional home gardens located in the humid tropics.[47] Most impressive are the thousands of hectares devoted to coffee cultivation (see Chapter 8). Traditional coffee production ranges from 'rustic' to polycultures of overstorey trees. In rustic production the original forest remains, except that the understorey has been replaced by coffee bushes. The entire ecosystem looks very much like a natural forest. Indeed, aerial photos invariably fail to distinguish rustic plantations from natural forests.

Agroforestry as a machine to absorb carbon

Since approximately the birth of the Judeo-Christian-Muslim tradition, the concentration of carbon dioxide in the atmosphere hovered around 270 parts per million (ppm), until about a century ago. Then, there was a precipitous increase, timed perfectly with the maturation of the industrial capitalist system. That pattern gave us our current climate crisis. It is sometimes not fully appreciated that the industrial agricultural system contributes a huge amount of greenhouse gases into the atmosphere.[48] Direct industrial process such as confined animal operations (CAFOs), the use of heavy machinery (and associated compaction of the soil), and the use of large quantities of nitrogen fertilizer, generate both methane and nitrous oxide, both greenhouse gases, and both more potent than CO_2. The fossil fuel consumption in the industrial system is evident in both transportation of food and other agricultural products and the energy needed to service large-scale, minimized labour, operations – that is, CO_2 also comes explicitly from industrial agriculture. And finally, tropical deforestation, although recognized as a direct and major contribution to greenhouse gases, needs to be recognized as part of the industrial agricultural system since most forest clearing is ultimately for the purposes of industrial agriculture.

Climate scientists today say that to avoid all sorts of disasters we need to keep that concentration below 350 ppm. Today's value is 410 ppm. Much concern has been voiced over this problem, and as much as 410 is greater than 350, there is little doubt that humanity faces a true crisis of proportions never faced before in the modern world. However, although less appreciated, we face two distinct problems. First, as generally acknowledged, we must move energy production to renewable sources and we must transform our agricultural system away from the industrial model toward agroecology. Both must be done to stop the continued CO_2 exhaust that the world capitalist system is exuding into the atmosphere. Yet, that represents solving only one half of the problem. Once we stop burning fossil fuels and adding nitrous oxide and methane from industrial agriculture to the atmosphere, we still face the daunting prospect of removing the excess greenhouse gases that already exist. We need not only to stop the increase, we must reverse the process. We need a metaphorical machine that takes CO_2 out of the environment.

That machine, or the many proposed machines, are frequently visualized as based on high-tech physico-chemical processes that, not incidentally, will require huge investments in profitable endeavours. Yet, as is evident to schoolchildren everywhere, there is already a machine that does the job, and does it remarkably efficiently. It is called a 'green plant' and more specifically, a tree. The process that converts CO_2 and water into energy and glucose (providing the atmosphere with oxygen as a side product) is contained within that machine. The forests being cut down to make way for yet more industrial agriculture in tropical Asia, Africa, and Latin America, contain millions of those machines, and we are destroying them at an alarming rate. So, part of the goal to stop the continual release of CO_2 is, and should be, curtailment of forest cutting. But part of the goal of reducing the current level down to its acceptable level of 350 ppm will, of necessity, involve planting trees. Lots of them.

However, the programme whereby we plant those large number of trees is not as evident as the laudable 'let's plant a tree' movements around the world.[49] Yes, we should all plant at least one tree per year if we can, but we evidently need to do much more. A vision is afforded through a recent review of the technical literature on what is already a common practice, albeit mainly in the Global South and thus sort of invisible to some scholars in the North. It is also directly related to the sort of agricultural transformation we advocate as a biodiversity conservation strategy. Small-scale farmers around the world commonly practice a system of agriculture long known as agroforestry. Any traveller to tropical climates has possibly noticed that farmers routinely use trees for a variety of perceived functions on farms, from the production of food (e.g. mangos, oranges), to the provisioning of other 'ecosystem services' (e.g. supplying nitrogen through their root systems, creating shade for livestock seeking refuge from the hot tropical sun, etc.). Using standard remote sensing technology, Zomer and colleagues estimate that currently there are many trees that exist as part of, mainly traditional, agroecosystems.[50] That is, the majority of agricultural production in many tropical regions of the world is actually in the form of agroforestry. Making standard assumptions of how much carbon is stored normally in trees, they estimate that 36.3 gigatons of carbon are already stored in trees in agroforestry systems around the world (as of 2010). It is currently estimated that approximately 80 per cent of agricultural land is dedicated to the industrial system, which is almost exclusively a treeless monoculture. There are obviously exceptions (e.g. almonds in California or palm oil in Malaysia), but as a first approximation, Brazil's or Illinois's mega maize and soybean farms represent a far greater portion of that 80 per cent. The approximately 20 per cent dedicated to small-scale agroecological production, especially those located in the world's tropical climates, are agroforestry systems. So, according to the calculations of Zomer and colleagues, 20 per cent of the land provides us with 36.29 giga-tons of carbon. Converting the other 80 per cent to agroforestry systems would yield 145.6 gigatons. That is a considerable amount of carbon, and a step toward the ultimate sequestration levels that will be needed. We, in no way, wish to imply that elimination of all emissions is not the main priority. Rather, we suggest that one of the agricultural transformations that would be positive for biodiversity will also move us to the goal of getting the atmo-sphere back down to 350 ppm. And, acknowledging that the prospects of a political challenge to the industrial agricultural system are already known to be enormous, the problem of reducing carbon from 410 to 350 ppm remains imperative. To this end, the transformation of agriculture, especially to agro-forestry systems, is perhaps the best, even essential, way we can get from where we are today down to what the scientific community calls non-catastrophic.

The politics of this proposed programme are complicated. It may not be possible to make the necessary changes in time to avoid truly catastrophic consequences, as many observers have already noted. But some analysts have already argued persuasively that the top-down approach taken by the Inter-governmental Panel on Climate Change (IPCC) and the climate change con-ferences are wholly inadequate. As much of history suggests, major societal changes (such as the emergence of democracy itself) gain their initial impulse and energy from the grass-roots actions. And here, a way seems to be emerg-ing from the growth of popular progressive politics and grass-roots popular movements like La Via Campesina. The video 'Together We Can Cool the Planet', produced by La Via Campesina and GRAIN, documents the benefits of peasant agriculture for reducing greenhouse gases,[51] an argument strongly supported by evidence.[52]

Biodiversity as it relates to agroecology

From the previous sections, two of our positions should be evident. First, that the industrial agricultural system was the natural child of an emerging capital-ist economic system and as such contributed to maintaining the structures of capitalism. And second, that there are many proven alternatives that are beginning to challenge the hegemony of the industrial agriculture model. Yet the specific relationship of the industrial model to biodiversity remains implicit: deducible, we feel, from a knowledge of both recent empirical work in agroecosystems themselves, and from recent theory in the science of ecology that emphasizes the importance of migration in metapopulation and metacommunity structure. In this section we summarize first the issue of bio-diversity in agroecosystems itself, and second, the important question of the agroecosystem as part of a high-quality matrix.

Biodiversity in agroecosystems

Contemporary literature increasingly recognizes that agroecosystems can be rich repositories of biodiversity.[53] Nevertheless, a common approach taken by those concerned with biodiversity conservation, has been to buy land for conservation, introduce fences and guards with guns, mainly metaphorical but sometimes real, around those areas, and hope for the better.[54] This so-called 'fortress approach'[55] not only ignores the importance of those areas for the livelihood of rural people, but most importantly for biodiversity conservation, it ignores the importance of the matrix in maintaining biological diversity. The extreme conservationist position tacitly assumes that when the original habitat (whatever that is) is converted to a managed system (e.g. an agroecosystem) the biodiversity is simply lost. Some would compromise, and admit that there is some biodiversity in managed eco-systems, but will qualify the term 'biodiversity' with adjectives such as 'trash', 'opportunistic', or 'non-forest species'. The literature now abounds with falsifications of that point of view, two examples of which are the coffee and cacao systems that we discuss in Chapters 8 and 9. Agroecosystems can be rich repositories of biodiversity, depending on how they are organized. A typical grape orchard in the Central Valley of California is reminiscent of the opening lines of *Silent Spring*, but a traditional coffee farm in the mountains of Chiapas contains as many ant species as an adjacent forest, and the mornings are filled with delightful birdsong. If you convert that traditional system to a pesticide-drenched coffee monoculture, many ant species will disappear and the mornings will be silent.

Agroecosystems as migration routes

From the point of view of ecological theory, there is every reason to expect that local populations go extinct on a very regular basis, only to be reinvigorated by migrations from other localities. This pattern is likely to occur even in habitats totally undisturbed by *Homo sapiens*. However, the principle becomes even more important when dealing with fragmented habitats. All of this was presented in detail in Chapter 2. Recall that the fundamental equation of metapopulation biology is $p = 1 - e/m$, where p is the fraction of potential habitats actually occupied by a species, e is the extinction rate, and m is the migration rate. Obviously, as long as m is larger than e, the ratio e/m will be less than 1 and the fraction of potential habitats occupied will be greater than 0. But if m begins to decline (or e increases), a larger problem arises: as m approaches e, according to the standard equation, the population becomes regionally extinct (p approaches 0). While local extinctions cannot be avoided, which is to say there is little we can do about e, regional and even global extinctions can indeed be averted by concentrating on m, the migration coefficient; which is to say, by constructing agroecosystems that are conducive to inter-fragment migration.

In summary, we have argued in this chapter that the post–World War II period saw the evolution of a worldwide system of agriculture that is historically unique – based on monocultures and the use of agrochemical inputs. This industrial system is young and, despite much trumpeting, its long-term survival or sustainability has been called into question.[56] In addition to the poisoning of the environment and the threat to human health, this system has been devastating for biodiversity. Understandably, a great many people have come to the conclusion that the industrial agricultural system is too damaging to both the natural and social environment, and needs to be transformed dramatically. Parallel with this analysis is the argument of this book – the industrial agricultural system must be transformed if we are to avoid massive extinctions of species in tropical areas. In the rest of this book we discuss the broad social context in which this transformation must occur (Chapters 6 and 7), present examples that illustrate the interconnection between the ecological theory, agroecology and the social movements (Chapters 8, 9, and 10), and finally weave this all together into a new conservation paradigm (Chapter 11).

Notes

1 Vandermeer (2010); Tscharntke *et al.* (2012); Chappell *et al.* (2013); Perfecto and Vandermeer (2015); Vandermeer and Perfecto (2017a, 2017b); Vandermeer *et al.* (2018).
2 Conford (2001); Heckman (2006).
3 Darwin (1881).
4 Howard (1940); Niggli and Lockeretz (1996).
5 Conford (2001).
6 St. Barber Baker (1985).
7 Balfour (1950).
8 An example if this can be found in the practice of the so-called 'biodynamic agriculture', where manure is placed in cow horns and buried for a year during which period the soil is supposed to recuperate its astral essence. The biodynamic agriculture movement was initiated by a series of lectures given by Rudolph Steiner in the 1920s. It is frequently described as 'organic plus' and involves most of the methods used in standard organic agriculture, plus a host of semi-mystical procedures based on concepts such as 'astral power' and 'the inherent energy of the earth'. While still practised extensively in Europe, it is a movement that is not well substantiated using standard scientific methods.
9 Altieri (1987, 1990); Vandermeer and Perfecto (2013, 2017a, 2017b).
10 Altieri (1990).
11 Fish (2000).
12 Jacobsen and Adam (1958).
13 Runnels (1995).
14 Altieri (1990); Altieri and Toledo (2011).
15 IAASTD (2009b).
16 Méndez *et al.* (2015).
17 Rosset *et al.* (2011); Boillat *et al.* (2012); Palma *et al.* (2015).
18 Levins (1990).
19 Vandermeer and Perfecto (2013).
20 Brush (2007); Koohafkan and Altieri (2011).
21 Rosset and Altieri (2017).

22 E.g. Landell Mills (1992).
23 Willer and Lernoud (2017).
24 Altieri and Toledo (2011); Holt-Giménez and Altieri (2013); Khadse *et al.* (2018).
25 FAO (2002, 2017).
26 Badgley *et al.* (2007).
27 De Ponti *et al.* (2012); Seufert *et al.* (2012); Ponisio *et al.* (2015).
28 Rosset and Altieri (2017).
29 E.g. van Mansvelt and Mulder (1993).
30 Pimentel *et al.* (2003).
31 FAO (2014).
32 ETC Group (2017).
33 Ricciardi *et al.* (2018).
34 Graeub *et al.* (2016).
35 Fernandes (2013).
36 Howard (2009).
37 Roberts (2008).
38 Graeub *et al.* (2016).
39 Foster *et al.* (2010).
40 Foster (2017).
41 Foster (2017).
42 Donahue *et al.* (2014).
43 E.g. Soule and Piper (1991); Ewel (1999).
44 Jackson (1980, 1985, 1996).
45 Cox *et al.* (2010); Hayes *et al.* (2012); Jaikumar *et al.* (2012); Marquardt *et al.* (2016).
46 Conford (2001).
47 Altieri (1990); Ewel (1999); Gliessman (2006).
48 Smith *et al.* (2014).
49 We do not wish to denigrate the many efforts to get people to plant trees, for whatever purpose. We support all of those efforts, whether we agree with their underlying logic or not. See, for example, the activities of the Green Belt movement (www.greenbeltmovement.org/).
50 Zomer *et al.* (2016).
51 La Via Campesina is an umbrella organization that groups small-scale farmer organizations throughout the world and is part of a global peasant movement that supports agroecological transformation and food sovereignty. GRAIN is a small non-profit organization that works to support small farmers and social movements in their struggles for community-controlled and biodiversity-based food systems. The video 'Together We Can Cool the Planet' can be viewed at https://tv.viacampesina.org/Together-we-can-cool-the-planet?lang=en, accessed 9 December 2018.
52 Lin *et al.* (2011).
53 Harvey *et al.* (2008); Perfecto and Vandermeer (2008a); Paoletti *et al.* (2012).
54 Oates (1999); Terborgh (2004).
55 Brockington (2001).
56 IAASTD (2009b).

6 The broad social context

Theoretical considerations

The importance of the deep historical context

Conservationists are often very impatient people. The race to save species and habitats imposes a sense of urgency because of how rapidly biodiversity is disappearing and how quickly habitats are being made unfit to support species. But if a key element in the new conservation paradigm is the type of agroecosystem generated in the matrix, then the socio-political forces operative in that matrix are key to the entire conservation programme since they form the foundation upon which the agroecosystem evolves. For example, a rural development policy based on the industrial model is likely to reproduce what we today see dominating large parts of the San Joaquin Valley of California or of northern Iowa – a landscape that allows little room or opportunity for the survival of anything but the crops being produced and their pests. These areas have become harmful even to humans. It is precisely the industrial model of agriculture that made the San Joaquin Valley America's toxic agricultural capital,[1] where, according to state data, more than a million people drink water from sources that are among the most contaminated in the United States.[2]

Much of the current agricultural development for soybeans and sugar cane (grown largely for ethanol fuel production) in Brazil follows this rigid industrial model, that tends to obliterate biodiversity and provides a very low-quality matrix for movement and reproduction of organisms (Figure 6.1). As an alternative example, the exigencies created by the fall of the Soviet Union generated socio-political and economic forces in Cuba that forced a totally different model resulting in a rural environment on its way to the sort of high-quality matrix we advocate.[3] Today there are a large number of rural social movements challenging the industrial agrarian system that we identify as beginning with the spread and dominance of the European powers. While it is understandable that conservationists are impatient to get on with the task, they cannot afford to ignore either the general conditions established in the process by which European nations and their heirs (e.g. the US) transformed the surface of a large share of the planet, or the specific conflicts and problems generated.

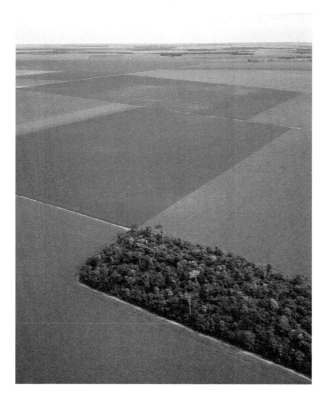

Figure 6.1 Soybean fields surrounding a forest fragment in Brazil.
Source: John Lee.

There has been a disturbing tendency for Western conservationists to see landscapes in the tropics, as elsewhere, as representing, in some sense, an original condition. For example, some European conservationists have seen tribal peoples in Africa in a certain relationship to the landscape, plants, and animals, and take what they see as the given condition of the place over an extended period. What they miss in this is that very few landscapes in Africa or anywhere else have escaped processes of major change over the last 500 years, influenced by many factors internal and external, but almost always transformed strongly by the history of European discovery and dominance.[4] In the case of Africa, conservationists eager to throw up a boundary around wildlife have often given little attention to the rather recent dispossession of African peoples by European settlers and by mining, timber, and agricultural corporations.[5] Thus, too little thought was given to the problem of both human and species survival within and around reserves on land that remained available after the richest land for both people and animals had already been seized for European agriculture.

But the problem is much deeper still. The massive taking of slaves out of Africa radically changed the relationship of various tribes and nations of Africa to one another, as some were largely the victims of slave raiding, and others engaged in slave raiding to capture slaves to sell to the trade that would take them across the Atlantic. Other critical changes were brought about by the imposition of taxes that had to be paid in currency, the creation of artificial boundaries that disrupted traditional methods of livelihood (e.g. pastoralism, hunting, seed gathering, agriculture, and trade), the creation of armies equipped with modern weapons, and indeed, the entire shock brought by foreign occupiers to a varied landscape and highly diverse peoples. All of these factors have played a significant role in the diverse human and environmental problems that continue to emerge: for example, the expansion of the Sahara desert southward and the periodic famines and armed conflicts in the region. Most of these factors have had major implications for the plants and animals throughout Africa, as well as for the diverse human communities.[6]

In the African context, the effort to preserve large mammals in extensive reserves came to Africa as an activity that was as strange and disruptive in many ways as were the very European influences that had brought those mammals under threat. A large literature now exists based on the relatively recent recognition of these factors and the enormous complications and difficulties set in motion by conservation activities in Africa. This literature attempts to identify alternative methods and policies that might create a new and more satisfying balance between human cultures and the legitimate desire to protect species. There is a relatively widespread recognition that failures to take into account the human historical and cultural context have undermined both human welfare and prospects for species survival in a distressingly large number of cases.[7]

The dilemmas of African wildlife conservation have become relatively widely known both because of the extremely high interest in the survival of African mammals and the severity of the problems associated with and often created by the attempt to protect them. It must also be admitted that Americans and Europeans have taken a larger interest than they might have in these problems because of, for better and worse, their tendency to see Africans, as well as African wildlife, as exotic and therefore interesting. The problem, however, is global, though in many cases less subject to transformation into handsome and captivating films. Nearly everywhere, the problem of conservation must deal with a long history of the interaction between local peoples, their landscapes, and the project of European political, economic, and cultural dominance.[8]

European colonialism

Sorting out the effects of European colonialism and domination over trade and economies in terms of the implications for natural communities is no easy task. First, we must recognize that by 1492 human beings were transforming the landscapes and biota of the planet in an enormous variety of ways. The planet was not a clean slate upon which Europe wrote the world's destiny.[9] China and India were civilizations that were in many, if not most respects more 'advanced' in industrial enterprise, technology and governance than Europe. Intense and highly productive agriculture covered large parts of Asia, especially where rice cultivation had been developed, transforming the physical landscape with drainage and irrigation works, terraces, and paddy construction, as well as creating major shifts in plant and animal populations. Rulers, including Chinese emperors and Indian Moguls, had already responded in some areas by establishing reserves and rules regarding wildlife, forests, and waterways, meant to maintain what we would now call a healthy and sustainable environment, or at least rich royal hunting grounds.[10] In Mesoamerica, relatively dense populations maintained themselves by a combination of extensive landscape modification and rigidly enforced rules governing agriculture, hunting, and fishing.[11]

One of the most remarkable of these was the *chinampa* system that, at European Conquest, supported Aztec and Toltec societies in what is now the Mexico City Basin.[12] Artificially created islands in the shallow bed of Lake Texcoco not only provided most of the food for dense human populations of hundreds of thousands if not millions but also attracted waterfowl and supported fisheries. The Spanish did not understand this system, and deliberately drained most of it in the early 17th century, though the remnants of the system were still able to provide most of the vegetables and a fabulous abundance of commercial flowers for Mexico City's population of about a million people in the late 1930s (Figure 6.2). In the 1980s, Mexico created an ecological reserve in the remaining *chinampas*, very belatedly recognizing the system's unique contributions to agriculture and conservation;[13] though by then the water had been greatly diminished in both quality and quantity. In North America, Native Americans used fire to increase the annual productivity of prairies for bison.[14] Along the Amazon, dense populations lived along the river courses governed by traditions that maintained precious fisheries, with soil management systems that, to this day, modern science fails to understand.[15]

Whether by draining *chinampas*, systematically killing off the bison population, or destroying the traditions and ways of governance that had maintained mostly sustainable systems of human subsistence alongside relatively high levels of biodiversity, Europeans dramatically disrupted biological systems. In most of the places the Europeans encountered, these systems had become co-creations of nature and human society long before the Europeans arrived.

One of the most powerful effects of the outburst of European trade, discovery, conquest, and colonization came about inadvertently. Europeans had long been exposed to a vast array of Old World diseases, some of which had had drastic effects among Europeans, such as the epidemic of bubonic plague in the mid-14th century that wiped out a third to a half of the Eurasian population, from Ireland to the Ganges plain. By the 15th century, European

(a)

(b)

Figure 6.2 Modern *chinampas* in the outskirts of Mexico City (a) and a current-day *chi-nampero*, or farmer of the *chiunampas*. (b) This is an ancient method of agri-culture that was practised by the indigenous people of Mesoamerica. Raised beds are constructed with the muck collected from the construction of the canals in shallow lakes. The system is semi-closed in terms of nutrient cycling because the aquatic sector is integrated with the terrestrial sector. The organic matter collected from the aquatic sector is used to fertilize the crops growing on the beds. Nutrients are eventually leached into the canals and are captured by the vegetation and aquatic life living in the canals.

Source: John Vandermeer (a) and (b).

exposure to the epidemic diseases of Africa, Asia, and Europe itself had conferred relatively high levels of resistance to these diseases among Europeans. This was not the case for the peoples of the New World who lived in an environment with few if any major epidemic diseases. Europeans took their epidemic diseases to the Americas, resulting in the loss of something like 85 per cent of the indigenous population within a century or so of initial contact.[16] This demographic catastrophe was obviously aided by the social dislocation, exclusion from access to land and resources, and outright slaughter that occurred along with it. Mexico would not recover its pre-Conquest population until about 1950 (although the population would quickly double and more due to drastically falling mortality rates resulting from better sanitation, vaccination campaigns, and other factors).[17]

This seldom meant, however, that plant and animal life would flourish because of the collapse of the human population. As well as transporting human disease, Europeans brought domesticated grazing animals to the New World for the first time, with the exception of llamas and alpacas, which had long been domesticated in the Andes. They also brought the plough. Ploughs and domesticated livestock wreaked havoc on soils throughout the Americas.[18]

Europeans also disrupted local traditions and governance. In most areas, they approached the conquered people and lands with a looter's mentality. Their recently developed advanced navigational skills and vast fleets of ships made it possible to ship the loot and produce of the New World on a scale never before imagined. With the corresponding rise of industrial technologies, all of this could be achieved with ever-greater speed and efficiency, though at the cost of enormous environmental and human damage in both Europe and the colonial areas. The creation of state-ruled trade organizations and vast private trade enterprises, such as the British East India Company and the Dutch East India Company, established the model for the international corporate organizations that would come to dominate world trade and finance to the present day.[19] This vast technological and organizational machinery created the world in which a small fraction of the world's population, largely in the US and Europe, continue to consume the majority of the world's resources. Part of the same structure arising out of colonialism determines that a majority of the world continues to have access to a minority share of the resources and in economies that are critically and often disastrously dependent on the world's dominant nations.[20]

The combination of domesticated grazing animals, the plough, disrupted traditions and governance, and new organizations driven by greed was devastating to diverse environments of the New World. In many regions, the landscapes encountered today bear little resemblance to those first encountered by Europeans. In much of highland Mexico, currently barren, unproductive, arid environments were once areas of abundant agriculture and diverse, lush landscapes. Such is the case with the Mezquital, for example, north of Mexico City, famously subjected to what environmental historian, Elinor Melville, called 'the plague of sheep', using a term that observers contemporary to the ruin of so much of Mexico's soils had used.[21] Geographer Carl Sauer long ago pointed out that the American Southwest was a far richer environment before the arrival of Spanish agriculture and grazing and the subsequent reshaping wrought by North American pioneers.[22] Vast forests throughout the Americas fell to European exploitation and settlement. Few regions of the world escaped drastic biological and landscape modification due, at least to a significant degree, to European influence.[23]

It would not have been possible for Europeans to transform the world without substituting new forms of governance for those they modified or destroyed. In much of the world this meant direct colonial government. In some areas, such as India, it was actually private enterprises, in this case the British East India Company, that exercised direct governing power for long periods of time, eventually to be replaced by direct colonial rule by the British government. The Dutch East India Company long held sway in much of the rest of Asia. In China, political and economic dominance did not depend on direct colonial government but on a variety of treaties and formal and informal arrangements unfavourable to China's interests.

After overthrowing colonial rule by Spain and Portugal, Latin America was mostly dominated by informal political rule and economic dominance by Britain into the 20th century. The same kind of informal power then shifted largely to the US. Until roughly the mid-20th century, this rule was focused on maintaining the conditions to facilitate low-cost economic exploitation of agricultural commodities, minerals and other raw materials exported from Latin America. More recently, the goals have shifted to include maintenance of secure conditions and favourable terms for investment in industrial and service enterprises, including mining and tourism. In Africa, formal colonial rule was largely established in the 19th century to give way to independence in the mid-20th, followed by very troubled efforts to establish effective national governments under an overarching framework of neo-colonial exploitation of natural resources. The earth-shaking events of Russia, India, China, and Vietnam have been, to a significant degree, the result of the attempt to shake off control by the centres of power in Europe and the US. In the first half of the 20th century, Japan and Germany vied to attain the power over the world's people and landscapes that had been the privilege largely of the US and Great Britain. The entire history of colonialism and its post-colonial consequences have played a major role in shaping landscapes, from the creation of large mining complexes to the destruction of vast stretches of forest, to the creation of plantations and the displacement of indigenous people onto marginal and fragile lands. The enormous waste of war has been integral to the process as well.[24] The point here is not to take sides or to make retroactive moral judgements about all the implications of these events. Rather, it is to insist that conservationists recognize that their work is ultimately governed by massive shifts in power and interests that they cannot control but must attempt to understand if their work is to endure. Conservationism without political and historical consciousness cannot hope to be well designed or well positioned.

While Europe and the US imposed exploitive terms on the rest of the world, they also brought ideas that would challenge the regimes of exploitation. In Latin America, early anti-colonial revolutionaries were inspired by the American and French revolutions and the doctrines of democracy and human rights that underlay them. Movements for an expanded suffrage and democratically accountable government, the abolition of slavery, labour rights, women's rights, socialism, communism, and a host of other ideas inevitably influenced the ruled as well as the rulers around the world. As people struggled to gain a degree of autonomy from outside rulers, they were also inspired to look deeply into their own traditions of governance and political struggle, formulating new ideas that combined the European with the indigenous. Gandhi's particular philosophy of militant non-violent struggle is one of the most famous examples, but there are many others including the variants of communist revolutionary struggle developed throughout colonial and post-colonial rule. In India, Gandhi's non-violence, his rejection of the moral character of industrial capitalism, and his reverence for nature were powerful forces in the Chipko movement, known as 'the tree huggers', in which villagers and allies protected trees from loggers through direct confrontation.[25] Beyond Chipko, Gandhi's philosophy and methods have been widely influential around the world in the environmental movement and the movement for land rights for family farms. Chico Mendes in Brazil learned his politics from a formerly prominent socialist in internal exile in the Amazon forest.[26] Wangari Maathai, Nobel prize-winner for her efforts in promoting tree planting and land rights for Kenya's poor, and a worldwide Green Belt movement, worked within the Kenyan women's and labour movements, built from a combination of American, English, and Kenyan indigenous roots.[27] The Zapatistas of Mexico take much of their inspiration and guidance from indigenous traditions of southern Mexico.[28] Conservationists who want to do more than draw boundaries on maps and throw up fences patrolled by armed guards have to try to understand the complex intellectual and political terrain in which they move, just as they must understand the physical and biological landscape.

The Bretton Woods institutions

As World War II drew to a close, the victorious powers began to draw up rules and institutions to govern a new world order when most of the former industrial powers of Europe and Asia lay in ruins. Those among the victors whose economies had been devastated would need financing and other forms of support to rebuild their economies, and it was quickly obvious that the system would be highly unstable and subject to communist revolution if Germany, Italy, and Japan were not rebuilt as well. All of the former industrial powers would insist on enjoying some of the fruits of victory along with the US. It was also clear that control over world resources and investment opportunities would no longer be able to depend primarily on direct colonial rule.

At the Bretton Woods conference in New Hampshire in 1944, the Allied nations established a new system for global financial stability that, it was hoped, would avoid a repeat of the disastrous period between the two world wars. There was a general realization that the origins of the war were mainly economic, including the Great Depression, the complete collapse of the German economy following World War I, and the expansionist desires of both Japan and Germany. To rebuild the industrial nations that lay in ruin, the government representatives at Bretton Woods established the International Bank for Reconstruction and Development (IBRD), known popularly as the World Bank. To avoid the violently disruptive cycles of inflation and deflation that had plagued the interwar period, the Bretton Woods economists designed a system meant to provide reasonably stable exchange rates. This would be achieved and complemented by the need to promote international monetary cooperation and international trade. The International Monetary Fund (IMF) created at Bretton Woods would be the institution primarily responsible for carrying out these objectives. The World Bank and the IMF were to work closely together to promote economic recovery and stability.[29]

The Bretton Woods institutions became virtual arms of US foreign policy and played a major role in the strategy of the Cold War waged with the Soviet Union and China. The World Bank was, by most accounts, strikingly successful at post-war reconstruction in Europe and Japan, and on the basis of this success, set out to promote economic development in the world's poorer nations, many of them just emerging from colonial rule. These poorer nations, described at the time as the Third World (neither First World industrial capitalist nor Second World communist), were often attracted to socialism and communism. The US used the IMF and the World Bank in conjunction with a variety of other institutions and programmes to counter what it saw as the threat of communist advance in the Third World. The US built up the largest and most widely dispersed military force in the world, carried out overt and covert interventions to avoid the ascendancy or success of left-leaning governments, offered food and development assistance as political leverage, and sold and gave away massive quantities of armaments along with supportive military and police training to friendly governments. An aggressive diplomacy accompanied all these efforts. The IMF and World Bank were an integral part of this Cold War strategy. By the end of the 1950s their role had shifted almost entirely from post-war reconstruction to Cold War foreign policy implementation.[30]

The weak economies of the poorer, less-industrialized nations were subject to crisis. The IMF stepped in during many of these crises with a combined programme of bailout loans to stabilize the currency and 'Structural Adjustment Programs' (SAPs). As a condition of bailout loans, the IMF nearly always demanded that nations accept significant deflation. This had the effect, among other things, of making raw material and agricultural exports available at cheaper prices to other nations.[31] The SAPs combined deflationary policies with sharp cuts in government spending, lower taxes, reductions in tariff and other trade barriers, increased user fees for basic public services such as public transportation, stagnant or lowered wage rates, and reduced regulatory pressure on investors. The stated goal of currency stabilization was to be achieved through making the countries more attractive to foreign investors and making export products more attractive to foreign buyers. This meant opening the mines and running them at full speed, cutting forests, and converting farmland from local food production to commodity production for international markets. Reduced barriers to trade encouraged consumption of foreign goods by local populations. The essential idea was that the nations in crisis were suffering from an inadequate commitment to developing their resources and an overly protective attitude toward local businesses, farmers, and workers. They weren't as eager as they needed to be to sell their resources and labour. They needed to run faster. Were they to run faster, they would earn more and the already wealthy nations would also grow faster as a consequence of access to cheaper resources and more foreign sales to Third World consumers. Faster growth everywhere would be good for all – 'a rising tide lifts all boats'.

The World Bank worked closely with specific IMF programmes during crises, and, consistently with IMF philosophy, between crises. The World Bank concentrated on financing the infrastructure investments that would facilitate the IMF programme. They helped to build roads, dams, electricity networks, and other basic infrastructure to make it possible to gain access to and commercialize the nations' resources. The World Bank also planned its investments in close coordination with private investors, attempting wherever possible to use public financing (based on taxpayer contributions in the US and Europe) from the Bank to leverage larger amounts in private capital. To work as planned, the avowedly public-interest purposes of the World Bank and the IMF, had to be in concert with the purposes of bank and corporate finance.[32]

This combined programme did encourage a certain pattern of economic growth, but at the cost of widening inequalities between rich and poor. Whatever its direct economic consequences, however, it was clearly a recipe for creating severe environmental problems. Much of the programme was directly aimed at maximum resource exploitation at cheap prices with a minimum of regulation supported by infrastructure investments that guaranteed wholesale landscape transformation. The flow of resources maintained a downward pressure on their prices and created periodic price volatility while tending in the long run toward resource exhaustion. The work was to be done at as low a wage as possible, and most of the contracted goods and services were to be purchased from foreign firms who would gain a lion's share of the direct investment, which meant that local labour forces would remain relatively untrained and dependent on any kind of economic activity available, including pioneer settlement and forest clearance. The whole pro-gramme was carried out in a way that strengthened specifically those Third World elites who were most interested in maximum immediate profits. Influ-ential capitalists and public officials who were more interested in building stable local markets and more skilled labour forces in a reasonably regulated business environment, taking a longer view of what would serve the interests of their countries, were systematically pushed aside. The political effect was to set up and maintain a vicious economic cycle that required repeated visits of World Bank and IMF officials to keep a fundamentally flawed model of development afloat – all part of what disenchanted World Bank and IMF employees were beginning to call 'the debt trap'. When conservationists finally began to understand the implications, some international environ-mental and social justice organizations launched determined campaigns in the late 1980s to reform or eliminate the World Bank and the IMF. The critique has been so successful that today, the World Bank (and to a much lesser extent the IMF) have internalized much of the environmentalist criticism of their former policies, at least rhetorically. However, it remains very much an open question whether they will continue over the longer run to put their money where their mouth is.[33]

The IMF/World Bank/US foreign policy programme pursued during the last half of the 20th century inspired determined opposition in many nations. One of the central tasks of US foreign policy and military apparatus was to dampen or eliminate such opposition. For most of that time, the justification for US pressure, intimidation, and covert and overt military intervention abroad was the Cold War. However painful the path of capitalist develop-ment might be, it was argued, it was certainly preferable in this perspective to socialist and communist approaches, and even to nationalistic capitalist pol-icies. The US carried out violent interventions to enforce this view in one nation after another, including Greece, Iran, Guatemala, Nicaragua, Chile, Argentina, the Dominican Republic, Grenada, Indonesia, the Philippines, Vietnam, the Congo, Angola, Mozambique, and many others. In virtually the entire Third World, including Brazil (as discussed below), the US frequently worked closely with selected elements of local elites to subvert democratic processes, whether through secret contributions to friendly parties and can-didates, fixed elections, or support for military coups.[34]

After the Cold War

The end of the Cold War eliminated the possibility that opponents of US policy might enjoy support from the Soviet Union, and thus also seemingly eliminated much of the justification for intervention. As a consequence, old wine is packaged in new bottles to pursue familiar aims. A programme to promote 'free' or 'liberalized' trade, and thus generalized 'globalization', has been vigorously promoted from Washington since the late 1980s. Putting more emphasis on the importance of eliminating trade barriers, the 'neo-liberal' programme, often styled the Washington Consensus, contains very familiar elements. Government regulation of business activities and a variety of subsidies and policies to support activities considered in the public interest are redefined as barriers to trade to be eliminated or reduced. These can include restrictions on export of capital from poor nations, assistance to farmers, and many kinds of taxes on goods and investments levied to support social spending on education, health care, and the environment. The US and sometimes other rich nations exert severe pressures to eliminate such barriers, arguing that generalized economic growth supported by liberalizing trade and investment will eventually provide the right economic and social conditions for universal prosperity and appropriate environmental programmes. At the same time, richer countries, including the US, have hypocritically maintained barriers to trade of all sorts that serve their own interests.[35]

Much of the 'Washington Consensus' is carried forward by the World Trade Organization (WTO), which replaced an older and more limited arrangement known as the General Agreement on Tariffs and Trade (GATT). The WTO is the organ intended to supply the overall supervision and legitimization of the new global order. Unfortunately, the WTO has little transparency or accountability to democratic institutions. Much of its operations are conducted in secret and there is little obligation to explain or justify the organization's actions. It has been largely run by economists allied with large corporate interests and it attends to the neo-liberal model as though it were a law of nature. It has thus far acted to smooth the way for corporate activities carried on without disruption. Most vociferous complaints about the WTO have come from environmentalists and labour groups, and more recently from small-farmer organizations which have suffered the most from the brunt of the neo-liberal policies for agriculture.[36]

The fly in the ointment emerged in the 1999 meetings of the WTO in Seattle, USA, and then in Cancun, Mexico, in 2003. An international coalition of environmentalists and unionists came out in force to effectively shut down the meetings. Especially at Cancun, from within the meetings a coalition of nations led by China, India, South Africa, and Brazil launched a critique of WTO policy that was similar in some respects to the one coming from dissidents outside the hall, pointing out the systematic inconsistencies and contradictions in WTO policies that had the effect of favouring the interests of the rich over the poor. Since these meetings, the WTO has been unable to meet without a retinue of security that makes the whole operation look at best a bit suspicious, at worst like a new brand of fascism. For a decade now it has been impossible for the WTO and the Washington Consensus that energized it to shape any broad new initiatives or agreements. The world economic crisis of late 2008 cast further doubt on the likelihood of advancing the Washington Consensus or of forming a new consensus on anything like similar principles.

Inequalities have widened under the neo-liberal programme.[37] At the time of writing the first edition of this book (2009), the wealthier nations, which accounted for about 17 per cent of the world's population, used about 70 per cent of the world's energy resources. That left about 30 per cent of the available energy for the other 83 per cent of the world. The figures are similar for most other resources.[38] As of the current writing, the per adult average income of an average citizen of the US/Canadian sphere is 50,000 euros, of the European Union, 31,400. Comparable figures for Latin America are 15,400, for sub-Saharan Africa, 5,100, and for India, 7,000.[39] Ignoring the well-known inequalities within countries and regions, on average, a person in the Global North earns more than four times what a person in the Global South earns. Some suggest that leaving 83 per cent of the world with only 30 per cent of its goodies is immoral and unfair, as is having an entire region earning only ¼ of another; others argue that it is highly unstable and therefore self-defeating. Recent events suggest that the rest of the world will insist on significant changes in a context that is made even more complicated by the problem of global climate change and the eminent end of oil as a basic energy source. Conservationists in the Global North will find that many of their partners in the tropical world understand this situation and deeply resent it. The problem and the consciousness of the problem invariably condition all international conservation work.

The underlying dynamics of expansionist capitalism combined with the specific programmes for growth and development adopted by the ruling powers after World War II would seem to constitute a reasonable foundation for understanding both the ever-more rapid use of the world's resources and the relentless concentration of their use by a minority of the world's population. Careless habitat degradation and the decline of biodiversity follow the same logic. The problems, in a sense, are not accidental consequences of economic growth but are the accepted price of a well-developed overarching policy.[40] In this sense, they are simply part of the plan. Interestingly, much of the conservation community has been slow to see this.

Malthusianism rearing its ugly head again

One of the main factors that has caused confusion has been the lingering popularity of varieties of Malthusian thinking in the conservation community, though it has been largely rejected elsewhere. Many intergovernmental documents and reports about food start with something like 'In order to feed the growing population, we need to…'. One kind of traditional conservation agenda has its roots in this primitive Malthusian equation, albeit only tacitly so in most cases. It is taken as an unquestioned assumption that as population grows – frequently leaving it vague as to whether the variable of concern is global or some more regional population density – natural habitat will be 'destroyed', by which is meant, converted to agriculture. With its more humanitarian face, this viewpoint normally adds the idea that people have to be fed, or they will either die (usually regarded as a bad thing) or encroach yet further on natural habitat (sometimes regarded as even worse).

Consequently, the problem is presented as a zero-sum game in which every piece of land devoted to agriculture cannot be devoted to conservation, as was discussed earlier in the section about land sparing in Chapter 3. With this fundamental framing, conservation goals include population control, increasing agricultural intensity on already converted lands, and higher vigilance of areas under 'protected' status.[41] The paradigm could not be more at odds with our perspective.

A crude Malthusian foundation remains a mainstay of some elements of the conservation community, even though it is rarely articulated explicitly. Thus, for example, a recent debate about the nature of the deforestation crisis takes as a starting point the size of a local rural population as a driver of deforestation.[42] Other examples could be cited, almost all of which are characterized by a simple confusion between correlation and causation (e.g. a deforested area tends to attract more people, thus possibly reversing the implied causation). In all human endeavours there are two, not one, population variables that are relevant to any discussion: first, the size of the minimum population that would be necessary to sustain whatever productive activities are currently the base of the population's existence, whether the production of cars in Detroit or the production of vegetables in Las Lagunas, Mexico; second, the total number of people that can be supported by those productive activities. In the case of Las Lagunas, a complex agricultural system, based essentially on the maintenance of terraces, had long provided a comfortable and sustainable food supply to the local community, which is to say the productive activities sustained the population. In the late 1970s, with a new oil industry booming in a neighbouring state, a large fraction of the males in the community migrated to take jobs in the new industry. The population decline was substantial enough that the maintenance activities required to keep the terrace-based agricultural system operative could not be met. Because of migration, the actual population fell below the population necessary for maintenance of the productive system. Consequently, the agricultural system collapsed and the previously well-fed population suddenly went to bed hungry – the agricultural system could no longer sustain the extant population.[43] In the case of all human productive activity there is a minimal number of people required to maintain productive capacity and a critical number of people that the productive capacity can sustain. Any population, regardless of its actual per hectare population density, can be either relatively overpopulated or relatively under-populated, depending on what level of technology is under consideration. The current population of Japan could never sustain itself without its modern productive systems, all of which are based on a large number of people; while the island nation of Madagascar, about the same size as Japan, could never sustain the population of Japan with its current productive system. Yet, Madagascar could not build the sort of productive capacity of Japan with its comparatively lower population and thus is an under-populated island, from this point of view.

Beyond the crude assumptions of Malthus, the apparently humanitarian notion that we must feed the world's population, leads to the assertion that we need to intensify current agricultural areas. The first problem with this assertion is that the world produces an enormous quantity of food already, and has done so since the British withheld food stores from its starving client states in Ireland, India, and China, to name a few.[44] Indeed, there are few known cases of massive human famines that occurred without export of food away from the famine victims.[45] In the contemporary world, we already produce so much food that the main problem faced by farmers is lowered prices because of an excess of supply.[46] The reason people go to bed hungry every night has nothing to do with the food supply – rather they have been excluded from land to produce their own and cannot afford food offered on the market, or because of war and political disturbances that block access.

The assumption that the elementary structure of the assumed zero-sum game can be easily manipulated by the intensification of agriculture, especially in areas on agricultural frontiers, is perhaps the most dangerous and egregiously false of the conventional assumptions.[47] Angelsen and Kaimowitz provide a series of case studies based on their initial formulation of the contradictory nature of the claims of conservationists and the standard assumption of development economists. If anything, the intensification of agriculture, if not done in a cautious fashion, is likely to increase the local pressure on natural habitat.[48]

On the other hand, it is not crude Malthusianism to assert correctly that in a general sense growing human numbers do potentially put additional pressure on world resources and environments. But that assertion does not go very far in understanding the complex circumstances that drive population growth and decline, nor does it go far in understanding the highly variable effects of human numbers on agricultural production and nature. Human population growth is responsive to social factors, and its effect on species and habitats are fundamentally conditioned by social organization and technological choices. The conservation community needs to get beyond Malthusianism. For we will see in the section that follows how inaccurate and dangerous it can be.

Difficult socio-political issues in practical conservation work

It is not a simple matter to assess the relationship of an existing or potential protected area with respect to the influence of the agricultural landscapes that surround it. An ecologist can be exceptionally well qualified to evaluate the biological character of a relatively wild landscape while at the same time being poorly trained to understand agricultural landscapes in the same region. Just as importantly, conservation biologists and other experts may make important errors in understanding how the agricultural areas surrounding a protected area developed historically, and therefore may be unable to accurately judge whether the quality of the relevant agricultural matrix is likely to remain as it is, deteriorate, or improve.

Imagine a conservationist looking outside the borders of a protected area in the tropics, actual or potential, towards surrounding agricultural areas. She observes people living in roughly constructed houses working on the land. The houses, or groups of houses, are scattered through what the conservationist sees as a 'disturbed forest'. There is a patchwork of fields devoted to various crops. In some areas are stands of forest that resemble the undisturbed forest within the protected areas, and there are areas that have some of the trees and other plants of the undisturbed forest, but the biologist observes many plants seldom seen within relatively undisturbed areas. Other areas are characterized by some combination of weedy, invasive annual plants, some perennial legumes and grasses, and a few shrubs. Local farmers cut and burn in some places, and may clear new fields within what looks like undisturbed forest.

The first problem of interpretation in this example is to understand where the process of human use of the landscape has been and where it is going. There is a strong tendency to see such landscapes as frontiers in which undisturbed forests are being degraded by either new migrants into the region, localized population growth, or progressive environmental degradation over time by a resident population, or as a combination of all three. This may be correct. Or, in contrast, it may be that the people of the area have been here for a long time – anywhere from decades to millennia – in a stable relationship with the resources and species of the area. In this case, the pattern made up of a combination of land that is actively farmed, land of relatively undisturbed forest, and land in various stages of succession between human disturbance and recovery toward a high level of diversity resembling the original forest, is a pattern that could have been observed here for much or all of the period of human occupation of the region. Disturbance is a constant, but so is the cyclical process of recovery and various stages of succession from farming to forest and back again. It may be a cycle in which the net amount of land experiencing disturbance versus that in recovery is relatively stable. The cyclical pattern may be supportive of a fairly high level of biodiversity within the agricultural landscape and an even higher level than within neighbouring undisturbed areas.

In many regions it will be difficult to tell the difference between frontier and long-term occupation simply by a snap-shot observation. One will have to consult the historical record. In order to ensure reasonable accuracy, researchers will have to examine written archival sources, interview local people and government officials, and carefully examine the land and its biota in detail. Making sense of the record is often far more difficult than one might imagine. It can usually be taken for granted that the evidence will not only be complex but it will often appear to be contradictory. It will be necessary to have a high level of skill to incorporate historical, anthropological, agronomic, and biological knowledge into an accurate interpretation. Making a mistake in this regard could be fatal to conservation efforts.

If the region is undergoing active frontier occupation or progressive deterioration from the actions of a resident population, the observed pattern might mean that if things keep going as they are, forest remnants and areas of recovery will become successively more scarce and smaller in area. Biodiversity both outside and inside the protected zone is likely to suffer, perhaps drastically. This kind of situation is common in tropical rainforest regions throughout the humid tropics. It may be found in the Amazon, in Central America, Africa, India, and Southeast Asia. In most areas where frontiers of human settlement are pushing insistently into the forest, conservationists face one particular set of very difficult dilemmas. How will it be possible to stop or, more likely, limit the impact of frontier development on the agricultural landscape with all of its implications for any protected areas? Stated in its simplest form, conservationists will have to devise a way to sweep back or channel a powerful tide that is running against their own efforts. In its more complicated forms, any real hope for conservation within reserves will require that somehow there will be effective ways to deal with gut-wrenching issues involving social inequality, political power, exploitation, manipulation, corruption, and large-scale environmental and demographic change.[49]

It is also important to understand the motivation and organization of people in the region. This is likely to involve some quite complex considerations, but at the simplest level it is obvious that if frontier settlers are being given incentives by governments and/or businesses to move into an area, the problem is quite different than if the new settlers are simply individuals in search of new farmland. If, as is frequently the case in the Amazon and other tropical areas, frontier settlement is strongly driven by the eagerness of timber companies to enlist the settlers' labour in felling trees and clearing roads, then the conservation problem will not be solved if that fact is not appreciated and addressed. If further, as is also common in the Amazon, powerful and wealthy ranchers are eager to acquire land exhausted from crop production, another key factor is added. If the ranchers are willing, as they are in the Amazon and much of the rest of the tropics, to encourage settlers to clear land and then further willing and able to drive settlers off the newly cleared land by legal manipulations, corruption, intimidation, and violence, then the situation becomes yet more complicated and difficult. If mining or other extractive interests are also involved the situation becomes even more complex. This is the case in much of the Amazon and Africa, where people are lured to the region by the prospect of mineral prospecting or employment in mines and then are periodically laid off and forced to make a living from agriculture, sometimes only to be drawn back to mining in a cycle of employment alternating between mining and farming.

The character of agricultural activity may also be influenced strongly by whether settlers are acting completely as individuals or single families, or whether they are organized by informal groupings from prior associations in other locations (as towns and villages outside of the forest region may be more or less reconstituted in new agricultural areas or in urban slums), or whether they are organized by rural unions or land associations.[50] If the government encourages settlers through land distribution and titling, credit, extension services, and incentives for housing construction or crop production, the situation will be very different than if government does not provide support. Government may also be relatively effective or ineffective at enforcing the law, either in support of or against settlers' interests. Supportive services and law enforcement can be either positive or negative in terms of habitat and species protection, depending on how they are designed and implemented.

Some frontier settlement is done by settlers with virtually no experience with the ecological details of the region, as they often come, for example, from highland plateaus or more arid regions into humid forests. They may, however, bring with them a great deal of relevant knowledge and experience as prior forest dwellers or traditional agriculturalists. They may be relatively well equipped to experiment and learn, or they may be very poorly prepared to do so.

The likelihood of adaptation to the new environment will be strongly influenced by whether they are organized into rural unions, cooperatives, or land reform movements, and by the ideology and strength of those movements. Some rural organizations are strongly focused on advocacy for settlers with little or no interest in adaptation of agriculture towards sustainability in the frontier region. Other more sophisticated organizations are strongly aware of the difficulties of both new and established settlers and are actively engaged in a variety of ways to ensure that settlers are able to achieve sustainability in their new farms. These latter organizations are also likely to be more interested and effective in working with conservationists towards the compatibility of protected areas.[51] This is true, as we will argue below, for some both obvious and not-so-obvious reasons.

If the patterns observed in the region do not involve significant frontier settlement but rather are the established and stable ones of a long-term resident population, then human settlement and farming may prove to be supportive of the effort to maintain high levels of biodiversity inside and outside the protected area, as species move in and through the agricultural matrix as well as within the protected areas. In these cases, it may be possible to achieve a great deal simply through recognizing the positive contributions of existing patterns of settlement and agricultural activity, and it may prove fruitful and relatively easy to encourage initiatives that will go further to actually strengthen the relationship between the goals of conservationists and those of farmers or peasants in the region. These situations are also common in many areas of the tropics. Again, the degree of success in this regard is likely to be determined by many of the same cultural, organizational, economic, and legal factors as those that influence new settlers.

Sources of systematic bias in conservation practice

Unfortunately, whether involving new settlers or established residents, many situations are misunderstood for their conservation potential because of a powerful bias among conservationists, government policy makers, agricultural scientists, agronomists, and ecologists. From a synthesis of the research, much of it cited above, and from our own experience over four decades and in several countries, we can summarize this bias as it works at eight critical levels:

1   The powerful desire to maintain high levels of biodiversity in relatively undisturbed areas has led many people, including and sometimes especially those with high levels of biological expertise, to regard all kinds of agriculture as dangerous to conservation efforts. They simply do not appreciate the critical differences between various kinds of agriculture in terms of their potential to support conservation.

2   Many agricultural scientists and agronomists trained in conventional agricultural methods see the kind of agriculture that is often practised by forest dwellers, indigenous people, and small-scale farmers as hopelessly inefficient, unproductive, or unworthy of study or respect.

3   Conservationists are often best acquainted with the formally educated and socially prominent elites of a region both by personal preference and as a necessity for influencing government policy. These local elites may be strongly biased against poorer rural populations, their culture, their organizations, and their interests. The elites will pass on their biases to conservationists in myriad direct and indirect ways, and they will act vigorously to protect their interests against assertions and initiatives by the poor. Deep prejudices against the poor are also sometimes found embedded within conservation organizations themselves and find ready reinforcement from elites at various levels.

4   Lack of language skills and/or cultural understanding creates a serious gap between those involved in conservation and agricultural development and the people engaged in agriculture. This is likely to be especially true where the people farming belong to groups with well-established traditions and strong group identity: precisely those who are most likely to be farming with sophisticated traditions of local knowledge that are culturally evolved to adapt to the landscape in a sustainable way.

5   Policy makers and conservationists often lack an understanding of the dynamic nature of what they are observing – they are eager to solve problems as they appear in the present and are impatient with historical analysis that may be necessary to understand the forces and trends that are driving land use in particular directions. Analysis of the present situation is systematically privileged over attempts to understand the arc of historical change. This is likely to lead to poorly designed policies.

6   As the present is privileged over history, so may analysis of local circumstances be privileged over understanding the larger regional, national, and international context. Again, the results can be quite troublesome.

7   Conservationists may lack the knowledge and willingness to engage in teamwork across science and social science disciplines that are usually necessary to investigate the situation properly, draw accurate conclusions, and make useful recommendations.

8   Conservationists often apply some version of Malthusian analysis to environmental problems, often inappropriately. It is all too easy, and frequently severely misleading, to see biodiversity conservation as simply a function of insistently rising human numbers relentlessly destroying habitats and species.

Notes

1   Leslie (2010).
2   California Water Boards (2013).
3   Funes *et al.* (2002); Wright (2008); Rosset *et al.* (2011); Fernández *et al.* (2018); Goulart *et al.* (2018).
4   McNeill (1977); Hughes (2001); Richards (2003).
5   Anderson and Grove (1987); Neumann (1998); Adams and McShane (1997); Hardin and Bahuchet (2011).
6   Franke and Chasin (1980).
7   Escobar (1998); Brechin et al. (2002); Adams and Hutton (2007); Brockington *et al.* (2008); Büscher *et al.* (2012).
8   Tucker (2000).
9   Hughes (2001); Richards (2003).
10   Guha and Martinez-Alier (1997).
11   Rojas Rabiela (1983, 1994); Mann (2005); Wright (2005, Chapters 5–8); Miller (2007); Heckenberger *et al.* (2008).
12   Rojas Rabiela (1983).
13   The *chinampas* in Xochimilco, Mexico, is one of the last reminding refuges of the critically endangered Mexican salamander, the axolotls (Voss *et al.* 2015).
14   Pyne (1982); Williams (1994).
15   Heckenberger *et al.* (2008); Mann (2011); the so-called 'black soils' of the Amazon, laden with deep deposits of biochar, are thought to have supported very large pre-European populations of Native Americans through what must have been very sophisticated soil management (Mann 2008; De Gisi *et al.* 2014).
16   McNeill (1977); Hughes (2001); Mann (2011).
17   Wright (2005).
18   Melville (1994); Richards (2003); Wright (2005); Miller (2007).
19   Keay (1991); Prakash (1998).
20   Tucker (2000); Wallerstein (2004); Speth (2009).
21   Melville (1994); Miller (2007).
22   Sauer (1955).
23   McNeill (1977); Hughes (2001); Richards (2003).
24   McNeill (1977); Hughes (2001); Richards (2003); Tucker (2018).
25   Guha and Martinez-Alier (1997, Chapter 8).
26   Revkin (2004).
27   Maathai (2006).

28 Stahler-Sholk (2010).
29 Rich (1994); Stiglitz (2003, 2005).
30 Escobar (1995).
31 Stiglitz (2005, 2007).
32 Rich (1994); Stiglitz (2003, 2005).
33 Rich (1994); Stglitz (2003, 2005).
34 Johnson (2004).
35 Stiglitz (2003, 2007); Harvey (2005).
36 Jones (2004); Harvey (2005); Rosset (2006).
37 World Inequality Lab (2018).
38 Speth (2009); at the present time these figures have got worse, if China is excluded from the analysis. China has increased its per-capita wealth considerably during the intervening years since the publication of the first edition of this book, so that if China is included in the analysis the picture looks similar. Most poor countries just got poorer with less access to resources, while China got richer, but still not reaching a status that could be called wealthy, but using quite a bit more energy. In sum, the average figures for 2009 are not that different today.
39 World Inequality Lab (2018).
40 Harvey (2005); Klein (2008).
41 Oates (1999); Terborgh (2004).
42 Wright and Muller-Landau (2006); López-Carr and Burgdorfer (2013).
43 García-Barrios *et al.* (1991, 2009).
44 Davis (2001); Patel (2012).
45 Lappé *et al.* (1998).
46 The displacement of small-scale farmers from the countryside, the dependence on imported foodstuff, droughts, and other climate-related disruption to agricultural production, and an increase in consumption in rapidly developing countries like China and India have caused major disruptions to local food systems, especially in the Global South, and have generated a food crisis toward the end of the first decade of the 21st century. However, this food crisis is not based on the absolute amount of food, since the world as a whole still produces more food than is necessary to feed the world population.
47 Perfecto and Vandermeer (2010); Fisher *et al.* (2011); Tscharntke *et al.* (2012); Kremen (2015); Goulart *et al.* (2016).
48 Angelsen and Kaimowitz (2001a, 2001b).
49 Chappell *et al.* (2013).
50 Barraclough and Ghimire (2000); Yarnall and Price (2010); Hecht (2010, 2014).
51 Barraclough and Ghimire (2000); Wittman (2010); Shanley *et al.* (2011); Delgado and Rommetveit (2012).

7 The broad social context
Practice

The Brazilian Amazon: a case study in conservation, livelihood, and social movements

A case study that summarizes all of the material in the previous chapter can be seen in a review of major conservation efforts in the Brazilian Amazon over the last several decades – efforts meant to slow deforestation and establish effectively protected reserves for habitat and species conservation. The Brazilian Amazon has rightly been the focus of immense effort on the part of the international conservation community and a subject of intense controversy and debate within Brazil. The story is far too long and complicated to tell in detail here. A brief review, however, brings out the dangers of all of the sources of systematic bias identified at the end of Chapter 6.[1]

The variety of human activities

For many years (roughly from the 1960s to sometime in the 1980s, with some conservationists and their organizations still stuck in the 'systematic biases' discussed in the last chapter), most conservationists in the region failed to undertake any serious evaluation of the differences among types of human settlement in the region, essentially working to reduce activity in the Amazon forest, everywhere and in all circumstances. This was an impossible and self-defeating task. They failed to achieve their objective and made enemies of nearly everyone. The one significant exception was for long-established indigenous people, and even in this case, conservationists did relatively little at first to reach a profound understanding of the dynamic character of indigenous life and its effect on the forest, preferring instead to see the indigenous people of the Amazon as essentially frozen in time. Indigenous people were seen as static, having no history of change in regard to their effect on the forest and having no future other than a stark choice between elimination or preservation, essentially as living cultural museums. Indigenous people were too often seen as lacking the ability and sometimes even the right to make autonomous decisions. Such attitudes were radically at odds with what is known about Amazon indigenous groups, who have undergone dramatic changes over the last 500 years, and who continue to change. For example, recent research has increasingly supported the view (as reported by the first European observers in the region and later discounted) that the Amazon supported populations much larger than previously supposed, living essentially in long cities stretched along the rivers.[2] The population may have been as high as the present population of the Amazon region. In recent decades, researchers have discovered more compelling evidence for long-term settlement and human management of at least parts of the forest. Among these discoveries are the 'terra preta' (black earth) soils that occur in many Amazon regions.[3] The black earth soils were created by intensive farming over long periods of time, with deliberate additions of significant organic material into the soils in order to make them more suitable for long-term cultivation. Such soils indicate that original populations of the Amazon may have been much more actively engaged in manipulation of the forest and the soils, and that these populations were more concentrated and permanently settled than previously assumed. In any case, it is clear that the scourge of European disease and European enslavement forced indigenous people to adopt new and more difficult survival strategies. In one version of this view, Amazon indigenous people are still in the process of sorting out the consequences of a way of life that, prior to European contact, was much easier and more prosperous. In the present era, indigenous people are increasingly influenced by recent incursions of settlers and economic enterprises, and are challenged again to work out new means of survival. They are not 'museum' cultures, but rather cultures that have for 500 years been dealing with the challenge of living within a context that has been getting narrower, more complex, and more difficult over time. In most instances, they are not a set of peoples living in glorious isolation until recent encounter, but rather a set of people engaged in a very complex pattern of strategic retreat, periodic contact, partial assimilation, disastrous exploitation, and exposure to Old World disease.[4]

The consequences of failing to understand the enormous complexity were mostly bad for indigenous people and bad for the cause of conservation. Disregarding, idealizing, and attempting to freeze indigenous people within a particular way of life were all responses based on ignorance. This was true in spite of the fact that Brazilian and foreign anthropologists had for many decades been creating a rich literature on Brazilian indigenous people that, if understood, would have avoided such mistakes.

Perhaps even more serious for the cause of conservation, as well as for the welfare of Amazonian peoples, was the tendency to overlook the differences in subsistence strategies among non-indigenous inhabitants. Some of these were recent settlers and some had several generations of experience in the Amazon. The cultural, physical, and ethnic boundaries between many of these people and indigenous people were often so blurred as to be insignificant. Inter-marriage and two-way cultural assimilation have made the word 'Indian' or 'indigenous' difficult and slippery terms in the Amazon – when visitors say they want to 'visit the Indians', locals know that what the visitors want is a kind of artificial museum experience that in most instances is faintly related to cultural realities in the region, but locals will often attempt to oblige by creating a satisfying experience in the interests of courtesy or money. Some of these people moved across the landscape, relentlessly cutting forest and exhausting soils, providing trees to the timber industry as they cleared land, inadvertently or intentionally, for predatory and unsustainable ranching operations. In sharp contrast, many others established relatively stable communities, practising a livelihood that included agriculture, fishing, hunting and gathering in a way that was consistent with high levels of biodiversity conservation. Such communities often protested vigorously against wholesale forest destruction and sometimes stopped the timber and mining industries from further forest encroachment. A large share of Amazon people lived lives that alternated between extremes of destructive work on one hand and livelihoods based on conservation practices on the other, which also often involved moving back and forth between city and country. This is still the case. Most people who live either temporarily or permanently in rural areas are in search of permanent homes and communities and in need of sustainable practices that would support permanence.[5] It has to be noted here that the Amazon population as a whole is one of the most urbanized in Latin America, with more than 80 per cent living in cities. This also has very important implications that we will not try to explore here.[6]

In 1978, we heard one of the most passionate and knowledgeable talks we have ever heard on the need for forest protection and what was required to achieve it from a man employed as a bulldozer driver whose machine was fitted with long chains attached to another bulldozer for the purpose of literally mowing down trees in vast swathes. What made it remarkable was the machine operator's intimate and subtle understanding of the interplay of various economic and political interests at work in the region. At that time and to the present, even those making their living from wholesale forest destruction sometimes understood far better than conservationists the dynamics of forest destruction, and yearned for the conditions and policies that would make it possible to have livelihoods consistent with forest protection.

Major international conservation groups, most prominently what was then the World Wildlife Fund (WWF, now known as the World Wide Fund for Nature), demonstrated their lack of understanding of the nature of the problem in their collaboration with Brazil's military government in a vast development scheme along the Amazon's southern border. The project was planned in the late 1970s and initiated in 1981. The military, heavily financed by the World Bank, envisioned a typically technocratic and authoritarian approach to the problems that were being encountered in the Amazon in what they called the Polo Noroeste Project. The WWF and some other conservation groups and individual conservationists had decided that it was better to attempt to cooperate with the Brazilian government than merely to oppose it. The elaborate plan would theoretically protect a large part of the region's forest and its key biological resources. The plan called for strictly limiting road-building, the number of new settlers allowed into the region, the amount of timber cutting, and the extent of ranches. It would protect indigenous people from settlers, and forest reserves would be strictly delimited and guarded.[7]

The plan quickly went awry. Land speculators, settlers, ranchers, and the government's own officials virtually ignored the plan. The government continued to support infrastructure development, most importantly road-building, while failing to enforce conservation laws. This should not have been surprising, since the very interests that supported the military were the same ones that were so avid to have access to the Amazon's resources. Conservationists had been persuaded that the military's technocratic approach combined with its hard-fisted ability to stamp out opposition to its will would mean the plan would succeed. They did not understand that the application of the hard fist would be exercised to promote various financial and corporate interests and against the interests of the poor and the forest. Conservationists had been seduced by the military's claim to technocratic efficiency and consistency, not realizing, as one would expect of anyone with long experience in Latin America, that the power of military regimes is nearly always exercised on behalf of various national and international economic interest groups who are the military's base of support.[8]

Many conservationists at this point privately lamented that the problem had been the implacable force of the wave of desperately poor immigrants into the region. They were often made vulnerable to this position by their implicit or explicit Malthusianism – a view largely discredited in economics and the social sciences, but that still has a lively place in the minds of some biologically trained people. What they saw was a very rapid growth rate in human population in Brazil, that in the early 1970s had approached a 3.4 per cent annual growth rate, in itself certainly a cause for alarm. What they did not see was that rural poverty in Brazil, according to nearly every serious analyst of the problem, came from massively unequal access to land and resources combined with a corrupt and violent rule by major landowners and their political allies. As ecologist/anthropologist William Durham has put it, in a rigorously theoretical and mathematical analysis of the case of the much more densely settled tiny nation of El Salvador, this was a case of competitive exclusion and not a case of overpopulation.[9] Neither did they see that it was poverty, inequality, repression, and lack of education that kept birth rates high even when death rates had fallen dramatically. Most conservationists were largely unaware at this point that Brazil had agricultural land sufficient to support the rural population in dignified livelihoods, as it still does. And in the 1970s they were unaware, as were most people including many professional demographers, that birth rates were about to fall dramatically to below one child per woman by the end of the century, with population growth rates as a whole slowing. Overall positive population growth rates were maintained with fertility rates lower than replacement rates, only by the age structure that determined that young people near or in their reproductive years were a disproportionate share of the population. Birth rates were to fall as a consequence of loosened restrictions on birth control information, pills, and devices, along with positive campaigns for family planning that were put in place as the reactionary military regime lost influence. Even more importantly, they fell as a function of urbanization, education, and the growth of both rights and education for women in particular, as was the case in most of the world. The Malthusian view that the Amazon was being chewed up by the population's propensity to expand beyond resources available was not an accurate analysis at the time of the Polo Noroeste Project, and it is even less viable as an explanation of the Amazon dilemma today.[10]

Conservationists could not allow themselves to see that they had been suckered into participation in a fundamentally corrupt development scheme, just as the settlers were being suckered into the region by false promises of wealth, livelihood, and secure landholdings. The Polo Noroeste Project left many prominent conservationists demoralized and more confused than ever about what was actually going on in the region and, in particular, about the complicated role of both long-established settlers and pioneers. On the positive side, more people in the conservation community did begin to think less in terms of implementation through authoritarian government and more in terms of the need for popular support for their objectives. They became more willing to begin the task of understanding the people of the Amazon forest.[11]

Emergence of the ribeirinhos

What came to be known as 'the rubber-tappers' movement, led by Chico Mendes, eventually brought some conservationists to see that not all settlers in the Amazon were the same in terms of their impact on the forest. Mendes and his followers did not begin their fight as 'environmentalists' or 'conservationists', but rather as people seeking rights to land: the eternal quest of the Brazilian rural poor. Many of the rubber-tappers and their allies were only part-time or occasional rubber-tappers – a better term might have been '*ribeirinhos*', or river people. People living along rivers had worked out combinations of small-scale agriculture, hunting, fishing, and gathering of forest products, including latex for rubber, that were relatively compatible with forest conservation (Figure 7.1). While they had worked out viable livelihoods, most of these people were desperately poor and had been viciously exploited for a century and more by middlemen who sold their products downriver. In many cases, they were being newly pressured by an advancing frontier of new outside settlers, ranchers, timber firms, mining companies, and government officials, including, at times, conservation officers of the government and supportive conservation organizations. The argument for the rubber-tappers as environmentalists developed out of confrontation and eventual dialogue between rubber-tappers, new migrants, conservationists, and

Figure 7.1 Houses and settlements of Brazilian *ribeirinho* people, near Belém, Pará. Typically, this type of settlement incorporates a variety of perennial tree crops, such as the acai palm (seen here) along with piaçava palms (from which is harvested a high value fibre), bananas, various tropical fruits, and other native trees. Fishing and sometimes hunting and rubber gathering are important elements of household economies, and in this area, *ribeirinho* children receive free public schooling and health care.

Source: Angus Wright.

government officials. They had the help of some very capable Brazilian and North American anthropologists. It began to emerge that river dwellers and conservationists shared many goals in their desire to slow or stop the advance of both old and new forms of exploitation of people and land.[12]

The formation of an alliance among forest dwellers, mostly working under the loose rubric of 'rubber-tappers', constituted a great advance for the cause of Amazon forest conservation. It made it possible for conservationists to shed their image, and sometimes their practice, as enemies of the poor. As long as they were seen as the enemies of the poor they would be opposed by the majority of Brazilians. Worse, in terms of strategic losses, conservationists would also be subject to cynical and highly effective attacks by investor and business interests who were able to exploit the perceived opposition between mostly foreign conservation organizations and the majority of the Brazilian people. Tying the fight against deforestation to the fight by and on the behalf of the poor was successful to a considerable degree. The positive link between conservation and the welfare of the poor changed the nature of the debate and of political battles fought over conservation at the regional, national, and international level.

Beyond debates and battles, the alliance between those in the Amazon who had established relatively sustainable livelihoods in and on the edges of the forest made it possible at last to formulate a credible programme for battling deforestation. It was obvious to virtually everyone, including most conservationists, that protected areas would end up protecting little in the way of habitat and species if they were seen simply as excluding Brazilians from using their own resources for the improvement of human welfare. The poor would not respect boundaries where they saw little reason to do so, and where survival was at stake. The vast distances and difficulty of travel would make effective enforcement impossible. Even more importantly, the powerful tradition of legal impunity for the rich and powerful in Brazilian rural life meant that reserves would be violated systematically by ranchers, timber companies, and mining corporations working under the cover of the claim to be contributing to 'national development' in terms of creation of government revenues and jobs. The fact that the most powerful of these extractive resource companies were international corporations with no particular long-term interest in Brazilian development was glossed over successfully. Even where law enforcement personnel were determined, skilled, and incorruptible, they were always stretched extremely thin and relied on the active aid and support of the local population – support they would never have if the cause of conservation was identified as opposed to the interests of the majority. The alliance between conservationists and organizations representing the poor was absolutely critical to overcoming these obstacles.[13]

As this alliance was forming, a new appreciation for the role of poor people in forest conservation began to become better known. Conservationists began to recognize that it was almost exclusively the rubber-tappers, indigenous people, and small-scale agriculturalists who actively risked their lives in collective actions to oppose wholesale deforestation. It was they who were most aware of violations of forest laws, it was they who put pressure on police to take action, and it was they who put their lives literally on the line in blockades against logging operations. When conservationists began to see the value of these activities, they were able to bring national and international attention to them, imaginatively creating films, television programmes, and fund-raising campaigns. Anthropologists helped by teaching indigenous groups, such as the Kayapo, how to make their own films with hand-held cameras and how to create effective media events – including one in which a Kayapo leader symbolically placed a spear against the throat of a high official of the World Bank in a public meeting. Small bands of urban conservationists involved in forest blockades made little impression on anyone. It was only with the clear and active participation of poor peoples' organizations that these activities could gain traction.[14]

The idea of 'extractive reserves'

One of the fruits of the new alliance was the creation of 'extractive reserves' as a new category under Brazilian law, an example of one component of the

Integrated Conservation and Development Units we discussed in Chapter 3. These reserves would allow rubber-tapping, gathering of wild fruits and Brazil nuts, fishing, hunting, and other activities considered compatible with overall long-term viability of the forest and a high degree of species conservation. Small-scale agricultural activities that did not involve massive forest clearing would also be allowed. Considerable effort by conservation groups went into the support of such presumably sustainable activities as bee keeping. Enormous 'extractive reserves' created under Brazilian law allowed for creative partnership between conservationists and the poor, literally 'on the ground'. Without doubt, the creation of 'extractive reserves' constituted a quantum leap in conservationist thought and practice, internationally as well as in Brazil.

The 'extractive reserve' concept alerted conservationists to the necessity of overcoming many of the eight systematic biases identified in Chapter 6. It became clear that it was necessary to examine human activities in and around forest reserves with a more discriminating eye. Not all agriculture was equally destructive, and considered within the wider context of the biodiversity conservation project as a whole, some agriculture and other activities could be considered supportive of the broad purpose of conservation programmes. There was a necessary recognition that agriculture could not be judged simply by its overall productivity, but by its meaning within a context of multiple human purposes, including conservation. Poor people were seen with fresh eyes, not simply through the narrow perspective of local and international elites. The necessity for conservationists to work to overcome cultural and language obstacles became obvious. To some degree, the importance of interdisciplinary training and cross-disciplinary teamwork became clearer. Conservationists began to have enough experience in Brazil and enough knowledge of its resources and social dynamics to see that the problem of the Amazon was clearly not in its most important aspects a Malthusian case of human populations expanding beyond the limits of available resources. The Brazilian 'extractive reserve' experience did not do away with these systematic biases in conservation work, but it began a process that is still evolving in a positive direction.[15]

Unfortunately, the 'extractive reserves' also revealed critical problems in the specific strategy, which in turn made it clear that some of the systematic biases were still at work among those who had not foreseen these failings. Had the problems been more clearly seen, it would not necessarily have been easy to design a strategy that would have overcome them. It would, however, have broadened the conservationist agenda, particularly at the national and international level, and it would have made it possible to some degree to anticipate and reduce the impact of some of the problems at the local and regional levels in Brazil.

As experience with 'extractive reserves' accumulated, several problems with them began to emerge. The most basic was that rubber-tappers and river dwellers, *ribeirinhos*, had always been very poor and highly exploited people and would remain so after the creation of 'extractive reserves'. The activities they used to support themselves were economically marginal, which is to say, with very little surplus available beyond subsistence needs. But while marginal, these were livelihoods that were partially and critically dependent on national and international markets. It was essential to earn cash, though the amounts might be small, beyond subsistence needs. Rubber, Brazil nuts, palm nuts, various marketable fruits, *piassava* fibre, and other emblematic products of the Amazon river peoples' livelihoods were dependent on national and international market demand. This demand was both limited and fluctuating. As the markets for extractive products were promoted by conservationists and government, they had an additional tendency towards saturation, such that early successes in the 'extractive reserves' often brought disappointment over time. For example, as the 'extractive reserves' were successful in increasing the harvest of Brazil nuts, the markets began to offer lower prices that undermined the entire strategy. These markets were also operated by middlemen, many of them working under the infamous *avia-mento* system of usurious credit and company stores, who ensured that when market demand was high, most of the benefit went to them, and when it was low, the effects were largely felt by the *ribeirinhos*. This relationship between producers and merchants was frequently enforced by violence, as it had been for centuries. The government lacked either the will or the ability to put a stop to it. Similarly, the violence exercised by ranchers and timber and mining operators, responsible for the assassination of Chico Mendes in 1988, did not end with Mendes's death nor with the creation of 'extractive reserves'. Shifts in economic growth rates in China and India, and changes in technological choices made with respect to the ratio of natural to synthetic rubber used in products manufactured abroad could and did have serious negative effects on rubber-tappers. The creation of 'extractive reserves' did not put an end to such difficulties, and in some cases may have reinforced them through market saturation. Many extractive communities that were meant to be the heart of the conservation efforts struggled to maintain themselves and were forced to engage in timber cutting and exploitation of forest resources beyond sustainable limits. Such activities included ranching, which was often favoured because cattle represents a source of savings against such emergencies as serious medical crises, can be a low management way of exploiting available land, and carries a certain kind of prestige in Brazilian rural culture.[16]

It was also naïve to assume, as conservationists often did, that the *ribeirinhos* lived in isolation from the corruption rife in local and national politics. So, for example, grants and subsidies meant to support such activities as bee keeping and aquaculture, often ended up in the hands of local political parties and bosses. Another issue arose with regard to the high levels of responsibility and authority granted to municipal governments under the new Brazilian Constitution of 1988. This meant that the success of 'extractive reserves' would depend considerably on both state and municipal government backing

for the programmes, in a context where some of these more local governments were supportive of the extractive reserve programmes and others were
outright hostile. There were also serious complaints that the national and
international conservation organizations tended to control the policies dealing
with the reserves and that they were more interested in conservation and
favourable publicity than the need for livelihood. Many of the grass-roots
organizations, including Chico Mendes's rubber-tappers' union, were strongly
interested in acquiring land titles and not merely access to reserve land, while
conservation organizations were less interested in or were opposed to granting title to poor people in the Amazon. This conflict has created ongoing
tensions and misunderstanding. Also, since much of the loan and grant money
meant to support local livelihoods in innovative ways came from international
conservation organizations, *ribeirinhos* and rubber-tappers became subject to
such problems as 'donor fatigue', the tendency for donors to conservation
organizations to lose interest in particular issues after a time, seeking excite-
ment and stimulation in new strategies and causes.[17]

During the first decade of the extractive reserve programme, Brazil was
also struggling to stabilize its economy. This meant that money for all government programmes was subject to frequent cut-backs. The Presidential
administration of Fernando Henrique Cardoso was committed to working
within what was then called 'The Washington Consensus', which meant
reducing barriers to trade and investment, minimizing government regulation
of business activities, and exercising severe budget austerity. Specific policies
based on these broad commitments had a variety of implications that tended
to increase pressures for Amazon deforestation. 'Neo-liberal' policies promoted rapid growth in export income, most easily obtainable by simple
resource extraction because mining and lumber operations required relatively
little capital investment. Austere budgets meant that funds for conservation
planning and enforcement were scarce. Illegal deforestation continued apace
inside and outside of 'extractive reserves' and other protected areas. The traditional legal structure and law enforcement agencies continued to work
effectively to protect the middlemen, timber and mining companies, and
ranchers. In contrast, the new conservation laws and agencies meant to
protect the interests of communities of poor people in the 'extractive reserves'
were under-funded and relatively weak. These problems continued under a
more socially and conservation-minded government led by President Luis
Inácio (Lula) da Silva that remained committed throughout his presidential
tenure, if not quite so firmly, to neo-liberal policies.[18]

All these problems that would arise in the extractive reserve initiative
could have been and mostly were foreseen by those with a reasonably deep
understanding of Brazilian life and politics, and of national and international
economics and markets. Such people had less influence on the extractive
reserve programme than they might have had because of their impatience to
do something positive about Amazon deforestation. But history and large-
scale context did matter and needed to be understood to avoid undermining

the programme and seriously discrediting it in the eyes of many participants and observers.

In fairness it must be said that some of those involved did, to some extent, appreciate the potential for these difficulties. It was understandable that they would downplay them in order to gain three steps forward even if it might mean taking two steps back. However, the cost was not only felt in the two steps back, but in the general demoralization that has ensued in some places as problems arose. This problem may, in the long run, prove extremely serious in blunting the momentum of conservation politics in the Amazon specifically, and in deflating the potential for larger-scale political change both in Brazil and internationally. Overselling an idea always runs serious risks. It was also the case that many of the conservationists were caught off guard by the problems that arose in the 'extractive reserves'. Some still operate with their blinders on because they cannot bring themselves to face the truth. Many have jobs or CVs that will suffer if the matter is discussed more dispassionately. Such considerations aside, it is also clear that if the extractive reserve scheme had been examined with a greater attention to the implications of deeply rooted historical factors and the national and international context, there would probably have been ways of designing it better, and there would certainly have been more realistic expectations about its promise and limitations.

A positive result of the earlier failure in the Polo Noroeste Project was that some conservationists did learn that they needed the support of grass-roots movements in the regions in which they operated. This contributed to their willingness to join with the rubber-tappers in trying to create an alternative. However, as we have seen, they had still not entirely learned how complex and challenging the problems of conservation in the Amazon could be. They tended to romanticize the rubber-tappers and indigenous people and did not face up to how severely constrained these people were by economic and political realities in a context of a fundamentally corrupt legal system that had been designed to serve the interests of those who controlled national and international markets. Conservationists seem to have a difficult problem understanding how profoundly rooted are problems of poverty and corruption. As a consequence, a good but limited idea was promoted as a broad and comprehensive solution.

In the end, the fate of the Amazon forest may critically depend on social policies and events that seem to many conservationists to be only marginally related to conservation. For example, subsequent to the creation of the extractive reserve programmes, the Brazilian government began to experiment with direct grant programmes for poor people. These programmes began with municipal sponsorship, then moved to the national level, and were expanded greatly under President Luis Inácio (Lula) da Silva, who took office on the first day of 2003. The programmes include access to a 'basic food basket', and 'family grants' that are given to low-income families who keep their children in school, send them for their required inoculations and health clinic visits, and meet other basic criteria for responsible parenting. These programmes have made a significant difference to poor people throughout Brazil. To the surprise of many, it does not take a large portion of the national budget to add significantly to the income of a large group of desperately poor people, so the programmes can be maintained while sticking to budget criteria determined by the Lula administration's continued adherence to neo-liberal policies.

What is not well understood is how these programmes affect the dynamics of deforestation and the practice of sustainable livelihoods in the Amazon and other areas critical to biodiversity conservation. Did the grant programmes give families the margin they need for dignified survival and thus make it possible for them to refrain from activities that undermine conservation efforts? Or did they give poor people additional resources that allowed them to become more aggressive settlers, less tied to the need to create sustainable livelihoods? Has it become more difficult, or at least more expensive, for timber firms and mines to attract people to dangerous and low-paying jobs? Did relative economic security reduce the corrosive effect of corruption and lawlessness, as many believed it would, or would the municipally administered programmes for the poor end up simply as a tool of corruption and control? At this point, we still do not know the answers to these questions.

Upon reflection, it seems likely that the various support programmes for poor people could have exerted a larger influence over biodiversity conservation, for good or for ill, than the 'extractive reserve' programmes,[19] although under the new government neither seems very active. We noted in the previous edition of this book that 'National context may matter far more than "on-the-ground" programmes that conservation organizations often design with insufficient attention to the influence of the context itself.' It seems that subsequent activities in Brazil made our point well.

Under the governments of Lula and his successor, Dilma Rousseff, deforestation rates in the Brazilian Amazon fell significantly. This was due to improved satellite monitoring techniques that allowed for better enforcement, but it was the willingness of the Worker Party governments to pursue enforcement that was essential as well. The '*ruralista*' faction of the Brazilian Congress, made up of an alliance of powerful ranchers and entrepreneurial farmers with national and international agribusiness corporations, declined in influence under Worker Party governments. This alliance represented the most important and aggressive force carrying out and lobbying for wholesale forest destruction. The Workers Party considered the *ruralistas* enemies on the basis of ideological affinities – they were seen as among the most generally reactionary groups in Brazilian politics and culture – and conflicting specific policy goals such as slowing Amazon deforestation. The Workers Party worked to slow deforestation based on support for the Amazon poor, including the Rubber Tappers, and urban environmental allies. The *ruralistas* were among the most powerful – if not the most powerful – of the groups that first weakened forest protection rules passed by Congress and then toppled the

Rousseff government altogether. Not surprisingly, Amazon deforestation rates began to climb rapidly again, as well-capitalized farmers and ranchers cleared the land for massive operations.[20]

Similar issues arise throughout the tropics. In the tropical forests of Africa, conservation efforts are severely threatened by war and failed states. In much of Southeast Asia, forests continue to be devastated by companies operating under the sanction of very aggressive programmes of national economic growth within the forested nations and their main trading partners. In much of Mexico and Central America the coalition of local ranching land barons and national governments maintains a rapid pace of deforestation even when local people organize determined resistance movements. In every case, the pro-globalization, neo-liberal 'Washington Consensus' pushes relentlessly for the rapid and undiscriminating economic growth policies that help fuel these problems.[21]

Conservationists cannot be expected to be able to control these problems. Nor can they be expected always to be able to minimize the effects of all of these problems simply through greater understanding – it is difficult to do much about some stubborn realities. In order to achieve long-term success in preserving biodiversity, however, they need to do two things: first, they should realize that they have little chance of success if they do not recognize the importance of understanding the local, regional, national, and international context in which they work. Second, they can, so far as possible, ally themselves with the organizations and movements that are working for positive change.

Direct action land reform

During the time leading up to the assassination of Chico Mendes and in the years following, a well-organized movement for agrarian reform was making itself known throughout Brazil. The distribution of Brazil's vast and rich agricultural land originally took place amidst the violence and injustice inherent in colonial rule, but with particular aspects that created remarkably high levels of inequality, even among colonial regimes. The economy was driven by large plantations worked mostly by African slave labour. Land law was written by and for the plantation owners. The result was a society that was marked by enormous wealth for the few, and grinding poverty for the majority. Even after Brazil had become one of the world's largest industrial economies (varying between the eighth and twelfth largest during the latter part of the 20th century), with more than 85 per cent of its 210 million people living in cities, it remained among all countries one of the worst in terms of the gap between rich and poor.[22]

Since the late 19th century, there had been one attempt after another to try to address the inequalities in the countryside (which in a variety of ways helped to recreate and maintain inequalities in the cities) through a land redistribution programme. None made much significant progress. A military coup in 1964 established a severely repressive government, in considerable measure to destroy increasingly powerful political movements demanding land reform. By the mid-1970s, the military regime was beginning to weaken and a new movement for land reform was slowly taking shape.

One of us (Wright) has written extensively about the *Movimento dos Trabalhadores Rurais Sem Terra* (MST) – The Movement of Landless Rural Workers. What follows is a summary of the movement as it relates primarily to biodiversity conservation.[23]

The main method of the new land reform movement was to organize hundreds of land-hungry people to occupy land whose ownership could be effectively challenged. This was the Brazilian version of what has come to be known internationally as Direct Action Land Reform (DALR). Under Brazilian law, if land was not being used for productive purposes, the title to it could be challenged by a counter-claimant. This law was meant to encourage settlement and had come to be systematically manipulated by influential landowners in order to wrest land from the less powerful. In the US, similar provisions of law, such as water law in many states, are referred to by the phrase 'use it or lose it', or more formally, as 'effective use law'. Also, because Brazilian land law had become enormously complex, self-contradictory, and corrupt, most nominally legal titles were subject to court challenge as fraudulent. In addition, considerable land was held by the states and federal government and was subject to claims of private appropriation in the name of putting it into productive use. By occupying land and mounting a legal case for ownership, the land-hungry rural poor were using the contradictions of Brazilian land law that had always served the interests of the rich in dispossessing the poor.

The land occupations were usually met by long campaigns of legal manipulation, intimidation, and violence, carried out by a combination of police, military, and private gunmen hired by wealthy landowners. In the late 1970s and later decades, however, the land reform movement was able to force the government to grant land to the occupiers, although often at great cost in lives and trouble. The amount of territory in land reform units by the end of 2005 was officially about 60 million hectares and was allotted to about 667,000 families. This accounted for more than 16 per cent of all Brazil's farmland. The process continues, with land reform advocates committed to gaining more land, and the government committed to an emphasis on making current land reform settlements viable through technical and financial support,[24] although actions of the current conservative government threaten its viability. This development would come to have a variety of important implications for the future of Brazil's forests and other wildlands.

The Amazon played a major role in the drama of land reform, but not simply in the straightforward sense of being a source for land reform appropriations. One of the chief tactics of the national government to try to forestall the development of a new land reform movement was to offer land grants in the Amazon. In 1970, Brazil's President announced that in the

Amazon, Brazil could provide 'a land without people for a people without land'. 'The land without people' being President Medici's quite inaccurate description of the Amazon. This was to serve the purpose of avoiding land redistribution in Brazil's established agricultural regions while providing labour for the ongoing development of the Amazon for economic purposes. The Amazon had a very negative image in the minds of Brazilians who had seen one settlement scheme after another fail in the face of harsh conditions in the region. Many of the poor who were there had been brought there under conditions of virtual slavery – this included many of the rubber-tappers who had been literally drafted into 'the rubber army' during World War II, received little or no compensation, and were left abandoned at the war's end. The Brazilian poor had often personally known families whose lives had been devastated by ill-advised attempts to settle in the Amazon, often succumbing to malaria and an array of other diseases endemic in the tropical forest. The poor were not clamouring to get to the Amazon. To the contrary, the military regime understood that it would have to attract them there by promises of land grants, schools, clinics, roads, and housing. The government hoped to lure people to the Amazon with the promise of land, not only, and perhaps not mainly, to clear new agricultural land, but also to use these people as labour for road and hydroelectric dam building, timber extraction, mining, and industrial enterprises, partly and precisely because agricultural settlement would in many cases fail. It was the military government's initiative in the Amazon that set off worldwide concern for the future of the world's largest tropical forest and, overwhelmingly, its most important reserve of biodiversity.

The government's programme to lure people to the Amazon posed a strong challenge to the land reform movement. The bitter historical experience of Brazil's poor was that frontier development had always been disappointing. For centuries they had been induced to clear forests and scrubland because they lived in desperate poverty in developed agriculture areas and were offered the hope of something better, including the possibility of owning land in the new frontier. On the frontier, they were often able to make and defend claims to new land for a time, but were nearly always displaced as the more powerful land barons moved in. The vast majority of those who had invested their labour and faced danger and disease in opening the frontier would lose whatever land they had gained. They would become, once again, hopelessly indebted dependents of wealthy landowners. Virtually all of Brazil's once enormous area of Atlantic Coast rainforest was felled in this manner, where it had not been cleared earlier by slaves. According to some estimates, more than 89 per cent of the Atlantic Coast rainforest has been cut.[25] In spite of this experience, it had proven possible to lure the poor to repeat it again and again, driven by desperation and a shred of hope. The land reform movement wanted to avoid leading people into this trap again, and they rightly foresaw that this was what would happen in the Amazon.

The leadership of the land reform movement also believed that for viable communities to grow from land redistribution, it would be critical for people to remain in the regions where the soils, climate, culture, sympathetic public figures, and neighbours were familiar to them. They also believed that the process of land reform would never be successful over the long term unless the political power of the rural oligarchies was successfully challenged. Land reform meant not only land redistribution but democratization of the countryside and of national politics. If people were continually lured into new frontiers none of this would happen – to the contrary, dispersed and unorganized families would be dependents of government colonization schemes and land barons rather than an organized political force in their own right. An authoritarian and reactionary regime of power would be extended over a larger territory. Following this logic, the land reform movement, and its most successful organized proponent, the MST, firmly rejected Amazon colonization and frontier development, consistent with their vision of what land reform was meant to accomplish, and consistent with the requirements of a viable organization.

In spite of this clear and consistent stand, conservationists tended to be wary of or hostile to the MST and other land reform organizations. They tended to see all small-scale agriculture in the same negative light. Often conservationists talked privately of the 'rapists and the ants' as the twin causes of deforestation, with the rapists being the large, well-financed landowners and corporations, and the ants being small-scale agriculturalists nibbling away furiously at the forest edge. Both were seen as the enemy, and only seldom was it remembered that if the ants were destroying the forest, they were encouraged or compelled to do so by the rapists. In any case, the objectives of conservation were largely defined as protecting forest from everyone and from all forms of encroachment.

Conservationists, to the degree that they knew anything at all about Brazilian land law, also worried that the 'use it or lose it' provisions of the law provided a powerful incentive for conversion of all land into agriculture as a requirement of ownership. When they saw the land reform movement making use of this provision of Brazilian law, they saw the movement as an implacable enemy, even after the law was definitively changed in this regard, with the cooperation of the land reform movement. A narrow focus on conservation has often resulted in a failure to appreciate or even acknowledge those aspects of the land reform movement that put land reform on the side of conservation.

The first and most obvious thing that conservationists seemed oddly blind to was the fact that the major land reform organization was among the most prominent and militant opponents of wholesale development of the Amazon, for the reasons cited above. They missed that this was not simply an opportunistic public relations stance by the MST, but a deeply self-interested position taken on behalf of the organization and its members. From the 1970s through most of 1990s, conservationists could not seem to see that if land were redistributed in already developed agricultural areas then the pressure on the Amazon would be eased by the reduction in the number of desperate rural people willing to take on the multiple challenges of life as Amazon settlers. Many still cannot see this. The majority of conservationists were taking for granted that the government's programmes for colonization of the Amazon were sincere but misguided attempts to provide land for the poor. Many Brazilian and foreign analysts were convincingly making the case that the government had little interest in the welfare of the poor, but rather was cynically manipulating them on behalf of Brazilian and international corporations interested in timber and mineral resources, and to protect what the Brazilian military saw as an imperative to settle the Amazon in the interests of national security.[26] Far too many conservationists for a long time failed to listen to this critique and made many mistakes as a consequence, wasting a lot of time trying to explain to the government the limitations of Amazon soils for agriculture. In fact, the government was far more knowledgeable and skilful at assessing the possibilities and limitations of Amazon soils than were conservationists, but had different objectives both to the conservationists, and to those the conservationists imagined were driving government policy.

By the early 1990s the land reform movement had to reassess its position with regard to the Amazon. They had not been very successful in slowing the pace of Amazon development, nor had anyone else. They also continued to struggle mightily with the problem of the legal impunity of influential landowners, a problem that undermined almost everything they accomplished. The Amazon frontier was a notoriously violent theatre of conflict between poor settlers and richer ranchers and agribusiness operations. In the most violent regions of the Amazon – which were also the areas where the most rapid deforestation was occurring – poor people were desperate to create or join organizations that could defend them and represent their interests. Older 'rural union' organizations had established a presence in the Amazon, but they were largely ineffectual. Many settlers hoped that the much more aggressive MST would better represent their interests. The MST began to move into the region. They organized themselves aggressively in selected key areas where they quickly moved into conflict with the dominant rural power structure. In 1994, a march of MST members in the southern part of the state of Pará was attacked by local police and state-controlled military forces, resulting in the deaths of at least 19 MST members. The incident shocked urban Brazilians who had been largely unaware of the struggles going on in the Amazon. Many felt compelled to protest and, perhaps more importantly in the long run, to try to understand what was happening. This led to a period of wider sympathy with the land reform movement and to a deeper understanding of the dynamics of Amazon deforestation.

In the Amazon, the MST must deal with difficult contradictions. It represents settlers who often have little experience or knowledge of farming in a humid tropical environment. Many of the settlements are unlikely to be sustainable for long. The MST is proud of always aggressively defending the interests of its members, but is it to conclude that this means commitment to a moving frontier of settlement? Does this mean that the organization will be abandoning its rejection of Amazon settlement as a solution to the problem of Brazil's landless rural people? We do not know the answer to these questions yet.

The MST has not always been clear and consistent in meeting this challenge. Vague statements about sustainability and respect for conservation efforts sometimes contrast with much more militant statements and actions that simply defend MST members in disregard of conservation questions. This is sometimes true whether or not their farms and their claims on new land represent sustainable enterprises. Fortunately, however, there are also more positive and significant elements in MST policies.

A significant positive development in MST policy is the product of another kind of inexorable logic about responsible land reform. The MST began with a rather simple-minded approach to the question of what kind of agriculture served the interests of its members. They were inspired by three sources: first, the kind of highly capitalized, highly mechanized, and chemical-dependent agriculture practised by those who were driving them off the land in Brazil, which they assumed had found the key to success since it was so successfully displacing them; second, the same kind of agriculture as seen in agricultural publications and taught to agronomists in Brazil; and third, the same kind of agriculture as seen in the US, Cuba, and the Soviet Union (places some of the membership had visited and others had admired from afar). This agriculture was mostly devoted to producing a limited range of basic commodity crops – maize, wheat, soy, and sugar – sold in highly competitive international markets.

The MST soon ran into several problems attempting to implement this vision of agriculture on land granted through the reform programmes. The first was that farmed as individual units, the farms were too small to support the level of capital investment required. While some in the MST wanted to establish large collective or cooperative farms, most of these ventures did not succeed and most members wanted to farm their own individual holdings. The second problem was that it was very difficult to get credit – government credit programmes were insufficient and private creditors had little faith in the land reform and its beneficiaries, and in any case, charged very high interest rates. Third, when MST farmers did succeed in surmounting these obstacles, they began to see that chemical-dependent agriculture based on one or a few crops with little crop rotation and diversification created a whole range of difficult problems (as discussed in Chapter 4) that required more capital to deal with and in the long run threatened irreparable harm to the land. Unlike the larger, more heavily capitalized farms, they could not compensate by leaving large amounts of land in fallow, nor by buying land in new frontiers of agricultural settlement, nor by seizing the land of other less powerful farmers. The land they had received through the reform would be the only foundation upon which they had to build their survival, which would be severely compromised if unsustainable methods were to render the land unproductive. The land reform law prohibited multiple serial assignments of land to the same family meaning that starting over was unlikely. Furthermore, they had little knowledge of, and no control over, the distant and volatile markets in which the major commodity crops had to be sold, threatening unsustainable local economic conditions also.

As this experience was accumulating, the MST also worked at getting advice from more progressive agricultural scientists and technicians in Brazil, and at building alliances with other groups in Brazil and abroad. Some of the more progressive people in agronomy were participating in a national and international dialogue and process of experimentation towards the creation of more sustainable forms of agriculture. Some conservation and social change groups they allied with were strongly committed to the cause of sustainable agriculture and agroecology, and many were eager to offer advice and training. The MST began to come into contact with the growing and lively community of people practising alternative agriculture in Brazil. A new vision began to form.[27]

The MST began to realize that they needed to practise techniques that preserved their hard-won land. They needed to minimize capital requirements, meaning that they needed to reduce reliance on chemical inputs; they needed to plant a diversity of crops; they needed to practise techniques to avoid damage to the soil and to control pests at minimum expense; and they needed to be able to sell in markets that were less volatile. They began to sell a diverse array of fruit and vegetable crops, fish from aquaculture ponds, poultry, honey, and herbs in local markets. This led to better appreciation of the advantages of capturing more profit by on-farm processing that added value to the product. They also made more profit from direct marketing in local farmers' markets. Participation in local farmers' markets brought them into more contact with neighbours and townspeople, improving the level of community understanding and support for what the MST was trying to achieve. Every aspect of this process was continually reinforced by people in many of the other groups with which the MST was allied, in Brazil and abroad.

By 2001 the MST was ready to announce that it was as an organization committed to practising 'agroecological farming', based on this new vision. Many members of the MST had little or no idea what this meant and others opposed the idea. However, there were many in the organization who had substantial traditional knowledge consistent with sustainable practice and had been putting that knowledge to work. Others were well ahead of the leadership in learning from the new local and international sustainable agriculture movement and from traditional practitioners in their regions. With some resisting and others out ahead of the central leadership, the MST has been unable to implement its vision consistently. It is, however, insistent on pushing forward with an agroecological ideal and with promoting and teaching the practice of it to its members. In addition, the MST has joined a coalition of environmental groups in consistently opposing the government's approval of genetic engineering in agriculture, where a convincing case has not been made for the value and safety of genetically engineered crops and livestock.[28] The organization also insists that the question of ownership of intellectual property rights in genetically modified organisms has to be resolved in favour of equitable access and use. The MST has also opposed a variety of environmentally damaging programmes supported by the government, including large eucalyptus plantations for cellulose and bio-fuel production, diversion of the scarce water resources of the Sao Francisco region, various dam projects, and incursions on indigenous territories. The MST has been among the sharpest critics of the enormous expansion of sugar cultivation for the production of ethanol, as proposed by the Brazilian government and an alliance of agribusiness interests. It consistently opposed the expansion of the large-scale soybean farms that have been devastating the savannah region (the *cerrado*) and, increasingly, the humid forests of the Amazon. Although in most of these cases they were working in collaboration with environmental groups, in almost all of the cases it was the MST, and very few of the environmental organizations, which was willing to mount aggressive and persistence actions that might make some difference. MST members have put their lives on the line in blockades, sit-ins, and other direct action with regard to important environmental issues over and over again, while in most cases environmentalists took little action beyond statements of protest and attempts to lobby politicians, putting little at risk. Still, some environmentalists are reluctant to give up long-held prejudices that make them wary of land reform in general and the MST in particular.[29]

In the Amazon, the MST, while jealously guarding its organizational independence, is eager to cooperate with environmental groups and conservationists. It tries to draw on the experience of those within the organization who have experience with sustainable agriculture in the Amazon. Many of these people are illiterate or semi-literate, so learning from them requires patience and close observation. Many MST settlers in the Amazon are anxious to minimize their impact on the neighbouring forest and are eager to find a way to do so while making a living. Fortunately, the MST has since its founding been based on the innovative educational ideas of Brazilian educator, Paulo Freire. Freire's methods encourage people to think in ways that are both relentlessly critical of existing realities and at the same time highly practical; they are ideally suited to the task of creating new kinds of livelihood when faced with new environmental contexts. One finds MST members in the Amazon practising a rich mixture of new and old techniques with an experimental attitude and a strong environmental consciousness (Figure 7.2). They represent an unusual opportunity for collaboration between farmers and conservationists.

At the level of national policy, the MST position remains sharply sceptical of plans to develop the Amazon and is opposed to continued deforestation in the Amazon and elsewhere in Brazil. However, it no longer attempts simply to stand in the way of Amazon settlement. Both the MST and conservation groups have come to realize that this cannot be done. What might be

Figure 7.2 Raimundo Pereira Galvão, a settler in an MST land reform community in southern Pará, Brazil. Pereira Galvão has planted 15 perennial tree and bush crops in order to achieve some of the agronomic advantages of a diverse mix of perennials in a rainforest environment. He also preserves a portion of his land in native rainforest, in accordance with Brazilian law.

Source: Angus Wright.

possible, however, is a set of new alliances between those fighting for social justice and those fighting for conservation: alliances that would go beyond the significant achievements of the rubber-tapper–conservationist alliance that led to the creation of the 'extractive reserves'.

Agrarian reform: a transcendent issue

In 133 BC, the Roman tribune Tiberius Sempronius Gracchus proposed the *Lex Sempronia Agraria* law, which would have provided land rights to landless peasants. During the final century of the Roman Republic wars in far off places required long deployments of legionaries, most of whom were farmers who left the farm with wife and children to defend the Republic. Frequently the understaffed farming unit would go broke and the land would be sold to another farmer, usually someone who already had a lot of land and worked the land with slaves rather than with his own hands. Returning legionaries, now landless, filled the streets of Rome, creating the sort of urban crisis so familiar (unfortunately) to those of us in the modern world. Tiberius sought to rescue some of that land and redistribute it to these homeless families – the basis of the *Lex Sempronia Agraria*. His reward was murder at the hands of senators with strong interests in protecting the land rights of another group: those who already had the land.

Ever since (and certainly before), the issue of who 'owns' the land[30] and who is without title has been an issue that drives historical change. Agriculture is obviously dependent on access to land, and the socio-political arrangements that dictate agrarian laws are seemingly under continuous dispute. Those with land normally also have political, as well as economic, power, and the legitimate argument that their way of life, and consequently the way of life of the country, nation, or state, is dependent on their owning large tracts of land. Those with no land have a much more direct argument: that they cannot feed themselves and their families without land, but frequently lack the economic and political power to confront those with the power. When the landless organize themselves into socio-political groups to exercise the obvious political power that emerges from group action, political power emerges. The political power of the landed rich then confronts the political power of the landless multitude – and sometimes agrarian laws are changed.

This process happens the world over, and has been a key element in political change. But the modern form, especially evident in Latin America, is what has sometimes been referred to as DALR, as introduced earlier in this chapter, in which the landless simply occupy land that is under-utilized by a large landowner. Armed with complicated legal structures that frequently enshrine the right to land only if it is utilized, such direct action basically challenges the state to intervene, as in the case of the MST as discussed in the previous section. The most common pattern is when an organized group of landless farmers occupies a piece of land that is evidently not being used for productive activities – frequently land that is being held for speculative rather than productive purposes. Governmental authorities are then challenged to adjudicate proper tenure. The arguments can become complex and politically loaded, but basically boil down to two contending ideas about who is the true owner: the landless family whose claim is the historical right to land, or the current owner, whose authority usually rests in a contentious claim that his or her ancestors somehow 'claimed' the land. Given the underlying construction of the problem, it is not surprising that such land takeovers are frequently referred to (by the landless engaged in the takeover) as land 'rescues', and why 'redistribution' is a word frequently employed. Likewise, not surprisingly, those with 'legal title' refer to such actions as 'invasions'.

From the perspective of this book, there is an obvious point that merits consideration: that 'under-utilized' land may very well be an old-growth forest or pristine savannah that contains huge amounts of biodiversity. The agrarian laws governing such land rescues have, in recent years, been significantly modified so as to avoid the claim that a biological reserve (say) is actually under-utilized, and most DALR efforts are in fact aimed at large, mostly absentee landowners who produce little, maintaining much of their land for speculative purposes. The matrix to which we have repeatedly referred in previous chapters is often the target of groups seeking land reform. It would seem obvious that our argument for a high-quality matrix as key to biodiversity conservation includes an argument to take into account the activities of the landless who are determined to be landed.

The dependency trap in biodiversity conservation

The essence of living and working within ex-colonies, with 'dependent' economies, is that one's livelihood is strongly affected by decisions and events that are taken elsewhere and that are unusually difficult to control. In one sense, this is clearly a matter of degree because nearly everyone's livelihood is affected to some extent by things that are remote and hard to influence. However, the degree to which it is true is absolutely critical to people's lives, and is dependent on particular kinds of relationships among national economies. A freeze in Brazil can result in higher coffee prices in New York or London, which may result in inconveniences for cash-strapped and caffeine-starved students and office workers. In contrast, if the price of coffee falls significantly on international markets, it is not simply a matter of inconvenience for those in coffee-producing regions and nations. A fall in coffee prices may result in tens of thousands of farm workers being laid off or tens of thousands of small-scale farmers loosing their livelihoods in a producer nation. Thus, low commodity prices have often resulted in generalized hunger and mass migration. Sometimes landowners are unable to pay off loans and therefore lose their land. Local people engaged in all the activities that depend on the coffee economy suffer as well. In most coffee-producing nations, economies are not highly diverse and domestic markets are relatively weak, so changes in commodity prices ramify strongly through the entire economy and political system. Brazil used to fit this description, but industrialization, a more diverse agriculture, and a better-developed local labour force and internal markets have greatly reduced the problem. Most of the coffee-producing countries of Central America and Africa still suffer drastically from coffee price fluctuations. Frequently, such price declines are quickly reflected in severe political crises as governments plunge into dramatic debt and are unable to protect citizens from the cascade of events that ensue.

The history of most tropical and sub-tropical countries can, to a large extent, be written in rough form as a reflection of the rise and fall of basic export commodity prices. Such crises can have dramatic effects on the survival of natural areas and biodiversity, sometimes leading to expansion of agriculture to compensate for low commodity prices, sometimes leading to timber cutting and other forms of raw resource extraction as the most readily available source of income when there is little money for capital investment. In some prolonged falls in commodity prices, there may be a benefit to biodiversity, as land is abandoned and allowed to return to natural vegetation. But more often, such abandoned agricultural land is turned over to extensive uses, such as ranching, with little if any gain for biodiversity, and often a loss.[31]

The world's developed economies are more buffered against these fluctuations in export commodity prices. First, they are highly diversified. Declines or disruptions in one part of the economy are usually cushioned by relatively normal conditions in other sectors.[32] Export dependence tends to be relatively low – most people engaged in making, buying, or selling goods and services are dealing with people within their own nation. Purchasing power is spread relatively broadly among the population so that there is a large and resilient domestic market. Political institutions are able to reduce the stress on people and firms in economic sectors that are suffering – most famously through commodity price support programmes in such areas as wheat, maize, and cotton agriculture. And governments have many other ways of taking the heat off distressed economic sectors. They create and maintain broad social safety net schemes, such as unemployment insurance and social security. Most financiers and corporate heads have at least some interest in and awareness of the need for maintaining the economic health of the majority of the population in order to provide a strong domestic market. They also see the importance of basic social services such as health and education that provide skilled and healthy workers. That this basic structure and set of perspectives in richer nations has been considerably shredded in recent decades clearly modifies this picture, increasingly so since the first edition of this book. But relative to most poorer nations, the contrasts still remain strong, if not stark.

The strength of the richer nations with respect to the poorer has been a fundamental fact of international relations for several centuries. Beginning in the colonial era and continuing to the present, richer kingdoms and later nations have been able to build diplomatic, military, trade, financial, and even cultural relationships that maintain and often reinforce the differences between rich and poor 'countries'. These relationships have been so meticulously built and defended for so long that they constitute a rather rigid structure of international power. Many are so ingrained that most people in richer nations tend to take them for granted and do not question their purpose or actions. They are more obviously troublesome to those in poorer nations who suffer from them. Indeed, one way to write much of the history of poorer countries is through the stories of leaders and political movements that have from time to time attempted to effect positive change in their own countries by negotiating better deals within the structure of international power, by trying to reform the structure, by trying to become more independent of it, and/or by direct diplomatic or military confrontation with it.

As discussed above, the neo-liberal model (also known as the 'Washington Consensus') which has been so influential in shaping international relations since the end of the Cold War, relied on persuading or pressuring poorer countries to accept a set of international agreements and policies that largely had the result of strongly reinforcing the older structure of power. The basic argument of this neo-liberal focus is that 'a rising tide lifts all boats' – that is, more economic activity results in better lives for rich and poor alike. The way to promote a rising tide in this view is by reducing barriers to trade, lowering taxes and lowering government expenditures on social services, and lightening the burden of regulatory measures on business. Freed to invest, buy, and sell as they like, businesses will be able ultimately to produce more goods and services and employ more people. Those who reject this argument, and we are among them, argue that such an approach at best maintains the inequality built into the old international relationships, and at worst, makes these inequalities much worse. When the neo–liberal model has its way, in the best of circumstances, the rich get richer and the poor get somewhat better off. In most real-world circumstances, the poor are helped very little or are harmed. In virtually all circumstances, the international structure of unequal power is strengthened and thus dependent economies become more dependent than ever.[33]

One of the most powerful effects of the structures of international economic relationships, even before neo–liberal dominance, is that poorer countries are chronically deprived of capital investment for economic growth. Apparently, the easiest solution to this problem is an unhealthy emphasis on raw commodity production – including timber and minerals – and cheaply produced agricultural products that require relatively little capital investment. Production of such commodities means that natural habitats and biodiversity pay a disproportionately high price. Furthermore, because of the relative vulnerability of dependent economies to economic slowdowns, cycles of severe debt are characteristic. Debt creates yet stronger incentives to rely on production of commodities that require little new capital investment, and once again, biodiversity suffers as a direct consequence of economic dependency.

It is essential to understand the importance of these issues when considering how to protect and prevent losses of biodiversity in the tropics. The situation of economic dependency is one that affects most of the countries that shelter most of the world's biodiversity. This creates several major problems. The first is obvious: most such nations are poor and are not inclined to favour conservation over opportunities for increasing economic activity. This aspect of the situation is understood by nearly everyone involved in the controversies over biodiversity protection. However, some are more responsive than others to the concerns it raises. Some are aware of the problem yet find it so overwhelming that they are forced into a posture of ignoring it. Some simply declare biodiversity to be so important that it cannot be allowed to play second fiddle to human welfare. In any case, the nature of this argument has long been clear and we can shed little new light on it here. We do take it as a given that no discussion of biodiversity protection that seeks to gain enthusiasm and consent within the biologically richest nations, most of them economically poorer nations, can long ignore the need for substantial improvement in human welfare. Conservation plans need to be at least compatible with measures taken to improve peoples' lives. Ideally, they should make a positive contribution to both conservation and human well-being. There are relatively few in the conservation community who openly disagree with this point of view.

A second and not quite so obvious problem created by structures of economic dependency is that many plans, projects, and schemes put forward by conservationists may fall disastrously into what some analysts have called 'the dependency trap'. Worse, in many cases conservationists may inadvertently be making the problems of economic dependency more severe than before they entered the scene, with plans and projects that recreate and reinforce the structures of dependency and poverty that are the chief threats to natural habitats and the biodiversity contained within.[34]

The most obvious form of dependency for conservation projects in nations of the Global South is the straightforward need for continual financing from other national governments, international organizations, and individuals. The national governments of tropical countries are often unable or unwilling to provide reliable financing and many have come to rely on outside funding for particular conservation efforts. This is often the case even when protected areas are designated as national parks or reserves, and administered by national governments: continual active administration and enforcement sometimes depends on continued external sources of money. When a funding cycle by, say, the UN, USAID, or the WWF comes to an end, the question may remain whether any new funds will be forthcoming. When they are not, or when they are reduced, the whole conservation project can languish or fail altogether.

Other forms of dependency can be built into plans to provide an ongoing local economic foundation for conservation projects. For example, some conservation projects rely heavily on tourism as a way of harmonizing conservation and economic development. To be compatible with conservation goals, tourism must be kept within strict rules and boundaries. Ecotourism of this type appeals only to a certain fraction of the tourist trade, while the majority of tourists would favour, say, a large new resort hotel with all the conveniences over a modest thatched building that is more compatible with conservation. Consequently, investors are generally more attracted to conventional than to genuinely conservation-based tourism. The experience of the last decades clearly shows that there is a healthy market for authentic ecotourism that lives within the limits of what the environment can bear, but it is a distinctly small portion of the total tourist trade.[35]

The kind of tourism that is based on luring visitors to healthy local ecosystems in the tropics is, sadly, another tropical luxury commodity, in the sense that – like coffee, tea, bananas, and sugar – it is subject to strong demand elasticity. Relatively minor downturns in richer countries can have disastrous effects on ecotourism, as it can on international tourism more generally. This problem is becoming more acute as there has been a rapid proliferation of ecotourism opportunities competing for the international tourist's dollar.

In addition, most of the people who are enlisted to work in the ecotourism industry have relatively low skills. In many instances, people with substantial and valuable local knowledge in farming and forest survival skills become baggage handlers, drivers, maids, and cooks. Some become guides who are given scripts to follow. Most guides are not recruited from the rural people living in and around the forest or other natural habitat, but rather are recruited among urban high school and college graduates, as with most of the relatively well-paid jobs. For the majority, the seasonal and vulnerable nature of the employment provides relatively low income, few opportunities for real advancement, and high levels of insecurity.

Conservation by dispossession

To make matters worse, many conservation schemes rely on the physical dispossession of people living in and around the conserved ecosystems, in other words, conservation by dispossession. Such people are literally displaced persons who are forbidden to use the local landscape in the ways they once did. Many examples of this sort of land grabbing can be found throughout the world.[36] This problem is acute in many of the East African wildlife parks, for instance, where widespread dispossession left whole peoples in a state of perpetual poverty.[37] In some of the more notorious cases, the best agricultural land went to white settlers during the colonial years, and more recently, to successful national politicians. And the best land remaining for hunting and grazing was converted to wildlife reserves. While not all wildlife reserves in East Africa have been so brutal to local people, many have, and it has made even the most conscientious conservationists much less welcome than they might otherwise have been. It is no wonder that, as a desperate means of survival, locals either turn a blind eye to wildlife poachers, or actively engage in poaching or supportive activities. Disreputable merchants, armed gangs, and corrupt politicians often find little resistance to the ruthless picking clean of the carcass of poorly designed conservation plans, as in Kenya during the 1990s, when prominent political leaders working through large international smuggling rings were found to be profiting handsomely from large-scale poaching in wildlife parks. Of course, wherever this kind of situation has prevailed, it has tended to undercut the primary purpose of wildlife reserves, that is, to conserve wildlife.

The charge is frequently made that those who promote compatibility of viable local economies with biodiversity conservation are blunting the desperate need for tough conservation measures because of an overly sentimental reaction to human distress. This charge is fatally flawed on two grounds. It is both morally bankrupt and contradictory to the conservation cause itself.

Many volumes have been written on this subject, and we do not intend to rake over the same issues again.[38] The key here is the urgent need to find ways to further biodiversity conservation without simply reproducing the situations of economic dependency that place both people and nature in danger. It is a general and most obvious principle that natural areas will be successfully protected over the long run only when they are embedded within local, regional, and national economies that provide for greater levels of economic diversity, resilience, security, and political participation. An essential tool for moving in that direction is the integration of productive and sustainable agriculture with conservation, which is the fundamental argument of this book.

The integration of sustainable agriculture with conservation cannot be accomplished, however, without strict attention to the legal and administrative framework and the macro-economic policies that must be in place to protect the larger purpose. For example, in Brazil, the national government placed enormous forested areas under various kinds of legal protection, amounting to an area of land roughly equal in size to Western Europe. The amount of land designated for conservation, dwarfs what richer nations of the North have ever put into protected areas. The problem is that much of this land amounts to what is commonly termed 'paper parks', that is, theoretical legal protection in a social and economic context that makes such legalities meaningless.[39]

For five centuries, Brazilian rural life has been dominated to a large extent by oligarchies of powerful families and firms who have shaped, twisted, bent, and mutilated the legal system to their benefit. They have relied on the use of police and the military for the purpose of furthering their private gain, and when this has not been sufficient, they have turned to private gunmen and gangs. The regions where the forest is being cut and burned in Brazil continue to be regions of frequent, endemic violence. Brazilian land law, as it is actually practised and enforced, is a quagmire of contradictions that provides not only opportunities but incentives for ruthless exploitation by timber and mining companies, ranchers, and large agribusiness firms. Furthermore, the chronic poverty and desperation of the majority of the rural population guarantees relentless deforestation by those with few alternatives than either to work for exploitive resource-extraction corporations or to cut down forest to plant crops for minimal food and income. The systemic inequalities and violence built into regions undergoing deforestation is also fertile ground for widespread corruption. The legally protected areas cannot reliably withstand the brutally destructive and corrosive context that surrounds and contains them.[40]

An example from Mexico

Mexico provides a somewhat different example. In some of the biologically richest areas of Mexico, villagers have done their best, under very difficult circumstances, to preserve forests and forest remnants over which they have exercised some degree of legal and/or customary control. They have struggled with many problems over centuries that have made it difficult to preserve forests, and often they have failed. But in spite of all the failures, there has been a degree of success – enough so that Mexico, virtually no inch of which has not seen some kind of human habitation and use over several millennia, has retained an unusual wealth of species diversity. Since the 1980s, however, and at an accelerating rate, the protection that villagers often exerted over forests has come under attack in several mutually reinforcing ways.[41]

Since the early 1980s, the Mexican government, partly due to its own internal problems and decisions, and partly in response to international pressures, began to reduce the already minimal amount of financial credit and other forms of support it offered to small-scale farmers and land reform communities. This deepened an already existing crisis for the majority of Mexico's farmers. More than 50 per cent of Mexico's agricultural land was held by *ejidos*, a form of landholding that had been created by the 1917 Mexican Constitution, written toward the end of the country's major social revolution of 1910–1920. The *ejido* was loosely modelled on ancient village forms of landholding that had prevailed in Mexico and Spain until large private landholders and the Catholic Church largely destroyed such forms. Under the terms of Article 27 of the 1917 Constitution, *ejido* land was owned by the state, controlled by a council of those who lived and worked it, and farmed, in most instances, by individual families. Various kinds of cooperatives, marketing co-ops and machine co-ops, for example, and work sharing among families were part of *ejido* life. Most *ejidos* had some amount of land that was treated as common land, including pasture and forest, and administered by the *ejido* council. The whole enterprise was almost totally dependent on government credit, because it was illegal to sell or mortgage *ejido* land. As government credit dried up, *ejiditarios* became desperate and many began to want to put an end to the legal restrictions that made it impossible for them to access most of the private credit market.

At the same time, the US government and the institutions it heavily influenced, including the World Bank, the Inter-American Development Bank, and the IMF, were pressuring governments all over the world to privatize state enterprises of all kinds, including state-owned rural credit banks. Private banks were eager to eliminate or take over such state financial institutions. Also, as part of the same international economic policy initiative, the US government was working hard to create a North American free trade zone, which would become known as the North American Free Trade Agreement (NAFTA). One of the key obstacles to NAFTA, from the point of view of multinational corporations and many officials in the US government, was the array of nationalist provisions in Mexican law, among which were the restrictions on land sales and mortgages that restricted the ability of companies to gain access to and own Mexican land and resources. These corporations, with the support of the US government, saw the elimination of such controls on land and resource sales as an important element in making a free trade area desirable and workable. In 1992, the Mexican government passed a Constitutional Amendment to Article 27 of the 1917 Constitution, making it possible for *ejiditarios* and other land reform beneficiaries to mortgage and sell land and such land-based resources as timber and minerals, without the kinds of restrictions that had applied previously. This paved the way for final negotiation of NAFTA itself, which came into effect in 1994.

The most important trade effect of NAFTA on rural Mexico was the elimination of the 50 per cent tariff on US and Canadian maize imported into Mexico. The agreement was founded on the promise of the US that it would eliminate virtually all price support payments and other subsidies to US farmers. However, after NAFTA came into effect, the US Congress proceeded to pass huge subsidy bills that meant that US grain farmers received an average of 40 to 50 per cent of all their income from government subsidy payments, while Mexican farmers were deprived of all significant subsidy payments. Along with the superb soils of the US Midwest and the very heavy capital investments made over decades in US grain farms, these events have resulted in a flood of cheap grain entering Mexico, severely undercutting Mexican farmers' already fragile economic viability.[42] The result has been an ever-more rapidly accelerating exodus from the Mexican countryside. It has also meant that those who do remain at home are frequently forced to sell their forests and minerals, if not the land itself, to companies that have rushed in to buy up everything of significant value. The negative consequences for forests and biodiversity are evident.[43]

There is probably some relief to native habitat and biodiversity in Mexico due to the reduction of the pressure of rural populations on local resources. However, in region after region of Mexico where villages are being virtually or entirely abandoned, a counter-effect is almost certainly more important for the future of the forests and biodiversity. When villagers attempt to defend rather than sell off their resources, the task has been made extremely difficult. The defence of Mexican forests in village hands has never been simply a legal matter. In hundreds of cases it has been necessary for villagers to defend their land and forests through determined organization, mobilization, and, often, resistance to physical violence perpetrated by local, national, and international timber, mineral, and land companies. Since the signing of NAFTA, there has been a sharp increase in the number of beatings, illegal incarceration, and murder of villagers trying to protect their forests from exploiters. One of the best-known incidents involved the machine-gunning of 17 villagers crowded into a truck on their way to a meeting to protest encroachments by a US timber firm. The incident is particularly well known because it was captured in all its brutality on video tape.[44] The most prominent lawyer working for communities in the state of Guerrero trying to defend forest resources was assassinated in her Mexico City office.[45] The future of Mexico's native habitats and biodiversity will almost certainly be more critically affected in a negative way by such pressures and events than it is affected positively by the creation of additional parks and reserves. At best, the programmes to create and protect reserves will be seriously compromised by the larger problems of rural Mexico and its villagers. Villagers who are now working in Mexico City, Tijuana, Los Angeles, or New York City can do little to help defend the forests and resources that their ancestors protected for centuries. Similar stories can be told in country after country.

What becomes increasingly evident is that biodiversity protection has long depended to a considerable degree on mostly poor farmers in relatively remote areas. While these same farmers have also often abused soils and forests badly, in the end, their physical presence in the areas and their self-interest in the protection of forests and watersheds have frequently made them a largely unrecognized force for conservation.[46] Their many failings and shortcomings in this regard should not be used to condemn them, but rather to call attention to the critical need to support them. Particularly, conserva-tionists should be active in seeking support for farmers who are ready and able to promote conservation goals. Such support cannot, in turn, be expected to be successful when national and international macro-economic and political forces make their very existence as rural people untenable. Nor is it justifiable for conservation policies, projects and strategies either to ignore the role such farmers can play in conservation or, worse, to promote or cooperate with economic policies that eliminate or weaken them.[47]

Grass-roots social movements

Resistance to the global system has been widespread, from Geronimo and before. Electing a government in Guatemala (in 1954) that promised to provide land to landless peasants was not an accident, but the product of grass-roots initiatives by farmers and labour unions. And the liberation strug-gles of the Vietnamese, the Nicaraguans, the Kenyans, the Chinese, and the Indians, were all the product of grass-roots social movements converging to form major political transformations in their respective countries (to say nothing of the US and France somewhat earlier). It is thus not surprising that we see the same thing today. It seems that the idea of democracy and justice are deep cultural rules that transcend the particularities of political and eco-nomic systems. Wherever there are oppressed peoples, there are grass-roots struggles to end that oppression.

Yet today these movements have become more important than ever. The crisis of neo-liberalism may be very similar to the previous crisis of the old capi-talist system (reflected in the Great Depression), but the solution, Keynesianism,[48] is no longer available to us, largely because of ecological fundamentals – biodi-versity will continue to be eroded and global warming may make the civilization that we know and live in unfeasible. Consequently, the solution to the previous crises of capitalism may not be available to us today. While a complete analysis of this possibility is beyond the scope of this book, we insist that any rational solu-tion to the global crisis created by capitalism and neo-liberalism will have to take into account the environmental background, which suggests an examination not only of neo-liberalism but of the capitalist system itself. If Keynesianism cannot be a solution, what are we left with? We do not have an answer to that ques-tion, but we are convinced that the transformation of the agricultural system in dramatic and radical ways will be an essential part of the answer. And in order to effect that transformation, the participation of the current and potential future rural sector will be essential. The good news is that the pathway to that future seems to be unfolding through myriad rural social movements right now. From the point of view of this book, the connection between those social movements and conservation is essential.

Much as traditional agriculturalists have given tremendous insights into the development of agroecological principles, there has been a recent, if under-reported, surge of interest on the part of small-scale farmers in the tropics in biodiversity conservation.[49] Indeed, our experience is that small-scale farmers in the tropics are frequently surprised to hear that some conservationists regard them as the enemies of conservation and the main cause of tropical deforestation. Although there is no doubt that landless peasants are partially responsible for the expansion of the agricultural frontier in areas like the Amazon[50] and the tropical rain forests of Central America,[51] blaming them for tropical deforestation is an oversimplification that ignores the political economy of agricultural development in tropical countries. We elaborated on this issue in our previous book, *Breakfast of Biodiversity: The Political Economy of Deforestation*.[52] Decades of environmentally destructive mega-projects, such as the colonization of the Amazon in Brazil, transmigration projects in Indonesia, and the establishment and expansion of banana plantations in Costa Rica and Ecuador, along with an uneven distribution of land, have been the principal causes of tropical deforestation.[53] Government policies that benefited a landed minority, owners of large rural estates, and in many cases foreign corporations, displaced small farmers from the best agricultural lands, effectively giving them two options: to move to the cities, or to migrate to the agricultural frontier.[54]

However, in many developing countries the small farmers and the landless are organizing and demanding access to land and their right to a decent livelihood. These farmers' organizations, increasingly organized under the banner of food sovereignty, sustainable agriculture, and biodiversity conservation, are an integral component of the discourse. Organized groups such as La Via Campesina, a coalition of over 180 small-farmer and peasant organizations representing over 200 million farmers from around the world, are not only actively opposing neo-liberal policies, but are now taking an active role in planning conservation activities and practising and promoting agroecology.[55] Four of the ten key points of the La Via Campesina platform are related to conservation and the environment. Brazil's MST, the largest rural social movement in the world, and a member of La Via Campesina, actively encourages and teaches agroecology, which includes protection of biodiversity and the use of ecological principles in agriculture.[56] At the 2003 meeting of the MesoAmerican Society for Biology and Conservation, a leader of La Via Campesina in Costa Rica, spoke of the acknowledged responsibility of small farmers to conserve the environment, and specifically, biodiversity, for future generations.

Another example is the autonomous territories of the Zapatistas, an indigenous movement in Chiapas, Mexico, that has created its own autonomous form of government called the *Junta de Buen Gobierno* (Board of Good Government). The word agroecology has been appropriated by the Zapatista communities, and agroecological projects are seen as part and parcel of community development (Figure 7.3a). Furthermore, their educational system

places nature and history at its the core, as prerequisites for all other disciplines (Figure 7.3b). In 2016 and 2017 the Zapatistas organized a conference,
ConCiencias (having a double meaning in Spanish – 'with science' and 'consciousness'), in which scientists from around the world were invited to
present their work directly to young representatives of the Zapatista

(a)

(b)

Figure 7.3 (a, b) Agroecological promoters within Zapatista communities promoting
the conservation of traditional seed varieties and soil conservation techniques; (c) Two Zapatista farmers building terraces and applying organic
fertilizer; (d) A picture of a mural painted in a Zapatista school in Oventic,
that states 'our philosophy is human beings as part of nature'.

Source: John Vandermeer.

(c)

(d)

Figure 7.3 (c, d)

communities. The 2017 conference included at least six presentations having to do with agroecology and conservation. Clearly, this grass-roots social movement is not only interested in striving to go beyond capitalism, but is invoking a scientific approach to agriculture and conservation as part of their fundamental programme.

In conjunction, all these movements and their initiatives represent not only an alternative pathway toward development but a dramatically different concept of sustainable development from that promoted by the neo-liberal model and its supporting institutions. In sum, it is a conceptualization that emerges from the grass roots, and as such, has social and environmental equity as a central principle, and grass-roots democracy as its methodological instrument. Although these so-called 'alterworlds' are in no way homogeneous, most of them have strong environmental ethics, and given that they are integrated by mostly small-scale farmers, peasants, and landless workers from the Global South, they are natural allies and should be regarded as integral to the efforts to conserve biodiversity.

Conservationists working with social movements

Realizing the need to work hand in hand with the small-scale farmers who manage the agricultural landscapes in the tropics, some environmental NGOs are beginning to pioneer conservation programmes in collaboration with these progressive social movements. An excellent example of the kind of work that incorporates social justice and conservation in a highly fragmented tropical landscape is the work that the Institute of Ecological Research (Instituto de Pesquisas Ecológicas, IPE) is doing in the region of the Atlantic forest in Brazil.[57] In the Pontal do Paranapanema, a large fragment and several smaller fragments of the Atlantic Coast rainforest form the Morro del Diablo reserve, surrounded by large cattle pastures with mainly absentee landlords, as well as settlements of small farmers established long ago by landless people, which are currently highly productive. These small farms are the result of a DALR programme that began in the 1950s with rural unions, was largely abandoned during the repressive phase of the military government in 1964, and was renewed in the late 1970s with the support of Liberation Theology Catholic priests and the organizations that would become the MST. Using the basic ideas of DALR, families of landless rural workers occupied (rescued) pieces of unused land, mainly located on the huge and highly unproductive cattle ranches. After intense legal battles (and some less than legal armed conflict with the ranchers), many of those participating in land rescues were given legal titles. And what we now see is the result of a 40-year experiment with this sort of agrarian reform.

A visit to these small farms today reveals a remarkable success story. These families have a dignified livelihood with what could be considered by most a middle-class economic standard of living. In a country like Brazil, with one of the highest levels of inequality in the world, this is quite an achievement. But more importantly for our argument, the farms themselves are complex agroforestry gardens that seem almost a perfect example of what we expect a high-quality matrix to be (Figure 7.4). Indeed, a study of birds notes that biodiversity in these gardens is almost what it is in local patches of forest.[58] For feeding, some of the birds even prefer the agroforests over the natural forests.

While the imposition of a military dictatorship in 1964 effectively ended the DALR operations throughout Brazil, the end of that dictatorship in 1984 saw a resurgence of those activities, and yet more unproductive cattle pastures

Figure 7.4 One of the agroforestry systems used by settlers in the Pontal do Paranapanema, São Paulo, Brazil. The picture shows Jefferson Lima, from IPE, and one of the settled farmers. IPE works closely with farmers in developing agroforestry systems to help maintain biodiversity in a region that is dominated by treeless and degraded cattle pastures. In 2002, Laury Cullen Jr, an IPE researcher, won the Whitley Gold Award, one of the most prestigious global conservation awards, for this project.

Source: John Vandermeer.

have seen land takeovers by landless people, some organized by the MST. Rather than seeing the MST as their enemy, IPE began collaborating with the MST to diversify and increase the tree cover on the farms in the settlements, as well as to improve the livelihoods of the families involved. Initiatives like this will, in our view, contribute to the creation of an agricultural matrix that is socially just, politically stable, and that will conserve biodiversity at the landscape level and in the long run. As of this writing (2018), the indirect parliamentary coup undertaken in 2016 followed by the election of an extremely right wing president, have changed the political landscape, and the immediate future of these types of conservation initiatives are uncertain in Brazil.

It is important to note that what we are proposing here is qualitatively different from the Integrated Conservation and Development Programmes (ICDP) that attracted so much investment by bilateral development agencies and the Global Environmental Facility (GEF) during the 1990s, and that generated debate in development conservation literature.[59] Although these programmes began by addressing the need to look beyond reserve boundaries and pay attention to the welfare of the local communities, they retained institutional top-down approaches and stopped short of recognizing the pivotal role of rural social movements as protagonists of the new conservation. But more importantly for our argument, the ecological processes involving extinction in the fragments and migration through the agricultural matrix (as explained in Chapter 2) were not fully appreciated. It does not matter how much the ICDP programmes may have met their short-term goals if the underlying ecology, especially the quality of the matrix, was not taken into account.

Social movements versus the traditional conservationist agenda

Conservationists in the past have focused on the purchase and protection of large tracts of land to be set aside as nature preserves, frequently with a 'fortress conservation' ideology (see Chapter 3), with surrounding areas acting as buffers. From what we now know about how biodiversity is structured ecologically, this is a doomed strategy if pursued in isolation, as we explained in detail in Chapter 2. A review of the implications for biodiversity protection of Amazon and African protected areas by scientists at Duke University (US) concluded, 'Size matters, and so, too, must the matrix.'[60] A recent study in Germany reports that insect biomass during the summer peak season has declined by 82 per cent, during the past 27 years, a decline unrelated to any observable variable and likely due to the increase in intensive agriculture surrounding the reserves (the biodiversity figures are from within the preserved areas only).[61] In the Americas, the iconic monarch butterfly seems on its way to extinction, principally due to the killing of its main food source in the pathway of its multigenerational migrations through the United States, a consequence of the industrial agricultural system.[62] It is difficult to see how any of these examples (and many others could be cited) could be reversed without acknowledging the importance of the whole landscape, including the quality of the matrix.

While there is no rational need to convert any more forests to agriculture, they are in fact being converted, and the future will almost certainly present us with mainly fragmented landscapes. It is in these fragmented landscapes that most of the world's biodiversity will be located. A long-term plan for biodiversity conservation needs to acknowledge this fact and work at the landscape level, not only to focus on preservation of the patches of native vegetation that remain, but also to construct a landscape that is 'migration-friendly'. Such a landscape is most likely to emerge from the application of agroecological principles. And these principles are most likely to be enacted by small farmers with land titles – a product of grass-roots social movements.[63]

Social movements will also be needed to change the overall conditions that determine the budget priorities and policies of governments, and the relationship of governments to banks and corporations. While agrarian reform movements are only a small part of the larger set of political initiatives that will have to be taken to achieve a more just and sustainable world, they are a necessary part and one that is particularly close to the problem of biodiversity conservation. The union of conservation with social justice in the countryside is one starting point for the wider union that will be necessary to protect habitats and species. While emphasizing that 'Most of the political and economic constraints on the sustainable use of natural resources by people in a rural locality originate elsewhere ... [and] can seldom be removed through local initiatives alone', Barraclough and Ghimire conclude, in a study sponsored by the UN and the WWF that:

> Without the mobilization and organization of peasants and other low-income residents with the aim of gaining greater control over resources and institutions, however, there seems little likelihood that wider subnational, national, and international political systems will seriously take their interests into account when formulating and executing policies affecting them. Local-level initiatives aimed at approaching more sustainable development through community-based resource management necessarily imply a struggle for greater control over resources and institutions in specific social contexts by those hitherto excluded from such control. Such struggles for self-empowerment are inevitably highly conflictive.[64]

We see a dichotomy of positions, with one side purchasing land in pristine areas to be protected by armed guards (i.e. land grabbing), and the other side marching with the poor in their struggle for revolutionary change. Naturally this is something of a caricature, but it does capture the two poles of thought that we encounter when we read and talk with people concerned with biodiversity conservation in the tropics. Beyond caricature, the two positions urgently need to be reconciled in order to achieve effective species protection, linked, as it must be, to improved human welfare.

If the phenomenal rise in rural grass-roots social movements continues, it is here that the future of rural landscapes will be determined. Since it is the agriculture in the matrix that is key to the overall conservation agenda, as we have argued, the social movements that form the basis for the future organization of that agriculture become key players in the overall conservation agenda. The Via Campesina, with its philosophy of food sovereignty, agroecology, and conservation of natural resources, offers an excellent example of where conservation energy should be placed at the present time.

Notes

1 The most recent and most analytically convincing review can be found in Hochstetler and Keck (2007: Chapter 4). Other useful overviews include: Ozorio de Almeida (1992); Moran (1996); Wright and Wolford (2003, Chapter 4); Brown (2006a, 2006b); Brown *et al.* (2007); Pacheco (2009); Rudel *et al.* (2009); Hecht (2010, 2012).
2 Mann (2005).
3 Woods *et al.* (2009), pp. 363–372.
4 Hemming (2004); Mann (2005); Heckenberg *et al.* (2008).
5 Hecht and Cockburn (1989); Ozorio de Almeida (1992); Browder and Godfrey (1997); Hall (2000a, 2000b); Wright and Wolford (2003); Hecht (2010).
6 Browder and Godfrey (1997); Padoch *et al.* (2008); Hecht (2010).
7 Bunker (1985); Hecht and Cockburn (1989); Rich (1994).
8 Foresta (1991).
9 Durham (1979).
10 Browder and Godfrey (1997); Moran *et al.* (2000); Carvalho *et al.* (2002).
11 Rich (1994).
12 Hecht and Cockburn (1989); Revkin (2004); Hochstetler and Keck (2007).
13 Bunker (1985); Hecht and Cockburn (1989).
14 Hochstetler and Keck (2007).
15 Hall (2000a); Moran *et al.* (2000); Carvalho *et al.* (2002); Hochstetler and Keck (2007).
16 Whitesell (1993, 1996); Brown and Purcell (2005); Brown (2006a); Brown *et al.* (2007); Gomes *et al.* (2012).
17 Brown (2006b); Brown *et al.* (2007).
18 Nepstad *et al.* (2002).
19 Boucher *et al.* (2013).
20 Mongabay (2018).
21 It has been argued that neo-liberal policies that can lead to some level of industrialization and massive migration from rural to urban areas or centres of employment (like maquilas in Mexico and Central America), can reduce deforestation, as they did in Puerto Rico during the industrialization period of the 1950s and 1960s (Aide and Grau, 2004). However, Puerto Rico is a very special case because of the colonial relationship of the island with the US that facilitated the migration of the 'excess' population to the US, and provided federal funding to support such a massive displacement without major political upheaval.
22 Some improvements were achieved under the government of Luis Inácio (Lula) da Silva. However, by most measurements of inequality, Brazil is still amongst the most unequal countries on the planet with a Gini coefficient of 51.3 (World Bank Data: https://data.worldbank.org/indicator/SI.POV.GINI?locations=BR, accessed 13 March 2019).
23 A longer, heavily annotated version of the summary that follows may be found in Wright and Wolford (2003, Chapters 1 and 3).
24 Ondetti (2008: pp. 229ff.); Wolford (2010).
25 Scarano and Ceotto (2015).
26 Bunker (1985); Hecht and Cockburn (1989); Wright and Wolford (2003).
27 Meek (2014, 2016); Pahnke (2015); Hoffman and Fox (2017); Meek and Simonian (2017); Schwendler and Thompson (2017).
28 Pellegrini (2009).
29 See websites of the MST (www.mst.org.br/mst/home.php) and Friends of the MST (www.mstbrazil.org) for a variety of statements on these issues over the last several years.
30 Ownership is not an evident category, but its full discussion goes beyond the intentions of this book.

31 Barraclough and Ghimire (2000); Wallerstein (2004).

32 There are obviously exceptions, such as the Great Depression of the 1930s.

33 Dreher and Gaston (2008); Picketty (2014); Bourguignon (2017).

34 Chappell *et al.* (2013).

35 Agrawal and Redford (2006); Ferraro *et al.* (2011); Coria and Calfucura (2012).

36 Benjaminsen and Bryceson (2012); Ojeda (2012); Galaty (2013); Holmes (2014); Rocheleau (2015). For various discussions on land grabbing see also the introduction to special issues in the *Journal of Peasant Studies* by Borras *et al.* (2011).

37 Bonner (1993); Adams and McShane (1997); Neumann (1998).

38 Cf. Anderson and Grove (1987); Johnston (1994); Sponsel *et al.* (1996); Guha and Martinez-Alier (1997); Zimmerer and Young (1998); Rothenberg and Ulvaeus (2001); Brockington *et al.* (2008).

39 Joppa *et al.* (2008).

40 Wright and Wolford (2003).

41 Wright (2005).

42 Relinger (2010).

43 De Janvry *et al.* (2014); Rocheleau (2015).

44 *El Universal*, 25 November 2008, www.el-universal.com.mx/nacion/163989. html, accessed 21 March 2015.

45 *La Jornada*, 11 December 2001, www.jornada.unam.mx/2001/12/11/005n1pol. html, accessed 21 March 2009.

46 García-Barrios *et al.* (2009).

47 Barraclough and Ghimire (2000).

48 Keynsianism is based on the ideas of 20th-century British economist John Maynard Keynes. According to Keynesian economics the state should stimulate economic growth and improve stability in the private sector – by, for example, adjusting interest rates and taxation, and funding public projects (http://en. wikipedia.org/wiki/Keynesian_economics).

49 Pretty and Smith (2004); Bacon *et al.* (2005); Cullen *et al.* (2005); Campos and Nepstad (2006); Bawa *et al.* (2007); Altieri and Toledo (2011); Vara-Sánchez and Padilla (2013).

50 Fernside (1993).

51 Kaimowitz (1996).

52 Vandermeer and Perfecto (2005a).

53 Hecht and Cockburn (1989); Ozorio de Almeida (1992); Sponsel *et al.* (1996); Tucker (2000); Vandermeer and Perfecto (2005a).

54 Wright and Wolford (2003); Davis (2006).

55 Desmarais (2007); Altieri and Toledo (2011); Rosset and Martínez-Torres (2012); Holt-Giménez and Altieri (2013).

56 Wright and Wolford (2003).

57 Cullen *et al.* (2005).

58 Goulart *et al.* (2007).

59 Oates (1999); Rabinowitz (1999); Wilshusen *et al.* (2000); Brechin *et al.* (2002); McShane and Wells (2004); Terborgh (2004).

60 Joppa *et al.* (2008).

61 Hallman *et al.* (2017).

62 Pleasants and Oberhauser (2013).

63 Rosset (2006).

64 Barraclough and Ghimire (2000).

8   Coffee production and biodiversity conservation

Biodiversity, agriculture, and social movements are the components of our argument, as discussed in previous chapters. With each of those issues, we peppered our discussion with particular examples. However, such peppering may have left the reader with a sense that the real world conforms to our perspective only in these anecdotal ways. In this and the next two chapters, we offer three extensive case studies. The first, the coffee agroecosystem, we discuss as an example of how an agroecosystem can be constructed in different ways and how that construction indeed does affect biodiversity. We choose this example partly because we have extensive experience doing research in the coffee agroecosystems of Central America, Mexico, and Puerto Rico, but mainly because, probably more than any other agroecosystem, a wealth of data are available on the consequences of management intensity for biodiversity.

The second case study is the cacao system, covered in Chapter 9. While less is known of the details of the biodiversity consequences of different management systems, a great deal is known of how socio-political forces have shaped the cacao landscape in Brazil. It thus represents a case study that reflects the need for a detailed knowledge of the history and socio-political structure of the agroecosystem of concern.

The third case study, recounted in Chapter 10, is not so much an example as an acknowledgement that food production rather than drugs (caffeine and theobromine) is the most important aspect of agriculture. An examination, even if superficial, of how the production of grains, in particular rice and maize, might relate to our story is thus warranted. In Chapter 10 we also discuss the role of social movements and diversified agroecosystems as an integral part of our argument.

Coffee and the technical side of biodiversity

As we explained in Chapter 2, there is now little doubt that isolating fragments of natural vegetation in a landscape of low-quality matrix, like a pesticide-drenched banana plantation, is a recipe for disaster from the point of view of conserving biodiversity. That isolation effectively reduces the migratory potential that is needed if the metapopulations of concern are to be conserved in the long run. Whatever arguments exist in favour of constructing a high-quality matrix, and there are many, the quality of the matrix is perhaps the critical issue from the point of view of biodiversity conservation.[1] The concept of 'the quality of the matrix' must obviously be related to the natural habitat that is being conserved, but most importantly, it involves at its core the management of agricultural ecosystems. The matrix matters!

One of the best-studied cases of a matrix that matters is the coffee agroecosystem. The importance of this ecosystem for biodiversity conservation is threefold. First, it is an important habitat for biodiversity per se. Second, it is important as a high-quality matrix through which organisms can migrate from patch to patch of natural habitat. Third, it represents a matrix ecosystem in which it can be demonstrated that biodiversity has a potentially important function with regard to the normative goals of management. We treat each of these three issues below, followed by some socio-economic speculations associated with the complicated issue of certification and coffee prices: topics intimately related to biodiversity conservation.

Coffee agroforests as habitats for biodiversity

In the northern part of Latin America, coffee has traditionally been cultivated under the shade of a diverse tree canopy. Having been domesticated from the understorey of forests in north eastern Africa, and therefore adapted to live in shaded environments, it was most naturally planted in the understorey of natural forests when first introduced in Latin America. Soon it became common to plant coffee in non-forested areas, while simultaneously planting trees to provide shade. These traditional shaded plantations are characterized by a high-planned biodiversity, including coffee, shade trees, and fruit trees. Most importantly, we now understand that they also include an enormous amount of associated biodiversity.[2]

This pattern of planting shade trees along with the coffee was characteristic of the major coffee-producing countries from Mexico to Peru. Elsewhere, notably in Brazil, the pattern was distinct. Due to historical accident, Brazil initiated its coffee history with 'sun' plantations and today has amongst the largest extensions of 'sun coffee' in the world. Our account in this particular chapter focuses on the experiences of northern Latin America, with only occasional reference to Brazil and other countries.

In order to quantify some of the associated biodiversity in traditional coffee farms, we sampled insects in the canopy of individual shade trees in a shaded coffee farm in Costa Rica.[3] The results of this study were astonishing. We were expecting high levels of diversity but nothing like what we found. From a single canopy of a small shade tree in a small traditional farm we collected 126 different species of beetles! This level of species richness was within the same order of magnitude as in the canopy of large trees in a Panamanian rainforest.[4] Even more surprisingly, a second tree yielded another 110 species of beetles, 86 per cent of which were different from those collected from the first tree. Subsequently many other researchers have documented high diversity of birds, mammals, plants, insects, and other organisms in these traditional coffee farms.[5]

There is no doubt that coffee agroforests provide habitat for a large number of organisms. Perhaps these managed systems are not able to maintain viable populations of jaguars or other charismatic megafauna, although there are several reports of tigers encountered in traditional coffee plantations in India.[6] However, they are certainly adequate for large numbers of birds, small mammals, amphibians, reptiles, epiphytic plants, and invertebrates. But what is most important, they provide habitat for these diverse organisms while at the same time providing a livelihood for farmers, as we discuss below.

Intensification and the loss of biodiversity

In the 1980s, a programme to intensify coffee production was launched in most of the northern Latin American countries. By the 1980s, these countries had accumulated a large external debt and were pressured by international lending agencies to increase their revenues in order to service it. USAID (the US development agency) provided US$81 million to finance a programme for the 'modernization' of coffee production.[7] The programme consisted of providing incentives to coffee farmers to 'intensify' coffee production by incorporating new varieties of coffee that could grow well without shade trees, thus increasing yield on a per area basis. The coffee intensification programme consisted of a technological package that included not only the new coffee varieties but also fertilizers and herbicides that could substitute for the most obvious functions of shade trees – provisioning of nitrogen through their association with nitrogen-fixing bacteria in their roots and reduction of weed growth through the shade they cast. The programme met with variable levels of success, with Colombia and Costa Rica experiencing far more success than other countries in convincing producers to shift to the more intensive system. As a result of the variable success of this programme we now have a variety of coffee production systems in northern Latin America.

Patricia Moguel and Victor Toledo, from Mexico, have developed a classification system that is now widely used to characterize coffee systems (Figure 8.1).[8] Rustic coffee is the most diverse and densely shaded and consists of coffee that is planted under the canopy of a previously existing forest. Traditional polycultures (or 'coffee gardens') have a high diversity of planted shade and fruit trees with a heterogeneous canopy. Although coffee is the main cash crop in this system, farmers produce many other products for sale or family consumption within these coffee gardens. Commercial polycultures are less diverse than traditional polycultures and focus on commercial crops, especially coffee and sometimes fruit trees. The last two systems are the most intensive and less diverse systems. Shaded monoculture is a shaded system in

Figure 8.1 Diagram of the coffee intensification gradient; the ordinate in each graph is the stature in metres.

Source: Modified from Moguel and Toledo (1999).

which the shade trees are sparsely planted and belong primarily to a single genus or species. Finally, the most intensive system is the unshaded mono-culture, frequently referred to as 'sun coffee'. Although this classification does not make any reference to the use of agrochemicals, the tendency is for the more intensive systems (shaded and unshaded monocultures) to use more her-bicides and synthetic fertilizers than the other systems.

We were in Costa Rica when many farmers were in the process of trans-forming their coffee farms from traditional polycultures (Figure 8.2a) to 'sun coffee' (Figure 8.2b). Knowing that the traditional coffee agroforests contain such a high biodiversity we could not help but ask where all that biodiversity would go when the process was completed. And, of course, we found it curious that most conservationists simply ignored this massive programme of what was effectively deforestation.

As introduced in Chapter 2, among the most important and least studied patterns of biodiversity is the relationship between agricultural intensification and biodiversity. We know that a natural forest contains more diversity than

(a)

(b)

Figure 8.2 Two types of coffee farms: (a) shaded polyculture and (b) monoculture, or 'sun coffee'.

Source: (a) and (b) Ivette Perfecto.

an intensive agricultural monoculture. We also know that the process of agricultural intensification results in a dramatic reduction in the planned biodiversity (the crops and animals that the farmer intentionally includes on his or her farm). However, we know little about the pattern of biodiversity loss along the intensification gradient for most agroecosystems. In the case of coffee production, the diversity of production systems enables us to examine this issue. A variety of studies comparing 'sun coffee' with 'shade coffee' or with coffee with different levels of shade have shown that the intensification of this agroecosystem results in a loss of biodiversity for most organisms that have been examined.[9]

Birds have been the focus of many of these studies. Indeed, the intensification of coffee has been implicated in the decline of several migratory species, since their winter feeding grounds were the shaded coffee farms that were deforested. Even so, the negative impact of intensification seems to be stronger for resident birds, presumably because they tend to have more specialized food and habitat requirements than most migrants. Coffee intensification not only reduces bird diversity but also changes the composition of the bird community by favouring birds that are more characteristic in open habitats – what bird-watchers refer to as 'tramp' species or 'weedy species'.

With a few exceptions, insects and other arthropods also decline with the intensification of coffee. For example, reviewing studies of the effect of coffee intensification on ant diversity, we encountered 21 published studies, 18 of which showed a significant decrease in ant diversity with intensification.[10] Since the intensification process results in a reduction of tree density and diversity, a sharp reduction of arboreal species should not be surprising. What is surprising is that the ants that live strictly on the ground are also affected by intensification. In seeking to understand why this might be the case, Colombian ecologist Inge Armbrecht conducted an important experiment with ants that normally nest in vegetation cavities and twigs on the ground on coffee farms. When given a choice, these ants select a habitat containing different species of hollow twigs, over one containing only one species. The surprising, and still enigmatic, aspect of this study is that a higher diversity of ants in the diverse habitat did not result from particular species of ants preferring distinct kinds of twigs, but rather it seems that ants select diversity itself – diversity in tree species begets diversity in ants. The specific details of how a diverse assemblage of shade trees (the source of the twigs) is more attractive than a less diverse assemblage remains perplexing.[11]

The few studies that have examined a gradient of shade suggest that the particular type of shade is also important. Some species of shade trees support higher diversity and abundance of arthropods than others. Furthermore, there is no reason to think that the pattern of 'species decline' with increased intensification should be the same for different kinds of organisms. Although comparative studies that include different taxa within the same sites are rare, our knowledge of the natural history of different groups suggests that some organisms should be more susceptible to intensification than others. This is evident for birds, where residents have been shown to be more susceptible to intensification than migratory birds. Our own data from a study in Mexico shows that ants and butterflies follow a very different trajectory of richness decline along an intensification gradient, with butterflies declining rapidly even with a low degree of intensification, whereas ants are far more resistant to the effects of intensification, declining in biodiversity only after the intensification process has reached a relatively high level. Such differences represent an important challenge to establishing criteria for certification of 'shade coffee' for conservation purposes, as will be discussed later in this chapter.

A recent quantitative meta-analysis demonstrates that the intensification of the coffee (and cacao) agroecosystem has a stronger impact on biodiversity than the conversion of forest to a diverse agroforestry system.[12] However, this same study reports a reduction in ecosystem services provided by biodiversity when forests are converted to coffee (or cacao) agroforests.

Coffee agroforests as a high-quality matrix in the landscape

In Latin America, coffee is produced mainly at middle elevations. Frequently the entire landscape of the coffee-growing regions is dominated by coffee farms (large and small) with interspersed fragments of forest in areas that are too steep for cultivation or in the riverbeds where they are protected by law. In southern Mexico, most of the areas designated by the government as high priority for biodiversity conservation are embedded within a matrix of coffee farms.[13]

As was discussed previously, most natural ecosystems in today's world are highly fragmented, and where coffee is produced in the Americas is no exception. In these fragmented landscapes, many organisms are likely to persist as metapopulations. Biodiversity conservation at the landscape level depends on the maintenance of that metapopulational structure, which is affected not only by rates of local extinction but, more importantly, by rates of migration through the matrix. If the matrix consists of 'sun coffee' monocultural plantations, it could deter the inter-fragment migration of many forest species. On the other hand, if the matrix is dominated by traditional shaded farms, it may facilitate migration from patch to patch, and therefore maintain the metapopulational structure of some species.[14]

It is important to note that the matrix does not need to provide a habitat of enough quality for the permanent establishment of a species. The matrix only needs to provide a habitat benign enough for individuals to move through. For example, in the site where we have been studying coffee systems in southern Mexico, there are two relatively large forest fragments that exist within a matrix of shaded coffee agroforests. Local bird-watchers have spotted the highland guan,[15] a large forest bird that reportedly needs a fairly old-age forest to survive and that is listed as a threatened species by the International Union for the Conservation of Nature.[16] However, it is almost certain that neither of these forest fragments is large enough (one is about 15 hectares, the other about 40) to maintain a highland guan population in the long run. Thus, the only chance for the species to survive is as a metapopulation within fragments of old-growth forest scattered amongst the coffee farms. But for this to be possible, the highland guan needs to be able to cross from one forest patch to another through the coffee matrix. The good news is that this species has been observed, very rarely, in shaded coffee farms. Given that this species needs fairly well-developed forest to survive, it was almost certainly just moving through the shaded farms when sighted there, perhaps in search of fruits or something else to eat. Once an individual highland guan ventures out of one of the forest fragments, there is a fixed probability that it may end up in another fragment where a new subpopulation can be established. This process, in effect, could maintain the metapopulation structure needed for the maintenance of this species in the region. Given what we know about the behaviour and ecology of the highland guan, it is very unlikely that an individual would venture into a 'sun coffee' plantation. So, if 'sun coffee' plantations surrounded the small subpopulations of the highland guan, the overall population would be doomed to extinction. Shaded coffee farms, we propose, provide the necessary high-quality matrix that allows migration of this species thus making its long-term survival more likely. Without that matrix, the highland guan would be expected to go extinct in successive forest patches, until it becomes extinct over the whole region.

Charismatic fauna, like the highland guan, is not all, or even the most important fauna to benefit from a high-quality matrix. Studies of ants suggest that shaded coffee farms form a high-quality matrix that can contribute to the maintenance of high diversity at the landscape level. Studying the species richness of ants within a forest fragment and at various distances from the forest edge into the coffee matrix we discovered that ant species richness declines sharply as you move into an intensively managed coffee monoculture. On the other hand, as you move away from the forest edge into a shaded polyculture or agroforest, ant diversity declines far less and far less rapidly, remaining relatively high at long distances from the forest (up to 800 metres).[17]

In the case of ants, as with any other organism, we repeat that it is not necessary for the matrix to provide a habitat that can sustain the population in perpetuity. The matrix could be only a transitory habitat for many species that may only have viable populations in the landscape as a whole, with the key habitat being the fragments of natural habitats, but the matrix being essential for migration. Thus, we suspect that many of the ant species that were collected in the diverse shaded farms were actually forest specialists. Ant colonies go through a process of colony building during which only unfertile workers (all female) are produced. Once the colony reaches maturity, it starts producing fertile males and females with wings, which then fly away from the colony, encounter flying individuals of the same species from other colonies, and mate. Fertilized females ('queens') then start searching for a place to establish a new colony. Depending on the inter-patch distance, the queen may be able to make it all the way to the next patch of forest. However, if the patches are too sparse, the queen may not be able to reach another forest patch and may land in the agricultural matrix. Once in the matrix, she will start looking for a nesting place. The chance that she will find a good nesting site, establish a colony and reach the point at which new queens are produced, is most certainly higher in a shaded farm than in 'sun coffee'. Even though the species may not be able to survive over the long term in a shaded farm, it is possible that individual colonies can survive to the point that they produce new queens that will mate and fly away in search of a good nesting site. Through this process, some queens will eventually land in another patch of forest where the colony can survive for a long period of time, probably in the presence of other colonies of the same species in a viable subpopulation, in effect completing the process of inter-patch migration that is necessary for maintaining a metapopulation structure. As in the case of the highland guan, the quality of the matrix is the main determinant for the maintenance of biodiversity, even with species that seemingly require natural forest for long-term survival. Without migration between patches, many of the forest species will be doomed to extinction at a larger regional level.

Perhaps even less charismatic than ants, is the forest tree *Miconia affinis*. It occurs in shaded coffee farms, as well as local forest patches. The species is dispersed by birds and pollinated by buzz-pollination. In other words, it cannot be pollinated by the invasive European honey bee. From a conservation perspective we are interested in both the dispersal of individual plants (seed dispersal) and the dispersal of genes (which is accomplished by the pollinating bees). Using polymorphic microsatellites (repeating sequences of DNA) it has been discovered that, (1) dispersal among trees within a forest patch is much lower than dispersal among trees in the shaded coffee surrounding the forest patch, and (2) pollination occurs over a far smaller area within the forest patch than in the shaded coffee plantation.[18] Clearly these results could not have been observed if the surrounding matrix had been a sun coffee monoculture – the presence of isolated individuals of the tree species in question was extremely important. Thus, it appears that the migration coefficient of the metapopulation (in other words, the dispersal of seeds) is significantly reduced by the fragmentation (dispersion out of the fragment is low), and the presence of individuals of the species within the matrix is essential to ensure true dispersal beyond the forest fragment boundary. Added to this are the genetic consequences of restricted reproductive activity, which may very well contribute to the timing of this species extinction with any forest fragment isolated in a low–quality matrix (i.e. sun coffee).

It is precisely this sort of patch dynamics that is ignored in land-sparing arguments, leading some to conclude that intensifying agriculture is a better conservation strategy because it spares land for conservation.[19] Since species-richness tends to be higher in natural forests than in agroforests for many organisms, the assumption that land will be spared is a seductive one. However, ignoring the extinctic and migration dynamics that occur in fragmented landscapes is a fatal error of those conservation biologists who advocate the land-sparing strategy.

In southern Mexico, the government has identified areas of priority for conservation and most of these areas are located in coffee-producing areas. From our discussion above we hope it is clear that the quality of the coffee matrix is exceedingly important for biodiversity conservation within the remaining fragments of natural forest. A matrix consisting of nothing more than 'sun coffee', with the almost inevitable use of pesticides therein, could easily result in the eventual extinction of forest species. Since many of the species in this region are endemic, their local extinction is the same as their global extinction. A conservation programme that focuses on preserving forest patches without paying attention to how coffee is produced in the surrounding areas will fail in the long run, probably resulting in the global extinction of many species. In this situation it is imperative that conservation policies are based on a landscape perspective that includes the coffee matrix as well as the patches of forest.

The 'function' of biodiversity in the coffee agroecosystem

A shaded coffee farm with many singing birds, large shade trees covered with different kinds of epiphytes, and a high diversity of other organisms can be a most aesthetically pleasing sight, and feed the soul of the farmer. But can it help her produce more and produce more sustainably? An important scientific question to ask about any managed ecosystem is 'what is the function of biodiversity?' It turns out that this is a surprisingly complicated question. Can a diverse array of organisms help increase the productivity and sustainability of the agroecosystem? Or, would it be better, as some agronomists propose, to select those species, and only those species, which are the most efficient, and eliminate the rest, which are only competitors? We propose that the high diversity of organisms found in coffee farms can play an important role in the functioning of that agroecosystem. Ecological debates on the role of biodiversity in ecosystem function suggest that we should be cautious in making general statements about biodiversity and ecosystem function.[20] Nevertheless, the question is an important one for farmers, since ecological function translates into benefits for humans. This is especially true in a managed system, like an agroecosystem, where the farmer manages the system for the purpose of obtaining a variety of provisioning and ecological services. Although the relationship between biodiversity and ecosystem services is equally as controversial as that of biodiversity and ecosystem function, the management implications are much more straightforward. When farmers diversify their systems, they do so with a particular goal in mind – for example, nitrogen-fixing trees are added to enhance nitrogen in the soil; flowering plants may be added to attract pollinators or predatory wasps, etc. A substantial literature now exists supporting the basic idea that biodiversity has positive effects on ecosystem services,[21] with substantial evidence that shaded coffee facilitates ecosystem services and consequently farmers' livelihoods.[22]

One ecosystem function that has long been associated with biodiversity is the regulation or control of insect pests. However, the relationship between pest management and biodiversity is complicated and requires a system-by-system analysis. In this section we present three examples of the role of biodiversity in the regulation of potential insect pests in coffee, even as we emphasize that control of pests is only one 'function' of biodiversity.

Ants and the coffee berry borer

One of the main coffee pests in the Americas is the coffee berry borer (CBB), a tiny beetle (about the size of a pinhead), accidentally introduced from Africa. The female adult beetle bores into the berry and enters the seed. Inside, she lays eggs, the eggs hatch, and the larvae eat the inside of the seed. They complete their life cycle within the seed, adults mate and the fertilized females leave the berry in search of new berries where they can lay their eggs and start a new cycle. The CBB significantly reduces coffee yields and is currently considered the worst insect pest of coffee in the world. Since this pest is not native to the Americas, it does not have any specialized natural enemies that can control it efficiently. Introductions of an African parasitic wasp and some general fungal pathogens seem to suppress the population but do not control it completely. However, it has been discovered that ants prey on the CBB. In Colombia, several ant species have been seen removing borer adults and larvae from the fruits that were sun-drying after harvest. Also, Moises Velez from the National Coffee Research Center (Cenicafé) observed how a relatively large ant actually 'hunted' the adult beetles as they alighted on a berry or when they were exiting the berry to infest another, an observation repeated by us in Mexico as well with the ant *Azteca sericeaseur*. Individuals of this ant were observed waiting on a berry for adult CBBs to emerge from the seeds, grabbing and immobilizing them with a sting, and taking them to their nest. It is worth noting that this species of ant has been collected from shaded farms and forests, but not from 'sun plantations'. The intensification of coffee most probably results in the ant's elimination and therefore the elimination of one of the important native natural enemies of the CBB.

In Mexico we discovered another ant, *Pheidole synanthropica*, that preys on the CBB quite efficiently.[23] This was one of the most common ants in the Soconusco region of Chiapas, Mexico. However, in response to the massive outbreak of the coffee rust disease in 2012–2013, farmers sprayed large quantities of calcium carbonate laced with copper sulphate over very large areas in an attempt to control the rust disease. This substance is known to affect ants negatively. After this massive application, and we suspect because of it, this ant species completely disappeared from the region.

It could be argued that since these two predatory ant species were numerically abundant they would be especially effective as a biological control system, that all other ant species are either irrelevant or redundant with respect to CBB control, and thus there would be no need to keep them, at least not with respect to this particular function. However, these two ants species are relatively large and cannot penetrate into the seeds through the small hole made by the CBB. This means that only the adult beetles can be susceptible to their predatory effect and only before they actually penetrate the seed. But there are many other ant species that are small enough to penetrate the seeds and eat the eggs and larvae of the CBB. However, these two dominant ant species reduced the foraging activity of all the smaller species,[24] arguably reducing their potential predatory efficiency. Indeed, surveys of the CBB in 2010 (before *Pheidole synanthropica* was eliminated) reported an average of 14 per cent infestation per coffee bush, whereas surveying the same general area in 2018 we found less than 1 per cent infestation. While we cannot demonstrate that the decline was due to the elimination of this one species, the coincidence is certainly suggestive.

In a similar vein, Colombian ecologist, Inge Armbrecht, reported 21 ant species preying on the CBB in coffee plantations in Colombia.[25] Some of these species are small enough to enter the holes made by the borers and prey on their brood. She also found that ant predation was a more important mortality factor in shaded farms than in 'sun coffee'. We also have evidence from Mexico, that several other species of ants, big and small, ground foragers and arboreal foragers, prey on the CBB.[26] Taken together, all these pieces of evidence suggest that a high diversity of ants may already be contributing to the regulation of the main coffee pest in Latin America. The problem may be that the extensive 'sun coffee' plantations or less shaded plantations are providing a refuge for the CBB, where it can escape the predatory effects of the diverse assemblage of predatory ants that live in the shaded farms.

Birds and the trophic structure of arthropods in coffee

Walter Peters, a coffee producer in Chiapas, Mexico, is a bird enthusiast. He plants specific kinds of trees to attract birds to his farm. He has been observing and recording birds on his property, Finca Irlanda, since he was young, when his father managed the farm. One day we were walking with him on his farm and noticed that a shade tree was being defoliated by a heavy infestation of a certain kind of caterpillar. We alerted Walter to this potential pest outbreak but he replied calmly, 'the migrant birds will soon arrive and will take care of that problem. No need to worry.' And indeed, two weeks later, the migratory birds arrived and took care of the problem. In addition to loving biodiversity for purely aesthetic reasons, he also understood the role of biodiversity in maintaining a healthy coffee farm.

With the exception of the introduced CBB, there are few other insects that have become pests in the coffee agroecosystem in the Americas. Although the caffeine contained within coffee plants is known to suppress herbivores (indeed, caffeine is undoubtedly the result of the co-evolutionary process between coffee plants and their herbivores), this alone does not explain the lack of insect pests in coffee, since there are more than 200 species of insects that are known to eat coffee.[27] A possible explanation for the lack of serious insect pests is that the system harbours a large density and diversity of natural enemies of herbivores, including insectivorous birds.

We decided to test the hypothesis that birds would be able to control insect pest outbreaks within diverse coffee farms, but not within intensive farms where bird diversity and density is lower. We conducted an experiment by constructing bird exclusion cages (with fish netting that prevented birds from entering the exclosure) and adding caterpillar larvae to plants inside and outside the exclosures, simulating a real pest outbreak in both a diverse and intensive farm.[28] The results of the experiment confirmed Walter's prediction that birds would 'take care' of this potential pest. After just 24 hours, more than 75 per cent of the larvae exposed to bird predation had disappeared from the diverse farm. Although some larvae also disappeared from the intensive farm, the numbers were the same outside and inside the exclosures (with and without birds), indicating that birds were not responsible for the disappearance of the larvae in the intensive coffee farm. This experiment suggests that the high number and diversity of birds within the shaded farms provides a sort of buffering mechanism against insect pest outbreaks, just as Walter suggested from his many years of observing birds on his farm.

With these results it is tempting to conclude that birds control all insect herbivores in coffee farms. However, Nature is seldom so simple. Through detailed studies of the trophic structure (who eats whom) of the coffee agroecosystem we have discovered that birds appear to have a positive effect on coffee plants, though through a complicated chain of indirect effects. Basically, it seems that under normal circumstances birds eat a large number of spiders (especially web-spinning spiders). In turn, the spiders normally capture a large number of the very small parasitic wasps that become entangled in their webs. But the parasitic wasps mainly attack the herbivores that eat the coffee. Through this five-level food chain, that includes 'intraguild predation' (carnivores eating other carnivores), the indirect effect of birds on the coffee plants is thus positive (Figure 8.3). The potential pests are controlled by the parasitic wasps, but the wasps are attacked by the spiders, which means the spiders actually have a negative effect on the coffee. But then the birds seem to prefer eating the spiders, thus releasing the wasps to prosper better and control the potential coffee pests. However, since birds seem to prefer larger prey and can also shift their diet according to what is most common, they are also able to respond to the occasional outbreak of specific pests. These sorts of complicated food webs have been identified in other studies as well and are likely a common feature of many diverse farming systems.[29]

In the tropics, there are also many insectivorous bats, and we have found that they have similar effects as the birds.[30] Using exclosures that were in place either during the day, during the night, both day and night, or neither day nor night (a control), we were able to demonstrate that bats play a com-

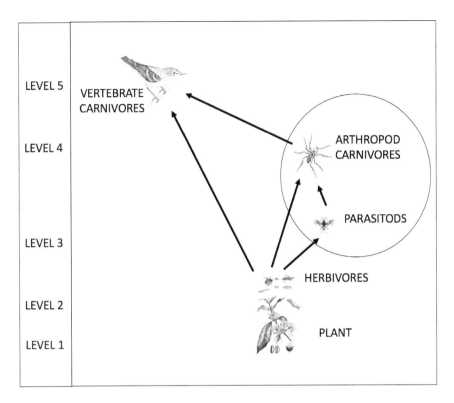

Figure 8.3 Proposed five trophic level food web in the coffee agroecosystems in Chiapas, Mexico.

Source: The authors.

plementary role to birds in terms of regulating insects in the coffee planta-tions. Furthermore, it appears that bats are even more important than birds during the wet season, when the migratory birds are in their northern breed-ing grounds. The pest control ecosystem service of bats and birds has also been demonstrated for other agroforestry systems.[31]

The ant, the scale, and the beetle

If the effects of birds and bats on potential coffee pests seems complicated, it is nothing compared to other interactions in the coffee agroecosystem. Con-sider the details of the *Azteca* ant, mentioned as a predator of the CBB above.[32] This species of ant nests in the shade trees in the coffee farms of Mexico and is well known to the coffee workers as a nasty stinging ant. When you disturb a nest, thousands of workers boil out of the tree and begin stinging. But far worse is the apparent effect these ants have on the coffee trees. While they do not build their nests in the coffee tree themselves, they frenetically move around tending scale insects on the coffee trees that are adjacent to the shade tree where they nest (Figure 8.4a and b). This is a remarkable relationship, well known around the world, in which the small, immobile scale insects produce a sweet secretion called honeydew which provides nourishment for the ants. While the ants are frenetically running around eating the secretions of the scale insects, they are also protecting the scales from the many small parasitic wasps and lady beetles that would otherwise attack and decimate the population of scale insects. The relationship is called 'mutualism' and, generally speaking, is the most common interaction of species we see in Nature, from insects that pollinate plants, to the bacteria that live in our stomach and help us digest food. In the specific case of ants and many different species of sap-sucking herbivores, like aphids and scale insects, the herbivores suck the juices out of plants, and part of what they do with these juices is produce the honey-like substance that they excrete to 'feed' the ants that protect them.

In the coffee farms there is a particular species of scale insect (its formal name is *Coccus viridis*, but we will continue referring to it simply as 'scale') that is well known to become, at times, a devastating pest of coffee and other agricultural plants, like oranges and grapefruit. There is no doubt that where it does occur on the coffee tree, it causes significant decline in yields of those particular bushes. So, if the ant population is very common on the farm, one might expect the scale insects to be a major pest. And, if there are no ants, one would expect the scales to disappear. So, a typical agronomist might see the ants as a potential pest of the coffee, worthy of all efforts necessary to eliminate them.

But the function of biodiversity is never simple, and this example is no exception. We take one particular coffee farm, Finca Irlanda, as an example here. On that farm, only approximately 5 per cent of the shade trees are occupied by *Azteca*, which means that on 5 per cent of the plantation, the scales dramatically affect the coffee trees. Naturally, there should be concerns about the ants expanding their area and possibly occupying more of the shade trees, such that, for example, if they ever got to a level of over 50 per cent, we would possibly be facing a massive decline in coffee yields. But this brings up the question as to why the *Aztecas* occur in such a small portion of the farm. One possible reason may have to do with a very small fly. The fly is a parasite of the ant. The female fly lays an egg on the ant and the fly larva develops inside the ant's head, eventually eating out all the tissue and causing the ant's head to drop off. For this reason, these flies are known as 'decapitating flies'. It seems that, in a shaded coffee farm, the ants have a more-or-less balanced relationship with the flies, such that the ants can increase the area where they live, but if they ever become too common, the decapitating flies start to build up their populations and, in classic predator–prey fashion, the prey (in this case the ants) have their populations controlled. So, the 5 per

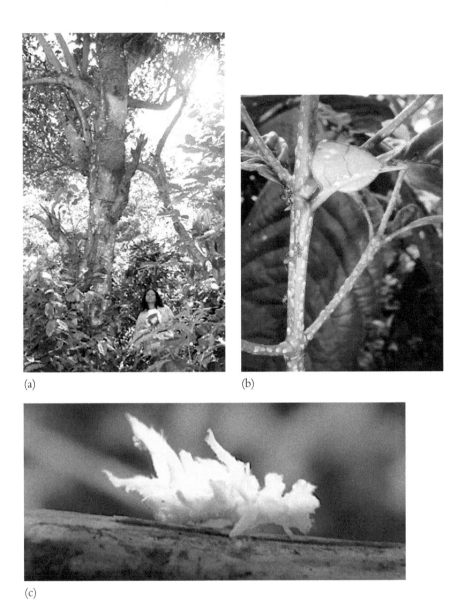

(a)

(b)

(c)

Figure 8.4 Azteca ants tending the green coffee scale on a coffee plant. (a) The ant, *Azteca sericeasur,* nesting in a shade tree and tending the scale insects, *Coccus viridis,* on coffee bushes (one of the authors, Perfecto, is standing next to the shade tree). (b) A close-up of the ants tending the scale insects on a coffee branch. The scale insects produce honey dew to feed the ants and in turn the ants protect the scale insects from their predators and parasites. (c) Predatory larvae of the ladybird beetle, *Azya orbigera*. The larval beetle preys on the scale insects and is immune to ant predation because of the white waxy filaments that cover its body.

Sources: (a) John Vandermeer; (b) and (c) Ivette Perfecto.

cent of the trees occupied seems to be a consequence of the controlling force of the decapitating flies.[33]

But the story does not end here. A species of lady beetle that normally eats scale insects also lives in the system. This sort of beetle is well known to be a devastating predator of the scales, which is obviously one of the reasons the ants are necessary, from the point of view of the scales. Close examination of the beetle larva reveals that it has waxy protrusions all over the surface of its body, which makes it immune to the attacks of the ants (Figure 8.4c). It thus can live on the scale insects, even though the ants are protecting those scale insects from other natural enemies that are trying to eat them, including the adult beetles. Furthermore, at least three different species of parasitic wasps also attack the beetle larvae. However, when the beetle larvae are in areas where the ants are protecting the scales, the harassing action of the ants also prevents these parasites from attacking the beetle larvae. Therefore, the beetles are found almost exclusively in areas where the ants are tending the scales, taking advantage of the non-intended protecting effect of the ants.[34] It is a well-known problem amongst biologists who try to cultivate this type of beetle that it is difficult to get populations of beetles surviving in the laboratory because almost all the beetle larvae you find in the field contain parasitic wasps inside them, and thus die shortly after being brought into the laboratory.

But the irony is that the adult beetles are not protected from the ants as the larval beetles are, with their waxy secretions. Thus, if they stay in an area where there are ants, they can't get enough to eat. So, the adult beetles have to travel outside of the ant areas in order to find food. Thus, we have the surprising situation where the beetle larvae cannot survive except when with the ants, and the adult beetles cannot survive except when not with the ants.

We now understand that the adult beetles, as efficient biological control agents, are responsible for keeping many of the potential scale pests under control over most of the farm. Yet if the ants did not exist in a small percentage area of the farm, all the beetle larvae would be parasitized and die, and their population could not be maintained. Since the ants only occupy about 5 per cent of the farm, their contribution to yield reduction is trivial compared to their contribution to biological control over 95 per cent of the farm.

So, what we have is a system that, on superficial examination, appears to be quite negative for the production goals of the farmer – a species of ant that protects a scale insect that is known to be a devastating pest of coffee. However, upon closer examination, it turns out to be a biodiverse system with complex interactions that actually functions to control this pest, and probably many more. A system of at least eight species of insects is involved: the ant, the parasitic fly, the scale insect, the parasitic wasp of the scale, the beetle and the three species of parasites of the beetle. And these species create this functional system. A superficial reading of the system (ants = scales = bad) could have led to a disastrous management recommendation – to eliminate the ants – which would have resulted in elimination of the main controlling agent of the scales (and perhaps other potential pests), the beetles. In this example, biodiversity has a function, and that function is achieved through complex ecological interactions.[35]

Coffee prices, biodiversity preservation, and free market fundamentalism

The island nation of Timor Leste (East Timor) was finally liberated from its imperial master, Indonesia, in 1999. The recent history of this country is an extremely sad account of the developed world either ignoring East Timor's plight or actively collaborating with a corrupt imperial power in Jakarta.[36] The massive looting and murder committed by the Indonesian military following East Timor's vote for independence shall forever blot the already soiled records of the world's developed countries. If ever there was justification for military intervention, the Indonesian military provided it with their genocidal activities in East Timor. Yet no such aid was forthcoming. East Timor produced coffee, not petroleum.

In a sad footnote, once the developed world did take action, they sent something that seems to have only exacerbated this poor nation's plight. They sent technical advisors. While it may have seemed obvious to those of us less well-trained in conventional economic theory, the technical advisors did not focus on the abundant small farmers who potentially could have produced food for the whole country. Instead, they focused on the export sector, as they always seem to do in the neo-liberal age. The major export of the country was coffee. The technical experts, rather than trying to improve the lives of small farmers, decided to recommend the technification of coffee production. Thus, with a focus on 'modernizing' coffee production, East Timor entered a world market already saturated by overproduction (the main current problem in world coffee markets), thanks to the international technical advisory community's advice to Vietnam, Indonesia and other countries. The glut of coffee beans on the market creates extremely low prices (something that, remarkably, seems to always surprise technocratic agronomists) and, while not filtering down to consumers, has caused a severe crisis for coffee farmers the world over. East Timor did not escape this devastation. In 2001, over 80 per cent of villages reported food shortages.

While there is substantial evidence that 'shade coffee' represents a potentially high-quality matrix with regard to biodiversity conservation, the current world politico-economic climate is hardly conducive to its promotion, as the situation in East Timor so clearly demonstrates. Yet the problem is not unsolvable if the world is willing to open up to a slightly different economic focus. The problem has been, since the end of the Cold War, a fundamentalist philosophy in both international economics and technological research. We recall an international meeting we attended in Costa Rica in 1990. Economists, sociologists, and development experts all completely agreed that the main, and perhaps the only, problem in the coffee sector was overproduction. There was simply too much coffee being produced, and this resulted in an oversupply, which resulted in low prices. There was no debate about this issue. Walking out of the meeting, which was held within the facilities of the main coffee research station in Costa Rica, we talked with the agronomists working in the station about their research projects. When we asked them what the main goal of the coffee research programme in Costa Rica was, they told us 'to increase production'! There is an enormous discrepancy between the recommendations of the experts – to increase coffee production – and their own analysis about the main problem – overproduction.

Biodiversity conservation and coffee technification are clearly at odds, as we have argued. Growing out of this contradiction has been a new niche in the specialty coffee market. 'Bird-friendly' certification is the latest coffee certification effort to be added to organic and fair trade coffee. In the mid-1990s, the Smithsonian Migratory Bird Centre (SMBC) developed certification criteria for certifying 'bird-friendly' coffee. Other organizations promote a similar message, such as the Rain Forest Alliance and the UTZ sustainable coffee certification programmes that cover agricultural practices, farm management, and environmental issues.[37] Later on, USAID, Conservation International, and the coffee company Starbucks launched the Conservation Coffee Alliance in Central America, to promote more of these efforts.

One of the more focused initiatives is the Bird Friendly® (BF) coffee programme, based on field-work conducted by researchers at the SMBC (Figure 8.5a). Coffee producers – whether estate or cooperative farmers – solicit an inspection by a Bird-Friendly-accredited inspection agency. Any farm evaluated for its shade management must be certified as US Department of Agriculture organic before being certified 'bird friendly' (BF). If the shade component of the coffee area meets the BF criteria, then coffee produced on that farm can be sold as BF in the marketplace. The actual inspection of the shade is carried out by inspectors who have taken a training workshop offered by SMBC staff. More than 20 coffee importers handle BF coffee in the US and Canada. They, in turn, sell to some 26 roasters who have written agreements with the SMBC to use the BF logo and the Smithsonian Bird Friendly trademark in marketing the coffee. Roasters selling labelled BF coffee remit a royalty fee to the SMBC for every pound of coffee roasted and sold as BF. This revenue funds the programme, as well as SMBC's research and education on coffee, conservation, and migratory bird issues.

While there is general agreement that shade promotes biodiversity, there is much debate over how much and what kind of shade should be recommended.[38] To make serious recommendations about the quality and quantity of shade will require more intensive research – how much shade, how diverse it should be, how high and heterogeneous the canopy should be, must epiphytes be present, should agrochemicals be allowed, etc. However, in spite of this uncertainty, a number of studies show that ecological certification of coffee (whether organic, Rain Forest, UTZ, Bird Friendly, or even Fair

Trade, which also includes some environmental criteria) has tangible environmental benefits.[39] But the majority of these studies have examined biodiversity at the scale of particular farms. As we have argued extensively in this book, long-term biodiversity conservation occurs at the landscape level, therefore, there is a mismatch between the scale at which most studies are conducted and the scale at which conservation actually occurs.[40]

Furthermore, the main question is whether shade coffee certification can also improve the livelihood of small-scale farmers, and that is a more complicated issue. While certification programmes are usually associated with a higher price of coffee, the low volume and additional cost of certification may counterbalance the economic benefits to the farmers.[41] Some studies have found positive environmental benefits but no economic benefits,[42] while others point out that the price premium is only one of many elements that could improve farmer's livelihood. Small-scale farmers need to form or join a cooperative or other type of association in order to afford and take advantage of the certification scheme. These associations frequently contribute to a higher level of social cohesion that can benefit farmers and their families in other ways besides higher income from coffee.[43]

Certainly, sustainable coffee certification is not the panacea for improving farmer's livelihoods and conserving biodiversity. Indeed, some analysts have argued that these kinds of market-based solutions make farmers even more vulnerable to the ups and downs of the market, especially for internationally traded commodities like coffee, and that price for certified coffee does not compensate for low productivity, nor for land and labour constrains.[44] Others have suggested that shade coffee certification programmes could encourage farmers to turn native forests into shaded coffee farms or simply cut more forest to expand production, annulling the potential environmental benefit of shaded coffee.[45] These are all valid concerns that should be taken into consideration every time a certification scheme is proposed.

Whatever the scheme developed for certification there is one issue that is of utmost importance – that of third-party inspection. Just as organic products need to have the production process and all handling post-harvest done by entities which have been inspected and certified as organic, other themes (like shade) need similar certification. Those with vested interests in reaping financial rewards from the certification cannot be the ones assuring the consumer of a coffee's environmental attributes. It is simply not a good model for instilling confidence in the marketplace.

Other possibilities for expanding on environmental and social concerns include various programmes of consumer–producer cooperation; that is, more direct links between producers and consumers. The basic aim is to incorporate the ideas of bio-regionalism and consumer–producer collaboration at an international level. For example, in Ann Arbor, Michigan, a group of university students made contact with a production cooperative in the highlands of Chiapas, Mexico, and promoted a direct connection between that cooperative of indigenous coffee producers and coffee drinkers in Ann Arbor. A Michigan

(a)   (b)

Figure 8.5 Two types of coffee initiative: (a) Smithsonian Migratory Bird Center's promotional poster for 'Bird Friendly' coffee developed to promote management practices that conserve bird diversity in coffee plantations; (b) 'Brewing Hope' initiative to establish more direct linkages between coffee producers and coffee consumer.

Sources: (a) Artwork by Julie Zickefoose; (b) logo design by Chelsea Wills, artwork by Beatriz Aurora.

roaster purchased the coffee at premium prices from the cooperative in Chiapas and marketed it under the trade name 'Brewing Hope' (Figure 8.5b). It was an 'international-local' affair, that unfortunately no longer exists. In the same way that some organic farmers talk of the importance of the connection between them and their customers, so was the Brewing Hope brand focused on developing that connection at an international level. The coffee drinkers of Ann Arbor were to know the lives of the coffee producers in the Yachil Xojobal Chulchan cooperative of the highlands of Chiapas, Mexico. Other similar initiatives are growing in other areas of the developed world. We see great potential in this fair direct-market approach.

Finally, it is worth returning to the beginning of this chapter to recall that, whether one thinks it is good or bad, governments will always be involved. NGOs can do a great deal, but they are ultimately only part of the mix of political structures that are needed to solve environmental problems generally. Elected bodies with accountability to their constituencies must also be part of the overall solution. Government budgets, enforcement policies, judicial systems, etc. can and should be brought to bear against land degradation, and should promote environmental protection and biodiversity maintenance. Governmental setting of coffee prices through market manipulations has a long, if sometimes less than noble, history. There is no reason why it cannot be used on behalf of the cause of biodiversity preservation and to support small-scale farmers. Funds set up for price premiums based on production processes could be used to reward good land stewardship. Just as petrol is taxed in many countries for government coffers, agrochemical taxes could fund programmes that are environmentally forward-looking. Just as many state fairs in the US award prizes to farmers with the largest hogs or biggest ears of maize, national contests for coffee production within the most ecologically responsible context could provide incentives to growers. There is no lack of ideas, but a great need for political will.

Notes

1 Perfecto and Vandermeer (2015, 2017).
2 Perfecto *et al.* (1996); Moguel and Toledo (1999); García *et al.* (2010); Méndez *et al.* (2010); De Beenhouwer *et al.* (2013); Jha *et al.* (2014); Perfecto and Vandermeer (2015).
3 Perfecto *et al.* (1997).
4 Erwin (1982).
5 This literature is summarized in Perfecto and Armbrecht (2003), Perfecto and Vandermeer (2015), and Perfecto *et al.* (2007).
6 Menon (2006); CoastalDigest.com (2014) 'Tigers on prowl in Chikmagalur coffee plantations', 12 September 2014, www.coastaldigest.com/news/68255-tigers-on-prowl-in-chikmagalur-coffee-plantation, accessed 23 December 2018.
7 Rice (1999).
8 Moguel and Toledo (1999).
9 De Beenhouwer *et al.* (2013).
10 Perfecto and Armbrecht (2003).
11 Armbrecht *et al.* (2004).
12 De Beenhouwer *et al.* (2013).
13 Moguel and Toledo (1999).
14 Perfecto and Vandermeer (2017).
15 Also known as the black penelopina (*Penelopina nigra*).
16 Bird Life International (2018).
17 Perfecto and Vandermeer (2002); Armbrecht and Perfecto (2003).
18 Jha and Dick (2008, 2010).
19 Chandler *et al.* (2013); Hulme *et al.* (2013).
20 Vandermeer *et al.* (2002); deLaplante *et al.* (2011).
21 Cardinale *et al.* (2012).
22 Jha *et al.* (2014).
23 Gonthier *et al.* (2013).
24 Ennis and Philpott (2017).
25 Armbrecht and Gallego (2007).
26 Morris *et al.* (2015, 2018); Morris and Perfecto (2016).

27 Le-Pelley (1973).
28 Perfecto *et al.* (2004). Although the initial intent was to exclude birds, the exclosures also excluded bats because they were left overnight. Therefore, the effect observed could have been due to bats or both bats and birds. Indeed, research conducted a few years later demonstrated that bats also have an effect on arthropods in coffee (Williams-Guillen *et al.*, 2008).
29 Karp and Daily (2014); Perfecto and Vandermeer (2015); Maas *et al.* (2016).
30 Williams-Guillen *et al.* (2008).
31 Maas *et al.* (2016); Vandermeer *et al.* (2010a); Perfecto and Vandermeer (2015).
32 Its full scientific name is *Azteca sericeasur* (mistakenly identified as *Azteca instabilis* in the previous edition of this book), but we will continue referring to it only as *Azteca* – it has no common name. For more details of this system see Perfecto and Vandermeer (2006, 2008b, 2015); Liere and Perfecto (2008); Vandermeer *et al.* (2008, 2010a); Hsieh *et al.* (2012).
33 Actually, this example is even more complicated than we indicate here. The effect of the fly is to change the behaviour of the ant so that its foraging activity is dramatically reduced. The colony, thus weakened because of the reduced foraging efficiency, is subject to any of a variety of mortality factors.
34 Liere and Perfecto (2008); Liere and Larsen (2010); Jha *et al.* (2009, 2012).
35 Perfecto and Vandermeer (2008b); Vandermeer *et al.* (2010a); Perfecto *et al.* (2014); Vandermeer and Perfecto (2017).
36 Chomsky *et al.* (2006).
37 In 2018 UTZ and Rain Forest Alliance announced that they would merge the two certification programmes into one under the Rain Forest Alliance label.
38 Philpott *et al.* (2007); Perfecto and Vandermeer (2015).
39 Blackman and Naranjo (2012); Rueda and Lambin (2013); Rueda *et al.* (2015); Tscharntke *et al.* (2015); Ibañez and Blackman (2016); Takahashi and Todo (2017).
40 Tscharntke *et al.* (2015).
41 Méndez *et al.* (2010).
42 Ibañez and Blackman (2016).
43 Martínez-Torres (2006, 2008); Méndez *et al.* (2010).
44 Raynolds (2002); Raynolds *et al.* (2007); Beuchelt and Zeller (2011).
45 Tejeda-Cruz *et al.* (2010).

9 Cacao and biodiversity

The historical development of a biodiversity landscape

Brazil's Atlantic Coast rainforest, or *Mata Atlântica*, harbours one of the highest numbers of species of any forest in the world.[1] Many hundreds of kilometres of arid and semi-arid country separate it from the Brazilian Amazon and about two-thirds of the plant and animal species that live in the *Mata Atlântica* are not native to the Amazon rainforest. Perhaps most remarkably, the *Mata Atlântica* still retains immense biodiversity comparable to that of any other terrestrial ecosystem in the world in spite of the fact that somewhere between 93 and 97 per cent of its forest is reported to have been destroyed over the last five centuries. One of the reasons it has been able to maintain such high diversity is that some, though not most, of the crops and cultivation techniques used in portions of the forest's range are themselves compatible with high species diversity. In particular, plantations of shade-grown cacao have recently been discovered to have sheltered high levels of biodiversity even during those periods when the plantations were a major source of raw product for the world's chocolate factories. The cacao-growing region has been designated by the International Union for the Conservation of Nature as one of the ten most important 'hotspots' for biodiversity conservation in the world because of the conjoined circumstances of enormously high and significant biodiversity and the powerful forces that could quickly destroy that diversity. How and why this came to be provides rich and complex insights into how agriculture can play a critical role in preserving biodiversity.

At the southern end of what is called 'the cacao zone' in the southern panhandle of the Brazilian state of Bahia lies the small town of Santa Cruz Cabrália. It is the spot where the Portuguese explorer, Pedro Álvares Cabral, blown off his course on a voyage to India, became the first European to lay eyes on Brazil. As it was Easter Day 1500, he named the mountain peak that was his first sight of Brazil, Monte Pascual (Easter Mountain). Monte Pascual is now a Brazilian National Park that contains one of the larger of the remaining remnants of Atlantic Coast forest that has never been clear cut, as far as is known. Cabral saw little of real economic interest here other than provisions easily harvested from the local forest and freely traded or given by the local inhabitants. He admired the strikingly healthy and beautiful indigenous people who extended their hospitality to his men. After writing some enthusiastic notes on the area and its people, he proceeded to India. As the Portuguese began to settle Brazil three decades later, they also turned their interest away from the low mountains of what would become the southern panhandle of the state of Bahia, and instead settled first in the coastal areas with rich alluvial soils, hundreds of miles to the north. There, they cut valuable dye-woods and timber out of the native forest, burned the rest, and planted sugar cane, and later, tobacco and other crops. These crops were the economic base for the colonial capital of Salvador da Bahia in the province, and later state, of Bahia, and its rival city, Recife, in Pernambuco. The Portuguese plantations virtually obliterated the *Mata Atlântica* in its northern ranges during the colonial period. The colonists did little, however, to destroy the *Mata Atlântica* in southern Bahia.

Early history

Over three and a half centuries, southern Bahia was slowly settled by *fazendeiros* (plantation owners) who established small plantings of sugar cane, fruit trees, and subsistence crops in the relatively few level areas near rivers. In the mid-19th century, things began to change, and by the end of the 19th century the region was one of the most profitable agricultural regions of Brazil. A newly lucrative cacao market was based on the increasing world demand for chocolate, as large urban populations in Europe and North America came to constitute a seemingly limitless market for tropical luxury goods: notably cacao, coffee, and a little later, bananas. So much money could be made from cacao in the boom years that competition for land and labour in southern Bahia was fierce, and the region became one of the most violent regions of Brazil, as it remains to the present day. Cacao planters clear cut a great part of the region. The deforestation was so closely associated with violence that Jorge Amado, Brazil's best-selling novelist, who grew up in the region, spoke of the process of settlement of the region in terms of the local phrase, 'derrubando mata, matando gente', that is, 'felling the forest, killing the people'. Not all of the forest was destroyed, however, and much of it was able to recover to a substantial extent in the form of a particular kind of cacao plantation. The cacao tree is a native of the rainforest, with one line of descent coming from the Amazon and another from the tropical forests of Central America and southern Mexico. It is an understorey tree, growing to about eight metres in height and flourishing naturally under a higher canopy of larger trees. Many planters first contracted workers (or earlier in the period, used slaves) to completely clear the forest, and then planted cacao, usually under a temporary canopy of bananas. Some rigorously cut back native trees as they tended to invade the plantation. Other planters allowed the regrowth of the native trees. In other cases, planters began by only cutting enough of the original forest cover to facilitate the planting of the cacao trees, and allowed the forest to continue to grow as the cacao trees matured and gave fruit over a period of decades. The plantations where a significant portion of the forest was left in place or allowed to regrow came to be called *cabruca* cacao systems (Figure 9.1). People outside the region often

(a)

(b)

Figure 9.1 The *cabruca* cacao system in southern Bahia: (a) A view from afar makes it impossible to distinguish between a *cabruca* cacao and a natural forest; (b) a farmer in a *cabruca* cacao farm.

Source: Angus Wright.

simply call it shade-grown, as they do with similar systems for coffee, as described in Chapter 8.

Beginning at least as early as the last years of the 19th century, when most of the region was yet to be cleared, planters argued over the advisability of *cabruca*. Those who tended to consider themselves more educated, 'advanced' and 'scientific' argued that *cabruca* meant low productivity because sunlight, nutrients, and water were captured by the trees shading the cacao. Some argued that the shade trees promoted pests and disease. On the other side of the argument, planters who preferred shade-grown plantations maintained that it was cheaper and that the cacao trees actually benefited from the shade of the other trees and probably benefited in other ways that they could not explain. In any case, they claimed, production was nearly as high as or even higher than in clear-cut plantations, and costs were lower.

One of the most influential enemies of *cabruca* was a Dutch agronomist, Leo Zehntner, with experience in the cacao plantations of Dutch East India (now Indonesia). He was hired to head an experiment station – which on arrival he lamented was nothing more than the caricature of such an institution – and to undertake a two-year research project on the state of cacao agriculture in the region. For the most part, his report is a model of serious observation and critical analysis. He was shocked by the low levels of education and sanitation in the region and the lack of public spirit among the planters. He was scandalized by pervasive violence and the general treatment of workers (which surely was not much better in the Dutch East Indies) and by the state of the roads and harbour facilities.

Zehntner's criticism of the so-called 'routine' or 'empirical' methods (terms used throughout Latin America at the time to describe agricultural techniques that were considered to be established practices, unaided by 'scientific' advice) of those farmers practising *cabruca* went along the standard lines: cacao would do best where the plantation was clean and orderly and where competing vegetation did not rob cacao trees of light, water, and nutrients. This appeared to be self-evident to him, despite the fact that his yield comparisons of *cabruca* and non-*cabruca* were inconclusive. Zehntner believed that *cabruca* resulted at least partly from what he saw as the general evil of Brazil's cacao economy, the predominance of relatively large-scale landholdings often in the hands of absentee or lackadaisical owners and/or managers. *Cabruca*, in this perspective, was simply one aspect of the social backwardness, or, as it was sometimes put, the decadence of the region. The credibility given to his opposition to *cabruca* was probably due more to the way it seemed to represent order and modernity than to its inherent logic. This perspective persisted among many planters, agronomists, cacao export firms, and others through most of the 20th century in the cacao region. It has only begun to wither since about 1990 under the force of dramatic events that have redefined the debate. Many still vigorously defend Zehntner's perspective in its 21st-century version, that adds the desirability of chemical fertilizers and pesticides applied to selected varieties grown in 'clean' plantings.

Other trained observers, including the Ukrainian agronomist Gregorio Bondar, disputed Zehntner's findings, at least as early as the post-World War I years and continued to do so into the 1930s. They argued that the cacao tree benefited from conditions that resembled the circumstances of the wild forest in which cacao had evolved. They produced statistical analyses, which while suffering from small sample sizes and vaguely defined parameters, seemed to demonstrate equal or superior yields and lower costs on the *cabruca* plantations.[2]

Later, we will see that the question of the viability and desirability of *cabruca* would have enormous implications on the fate of the native organisms of the cacao-growing region. These implications were not part of the debate over how cacao was to be grown until the last two decades of the 20th century. But to really understand how these implications would eventually play out, we will have to journey through a bit of the history of the region in the 20th century.

Brazilian cacao in world markets: the 20th century

Bondar, and others who shared his views, were as critical as Zehntner of the tendency toward absentee and inadequate management of the plantations. With price collapses during some years in the 1920s there also began to be increasing criticism of the nearly total domination of regional agriculture by cacao, leaving the region's fortunes totally dependent on volatile international markets for a single luxury crop with dramatic price elasticity. Price elasticity, the tendency of prices to be highly responsive to relatively minor changes in demand, is characteristic of luxury products in general and of such products as cacao, coffee, and bananas in particular. If peoples' incomes fall, luxuries are the first thing to go. Tropical luxury products depend on mass markets among people with marginal incomes, as well as those in more comfortable circumstances, and the tropical agricultural luxuries tend to be in chronic oversupply. In addition, Bahian cacao was of relatively low quality due to a combination of the variety that grew most successfully in Bahia and poor management, processing, and shipping practices. For example, it was known for its 'ham fat' smell because *fazendeiros* provided inadequate housing for workers, and so workers often lived and cooked their cheap cuts of meat under the cacao drying racks. It was also often contaminated with salt because the port city of Ilhéus had lost its harbour to silting up, due to the deforestation of clear-cut cacao plantations. The cacao had to be loaded onto freighters from 'lighters' (small barge-like boats), as they pitched and rolled in the open sea, with sacks of cacao beans taking on more than a sprinkling of salt. Bahian cacao thus suffered not only from market volatility but also from quality competition with better varieties grown under more careful management, usually on much smaller landholdings, in other regions of Latin America, Africa, and Asia. In general, smallholders, such as the Ashanti of Ghana, grew and processed their product with great care because the owners of the small farms were among the families who actually did the work, and the ability of the farm family to gain the maximum price determined whether the family prospered or starved. Holders of larger plantations in Brazil did little or none of the work, were often absent, and had sufficient income from cacao, combined with income from a variety of other investments, to prosper by emphasizing quantity rather than quality as a production strategy. Its quality deficiencies made Bahian cacao especially susceptible to market fluctuations.

The onset of the Great Depression beginning in 1929 sent tropical luxury products and the regions that produced them into a sharp and disastrous nose-dive. In Brazil, coffee and cacao beans in some years were so unsuccessful on the market that they were used to fuel steam locomotives. The economic crisis brought a reform-minded revolutionary regime to power in Brazil, as it did in much of the rest of Latin America. The new government was deeply hostile to the traditional domination of the Brazilian government by coffee planters and exporters and their allies among sugar, cacao, and other traditional landholding families. As a consequence, in the cacao region, the reformist Vargas regime created the Institute of Cacao, meant to resolve many of the region's problems. The most important function of the Institute of Cacao was as a marketing cooperative designed to gain better prices for Bahian cacao. However, the Institute carried forward a more general reform agenda. It planned to diversify regional production, improve technical research and extension, improve roads and bridges, tackle regional public health problems, and improve public education. The Institute nourished an undercurrent of criticism of regional domination by an alliance of foreign export houses with a faction of wealthy local planters, and promised a vaguely defined renewal and democratization of local life and institutions. Conditions in most of these areas were so bad that almost any action at all would have constituted progress. The landholders and exporters who had controlled the region saw little reason for public investment, and some even feared that improved social conditions would threaten their ability to hold wages down, resist challenges to their, often fraudulent, claims to land, and maintain political dominance. However, in the severe economic crisis of the 1930s, as a group, they were so weakened politically that they could not mount successful opposition to reform and were willing to grasp at nearly any straw that might bring them economic relief. The Vargas regime provided a sweetener in the form of a debt forgiveness programme for planters that overcame, for the time, any determined resistance. At least some of the powerful planters were willing to believe in the benefits of a marketing cooperative.

The story of the Institute and its partial successes and eventual destruction (which for all practical purposes occurred in 1942) is intriguing and complex.[3] With respect to *cabruca* cacao and biodiversity, however, the Institute's influence boils down to a rather simple matter. Endowed with a comprehensive reform agenda and associated with aggressive young politicians and public officials, the Institute came to represent a broad challenge to the regional power structure composed of conservative large landholders and export firms. Once the price of cacao rose again, partly due to the success of the Institute's cooperative, the conservative alliance was able to regain the economic strength it needed to launch a challenge to the Institute. The challenge was successful, and the groups that managed to restore their authority in the region came to a firm conclusion: no one should be allowed ever again to put forward an agenda of social reform in the region under the guise of aiding the recovery of the region's economy. Thus, when planters emerged from the glory days of high prices for cacao in the years immediately after World War II into a new overproduction crisis in the 1950s, they were determined that any proposed solutions be strictly technical with no aroma of social reform or unorthodox thinking.

The technification of cacao

In this technocratic state of mind, a new institution was created, *Commissão Executiva do Plano da Lavoura Cacaueira* (CEPLAC), financed by a direct tax on each sack of cacao exported from Brazil. This institution was not to get involved in cooperatives, social reform, or diversification of regional production. Its purpose, stripped of a certain amount of flowery rhetoric, was simple: to discover how to increase the yields of cacao such that Brazilian planters could out-compete cacao growers in the rest of the world. They were told by agricultural scientists that this could be done by the same kind of intensification that was occurring in the development of Green Revolution grain crops. As a consequence, as we shall see, the conservative alliance of planters and exporters, through CEPLAC, ended up mounting a major attack on shade-grown cacao as a technique while introducing a new kind of cacao agriculture that would represent a far more aggressive threat to biodiversity than even the older clear-cutting techniques as advocated by Leo Zehntner and other agronomists (Figure 9.2).

CEPLAC agronomists and other agricultural technicians selected and bred cacao varieties with higher yields than older varieties. These varieties proved their worth when heavily fertilized. They tended to be more pest-vulnerable and were thus dependent on synthetic chemical pesticides. And, CEPLAC specialists insisted, they would thrive only if competing vegetation and trees were rigorously eliminated, with exceptions under some circumstances for bananas and other temporary plantings to shade immature cacao trees. A powerful public propaganda and private advertising accompanied the programme, associating *cabruca* with ignorance, neglect, and decadence. New stores in the region opened up to sell farm chemicals and machinery and were seen as a sure sign of economic progress. Once the basic research for the programme was accomplished and its techniques took root in the shallow forest soils of the region, yields on the *fazendas* using the new package did indeed increase spectacularly. Regional production roughly tripled from the late 1960s to the mid-1980s.

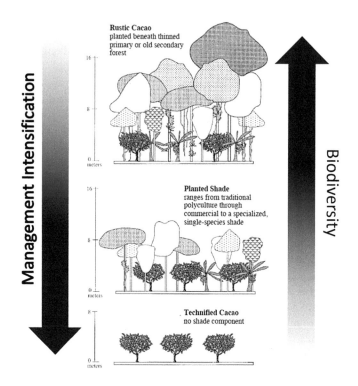

Figure 9.2 The range of cacao systems extant in the world. In southern Bahia it is possible to find all three systems and some intermediate ones. Studies have shown that with management intensification there is a loss of biodiversity (Rice and Greenberg, 2000).

Source: Modified from Rice and Greenberg (2000).

As has happened so frequently in the history of technological developments in agriculture, the production increases were a disaster for the region and for cacao growers around the world. Many new areas of cacao agriculture had entered the market during the period, and some other major producing countries – notably, Côte d'Ivoire – had pursued the same kind of research and technical programme as CEPLAC in Bahia. The global cacao glut led to a reduction of two-thirds and more in international prices in a few years. By 1990 southern Bahian cacao growers, whatever production techniques they were using, were facing ruin or had already given up farming in the region. The production package of new varieties, chemical fertilizers, and pesticides was nearly universally abandoned because the expense was impossible to recuperate in the depressed market. By default, plantations not in *cabruca* and not converted to cattle ranching or sugar cane were in general tending back towards something more like it as vegetation moved in. The final blow to the regional cacao economy was the appearance in 1989 of a fungal disease of cacao, witchbroom disease, or *vassoura de bruxa*. The disease was common in most cacao-growing regions of the world, but one advantage that Bahia had previously enjoyed was its absence in the *Mata Atlântica* region.

The conservation agenda faces the reality of land tenure

Ironically, it was just as the cacao economy was entering extreme crisis that conservationists began to appreciate the value of *cabruca* cacao plantations for the preservation of biodiversity. The International Union for Conservation of Nature (IUCN) had designated the region as a 'hotspot' for biodiversity conservation in 1988[4] and the WWF was investing money in the establishment of the Una Biological Reserve in the region. Conservationists were concerned about the broad range of species diversity in the region, but their campaign focused on the need to preserve the golden-headed lion tamarin, a small, endangered monkey species. Tamarins of several species were well known in Brazil. As cute, intelligent, and highly active primates, they held the kind of broad public appeal that helps conservationists raise money and gain public support and understanding of conservation needs. Golden-headed lion tamarins were thought to number no more than a few hundred, with the population concentrated in the cacao region. It was thought that the population owed its existence to the survival of relatively small stands of more or less undisturbed forest in southern Bahia. A young Brazilian biologist, Cristina Alves, wrote her Master's thesis at the University of Florida on the conservation of the golden-headed lion tamarin, showing that *cabruca* cacao was surprisingly effective as habitat and as a high-quality matrix between wild habitats.[5] In subsequent years, other researchers further validated her findings and were able to locate a much larger number of the monkeys in shade-grown cacao stands than had been imagined. It did not require much more thought for conservationists to realize, and begin to demonstrate, that what was true for the tamarin was true for a broad range of other plant and animal species.[6]

This was of added importance because the difficulties of buying and protecting wild forest remnants were becoming ever-more evident to the conservation organizations. Although the WWF had purchased 12,000 hectares of forest, this was at best a small parcel on which to base the survival of the golden-headed lion tamarin and other endemic species. Also, WWF was discovering that in southern Bahia, even more than in other regions around the world with similar problems, the purchase of land did not translate into being able to control what happened on it. First, land titles in the region were based on a system of property law that was shot through with legal contradictions, and that had been enforced, or unenforced, through political means that constituted pervasive fraud. The most popular Brazilian terms for land fraud were coined in the region. For example, the term *grilagem*, used generally for a variety of forms of chicanery in establishing land ownership, originated during the land rush in the cacao zone in the 19th century. It came from the word *grilo* (cricket) because the ancient character of land documents was commonly faked by enclosing newly forged documents for weeks or months in a box with crickets, which ate holes in the paper and yellowed it with waste products. Fraud was so common and establishment of legitimate land title rare enough in the region that little credibility was given to claims of ownership that could not otherwise be enforced than through influence-peddling and/or violence. When the WWF became aware of the likelihood that people who claimed to hold title to land in the area they were buying did not have clearly defensible legal title, they faced a difficult dilemma. How could they obtain the best claim to ownership when there were competing claims, and how could they deal with the counter-claimants? Privately, they told us that the solution adopted was one that had come from local advisors. Namely, the technique adopted was to cut through the, often insurmountable, difficulties of finding out who constituted the unique and legitimate owner of a given piece of land. This was done by promising to compensate the claimants who held the most political and economic power in the area on the condition that these claimants would receive the compensation only when they had physically removed the less-powerful counter-claimants. It was understood that they might do this through their influence over local police and/or through hiring from the large pool of gun-thugs who thrive in the region. The poorer counter-claimants were usually families or small groups of families who had established livelihoods in the forest and who usually based their title on the ancient and honourable Portuguese and Brazilian tradition of ownership as derived from effective use of the land – that is, the legitimate owner was the person who was giving 'social value' to the land by putting it to an economic use. Some of the counter-claimants had their own documents to pose against those of others. The WWF found, as is typical in many areas of Brazil, that the same piece of land might have multiple claimants to ownership, each with his or her own documentation and/or legally arguable rationale.

The dilemma created by the inability to discover who had the most legitimate claim to ownership is common. It is easy to understand why conservation organizations adopt rough-and-ready solutions to such difficulties, essentially because the legal system cannot be expected to provide a more just remedy. As understandable as the dilemma may be, the kind of tactic used in southern Bahia inevitably brings about a series of other very serious problems.

First, those removed from the area are human beings with their legitimate rights to livelihood and dignity, whatever the fine points of the legality in a profoundly flawed and corrupt legal system. Removal may mean that the families will be unable to find any other means of survival. This may lead to extreme deprivation, including malnutrition, disease, or starvation. Second, the families removed, and countless others in similarly difficult circumstances, will have powerful incentives to re-enter the area to poach or to attempt to re-establish residency. This poses a danger of further conflict and of continued threats to the forest and wildlife. Third, with the loss of the families who had been managing to live in a largely forested and species-rich environment, conservation organizations were likely to lose complex bodies of knowledge about how people were able to make a living within a substantially forested landscape. Fourth, the more powerful claimants who have been compensated sometimes remove valuable timber, or contract poachers either before they turn over ownership to the conservation organization or on an ongoing basis later. Since they are in effective control of the law in the area, this is easier and more likely than it might appear. Fifth, because of these problems and others, the conservation organizations and government agencies set up with the task of protecting the area will almost certainly suffer from a legitimacy problem in the region because what they have done continues to be based on the endemic contradictions, absurdities, and injustices with which people are all-too-familiar. The sense of support for the conservation projects (that virtually all experience shows is critical to successful protection of landscapes and species) will be weak or missing altogether. Many local people will be positively hostile to conservation. WWF and government officials privately admit to the persistence of such problems and to the partial control they have over the reserves established. Attempts to protect the forest remnants in southern Bahia have not been completely defeated, but they have been badly compromised.

With small reserves themselves lacking the full protection they need, the quality of the mosaic in and around the forest is of immense importance. This is where *cabruca* cacao became a vital potential partner in conservation efforts. For reasons that are not altogether clear, there had been substantial resistance to CEPLAC's attempt to eliminate *cabrucas*. The most important reason for the survival of the *cabrucas* was almost certainly that it was the lowest-cost way to produce cacao. This was attractive to owners of smaller operations with little capital, and to some holders of large plantations who had multiple economic interests and regarded their holdings as a cash cow but who had little interest in sinking money or management effort into their cacao farms. There has always been, in addition, a solid core of planters, large and small, who believed, and continue to believe, that *cabrucas* produce more cacao on average over the years. They also assert that in those years when production is low, the compensation of low costs makes up for yield losses. When CEPLAC's clear-cut, chemical-dependent cacao was made impractical by low cacao prices, *cabrucas* might have a better chance of gaining ground as a low-cost alternative. With additional research and financial support from conservation organizations, growing cacao in the shade of the forest might become yet more attractive. The great problem, however, is that cacao prices fell so low that few planters wanted to stay in cacao farming at all. The main alternatives in the very delicate soils that predominate in the region – a shallow layer of soil made up mostly of rapidly decomposing organic matter laid over sand or heavily mineralized thick clay – have been cattle ranching and sugar cane, both of which tend to be short-lived and highly destructive strategies (Figure 9.3). Ranching, sugar cane, and other cropping alternatives are so

Figure 9.3 Recently converted cattle pasture in the cacao-growing region of southern Bahia, Brazil. This farm used to be planted with cacao, but because of the low price and the introduction of the witch's broom disease, it was converted to a cattle ranch.

Source: Angus Wright.

destructive of the soils that they are capable of greatly retarding forest regrowth for a long time to come even if agriculture were subsequently to be abandoned altogether. The combination of soaring petroleum prices and a flourishing Brazilian programme producing the world's largest quantity of ethanol-based transportation fuels makes it particularly dangerous for sugar cane to become the most attractive expedient for local landowners.

All of this has been complicated by other factors. One is the weakness of the government's environmental enforcement agencies. While we were doing research in the area in March 1992, established logging mills were busy and three new sawmills opened during a time when officially all logging of primary forest was absolutely prohibited, and limited logging of secondary forest and *cabruca* stands required special permits, none of which had been issued and which would not be issued for months or even years after application. Local conservationists feared for their lives if they were seen to be documenting the logging. During the same month, the international renowned head of Brazil's environmental ministry, Jose Lutzemberger, resigned in a highly public gesture meant to express his complete frustration with his job, declaring the ministry to be 'nothing more than a den of thieves' and a 'one-hundred per cent owned subsidiary of international timber firms'. Logging firms operated in southern Bahia, as in most of Brazil, with complete impunity from the law. With cacao *fazendeiros* desperate for income, the logging of *cabruca* cacao stands and all other forested areas would speed up dramatically, even where no agriculture at all replaced them.

The conservation agenda faces a cry for social justice

Another complicating factor was the rise in Brazil and in southern Bahia of organizations of rural people pressing for land reform. These included traditional unions of rural workers enjoying something of a renaissance in the wake of Brazil's return to civilian rule, and, most importantly, the MST, a new, militant, aggressive land reform movement, as described in Chapter 7. A Roman Catholic organization, the *Comissão Pastoral da Terra* (Pastoral Land Commission) provided documentation and support for these organizations as they fought for access to land for the poor. During the late 1980s, these groups began to organize land occupations of poor families to claim land under the complicated provisions of Brazilian land law. The law allowed property claims by people who put land to use, even where those properties had been deeded to others who demonstrably were making use of the land.[7] Brazilian law also mandated a land reform that had never been seriously implemented. For the MST and some of the other organizations, their land occupations were meant not simply as an end in themselves, but also as a way of pressuring the government to undertake land reform as an aggressive, serious national policy, and to adopt other measures to improve the lives of the poor. We have written elsewhere on the complexities of Brazilian land law and land reform,[8] much of which, while fascinating, is beyond the scope of this book.

In southern Bahia, as throughout Brazil, individuals and small groups had long occupied land in the hope of getting either temporary or permanent use of it, and, perhaps, an ownership claim. Where land was of good quality, they would often be subsequently run off the land by more powerful people through legal manipulation, intimidation, and/or violence. What had changed by the 1980s was that the MST and others had begun more effectively to organize occupations and gain support for them, making it difficult to dislodge the occupiers. They were also succeeding to a significant degree in gaining the support of urban and rural people all over Brazil for a serious land reform. They were and are active in many regions of Brazil, and, with more than a million members, the MST is considered to be one of the largest social movements in the world (Figure 9.4).

Land occupations in southern Bahia took place in cacao plantations and in secondary or primary forest,[9] sometimes resulting in the deforestation of substantial tracts. It was just as the rhythm of occupations began to increase significantly that the bottom fell out of the cacao market, witchbroom disease arrived and conservationists who had begun to work more seriously to preserve biodiversity in the region started to understand the value of *cabruca* cacao

Figure 9.4 A march of 18,000 militants of the Landless Workers Movement (MST) in
    Brazil during its 5th Congress in 2007.

Source: Wilson Dias (licensed under the Creative Commons License Attribution 2.5: https://
commons.wikimedia.org/wiki/File:MST_06142007.jpg).

for conservation purposes. The convergence of all these elements raised a
variety of complicated and difficult questions for all involved.

Was it possible to protect or promote *cabruca* stands at a time of such low
cacao prices, and with the additional problem of a rampant disease outbreak?
If not, what would replace *cabruca* and what would the consequences be for
biodiversity conservation? Did the poor people occupying land represent a
threat to the forest, *cabrucas*, and biodiversity, or could they play a positive
role in conservation? With cacao agriculture in severe crisis, how would those
who occupied land and land reform beneficiaries make a living from the land
without further degradation of the soils and forests of the region? There are
still no definitive answers to these questions. However, one idea is for there
to be a more thorough, ongoing documentation of the positive role that
shade-grown cacao can play in biodiversity conservation. As a consequence,
international and local conservation groups have sought ways to promote
cabruca stands. In addition, the MST and other organizations of rural people
have sought to use *cabrucas* as part of a more comprehensive strategy of building rural livelihoods, as well as a way of gaining public sympathy and support.
These organizations and the communities affiliated with them have done a
great deal of experimentation of their own and have sought professional

technical advice on how to transform *cabrucas* into a more complex, stable, and profitable activity. Unfortunately, most agronomists and other professionals are poorly trained to be able to offer much help beyond recommending what seems to be the obvious.[10] The main method for doing so is diversification of economic crops within the forest stands, incorporating such crops as piassava (a palm that, when properly farmed, can produce a large and continuous crop of high-value fibre used for brooms, brushes and durable and attractive thatch roofs), citrus, cinnamon, black pepper, pineapple, *caju* (cashew, valued in Brazil for its fruit as well as the cashew nut), mango, banana, and other trees and perennial crops. The more conceptually sophisticated (though not necessarily the best or most productive) of these efforts focus on the use of a diverse combination of largely perennial plants, some of them producing an economic product, some not, to mimic the structure and ecological functions of the indigenous forest plants. The latest threat to the *cabruca* cacao system is climate change, and the drought associated with El Niño Southern Oscillations (ENSO), as climate models predict possible increases in the frequency of ENSO associated with global warming.[11]

Some planters and communities have found fish and shrimp aquaculture to be a good complement to forest-based production, making it possible to increase income while maintaining forest and *cabruca* stands that help to maintain water quality and stabilize drainage, reducing the dangers of droughts and floods to aquaculture ponds, as well as to other crops and to homes.

In this most species-rich nation on earth, the MST has officially committed itself to promoting agroecological models of development, with excellent examples such as those we have cited, although against considerable inertia and resistance from many of its more than a million members. Perhaps just as significant, the MST has worked for decades to develop a curriculum for rural life that has been instituted in a large share of public schools throughout Brazil and taught at the national college, Escola Florestan Fernandes. MST organizers and agronomists are trained to promote agroecological approaches. The curriculum is aimed at creating critical thinking and analysis at all levels, and also puts forward a specifically ecological vision of agricultural and rural economic activities of all kinds. Ecological ideas and practice, like the rest of the curriculum, are not seen as static 'lessons', but rather as departure points for thinking about, organizing, and creating socially just and ecologically sustainable rural communities.

These same ideas are reflected in the programme of the international organization La Via Campesina, an organization in which the MST plays a prominent role. The curriculum has also been influenced by the particularly Brazilian version of environmental justice referred to as 'socio-ambientalismo', which is built on the idea that social equity, democratic practice, and environmental protection and sustainability are mutually inter-dependent – none can be successfully built or maintained without the others.

Independently, some farmers and a few communities have earned further supplemental income through tourism, variously termed agro-tourism, ecotourism, adventure tourism, or just plain tourism. The very high rainfall and the endemic poverty and violence of the region tend to discourage tourists. However, the beauty of the hills and low mountains when substantially covered by forest and *cabrucas*, the countless spectacular flowering plants and the magnificent and seemingly unending beaches guarantee a certain modest flow of tourists to the region. Even the region's reputation as a kind of Brazilian Wild West, both excoriated and celebrated in the tens of millions of copies of Jorge Amado's novels in all of the world's major languages (a phenomenon much like that of Amado's imitator and friend, John Steinbeck, in Monterey, California), provides a stimulus to tourism.

In his older and more mellow years, Amado wrote of the region as a vast and beautiful cultivated garden. It certainly has the potential to be just that, incorporating both the preservation of a great array of wild species and reasonably prosperous livelihoods. Some *cabruca* cacao plantations and smallholder land-reform communities epitomize the possibilities. But the somewhat misty and sentimental vision of Amado's later years is very far from realization, and to the extent that it does exist, may be disastrously undermined in the future. Perhaps the most dangerous element is on the one hand, the failure to realize the region's enormous potential for a relatively harmonious balance between conservation and livelihood in a species-rich tropical forest environment, and on the other hand, the possibility of a steady slide into further extreme poverty and environmental destruction.

Notes

1 Conservation International (2007).
2 Johns (1999).
3 Wright (1976).
4 Mittermeier *et al.* (1998).
5 Alves (1990).
6 Rice and Greenberg (2000); Faria *et al.* (2006, 2007); Cassano *et al.* (2009); Oliveira *et al.* (2010); Sambuichi *et al.* (2012).
7 Because one of the most common ways of demonstrating economic use was to cut the forest, this provision of law has been devastating to Brazilian forests.
8 Wright and Wolford (2003).
9 The term 'primary forest' is probably a misnomer in almost all cases in Bahia, where there may be no forest that has not been cut over at some point.
10 Fernandes *et al.* (2018).
11 Gateau-Rey *et al.* (2018).

10 The production of food and the biodiversity connection

Both coffee and cacao are excellent examples of agroecosystems that can and do contain much biodiversity and form high-quality matrices. Furthermore, in many cases, especially in small landholdings, these systems contain many other species that contribute to the livelihood of rural families (other commercial crops, food, medicinal plants, ornamentals, honey, etc.).[1] For example, in 2008 we visited a 'coffee' farmer in the state of Kerala in India. His farm was one and a half hectares and his main crop was coffee, which he sold in the nearby town. However, in this small farm he also cultivated cardamom, vanilla, pepper, citrus of various kinds, bananas, coconuts, mangos, and a variety of other fruit trees and minor crops that were either sold locally or consumed by the family. In addition to the crops and trees, he had a few goats and a cow that provided fresh milk and cheese, and manure to fertilize his organic farm. According to him and other locals, it is not uncommon, especially for small-scale producers of coffee and cacao, to cultivate and promote many other species (either for their exchange or their use value) beside the main crop, a pattern we have seen elsewhere in the production of both coffee and cacao. Our preliminary work in Puerto Rico similarly shows that coffee is frequently intercropped with plantains and/or bananas, a variety of tubers, and citrus trees. Indeed, in our survey of more than 80 farms, we failed to find a single pure monoculture. Since many of these additional crops are used for household consumption and not sold in the market, they tend to be overlooked and not counted in formal economic valuations. Robert Rice, from the Smithsonian Migratory Bird Center, and director of the Bird Friendly® Coffee programme, estimated that the potential value of fruits produced in coffee farms in Guatemala and Peru could be between $95 and $270 per hectare, although little is consumed or sold.[2]

However, an argument in favour of the importance of the agricultural matrix for biodiversity conservation cannot be made based solely on these two cash crops, even when considering their substantial secondary crop production. Coffee and cacao have become iconic crops for the idea of wildlife-friendly agriculture, but people need to eat and most of the food we consume does not come from these highly diverse agroforestry systems. The vast majority of the calories we consume, whether directly or indirectly, come from just four or five plant species: maize, wheat, rice, soybeans, and cassava. None of these essential calorie factories can be grown in shaded conditions, so the sort of analysis of matrix quality so evident in the case of coffee and cacao may not apply in the case of these food crops. Indeed, one of the main victims of the industrialization of agriculture was diversity, the diversity of the plants, animals, and fungi that we consume.

Throughout human history, thousands of organisms have been regularly consumed. Even after agriculture became the main source of sustenance for the majority of human beings, the variety of crops and their genetic diversity remained astonishing. The industrialization of agriculture put an end to this vast diversity. Although visiting a supermarket in any big city today suggests that we have a wide selection of food products to choose from and that globalization has broadened our choices of foods, the reality is that the hundreds of different kinds of breakfast cereal, for example, are made mostly from maize or wheat, just two agricultural crops. According to the Food and Agriculture Organization of the UN (FAO), the world has lost 75 per cent of its food diversity over the last century. Indeed, today 95 per cent of the calories consumed by humans come from just 30 plant varieties.[3] Not only have we lost the diversity of types of plants and animals that are cultivated and reared, but also, within these main crops, we have lost much genetic diversity.

One of the main reasons for the decline in crop diversity has been the drive towards mechanization, and the consequent dependence on monocultures to facilitate the use of machines – all part of the historical, economic, and political roots of agricultural transformation discussed in Chapter 4. Later, in the 1960s, the Green Revolution, with its technological package that included high-yielding varieties (HYVs), pesticides, fertilizers, and irrigation technology, contributed further to the homogenization of rural landscapes. In this section we examine how food crops also relate to our main themes of how agriculture affects biodiversity conservation. We first focus on the rice agroecosystem as an important repository of biodiversity and second on maize and its traditional production system, with important lessons about how agroecosystems create landscape mosaics and thus promote high-quality matrices in fragmented landscapes. In both cases we will conclude that small-scale diversified grain farmers, who in some regions still produce a large amount of the food consumed in the contemporary world, play a crucial role in directly maintaining biodiversity and in the indirect construction of a diverse landscape mosaic, which, we argue, is essential for long-term conservation in a fragmented landscape.

Rice production and biodiversity

Rice cultivation is the main productive activity and source of income for more than 200 million farmers in Asia.[4] Rice is the staple food for half of the world's population, and often the main source of employment and income for many of the world's poorest people in the most food-insecure areas of the world. More than 3.5 billion people depend on rice as their main food. In 2009, rice provided 19 per cent of the world's human per-capita energy and 13 per cent of the per-capita protein.[5] Furthermore, hundreds of millions of people spend more than half of their incomes on rice to feed their families.[6] In the context of neo-liberal globalization, rice has followed a similar but distinct trend as maize (discussed later). Subsidies and technical improvements, mostly in developed countries, but also in some developing countries, have increased rice production, lowering production costs and resulting in overproduction and a predictable drop in price. The price plunge in the late 1990s had a significant impact on small-scale farmers by plunging them further into poverty, undermining household food security, and encouraging rural to urban migration.[7] The result has been an increase in the size of farms and levels of intensification of rice production systems and the displacement of small-scale rice farmers. Given the direct connection between rice markets and rural livelihoods, rice is a perfect subject with which to explore the interconnectedness of biodiversity, agriculture, and food sovereignty.

Wetland rice ecology and biodiversity

Rice cultivation covers 158 million hectares worldwide, producing 700 million tons annually, 90 per cent of it in Asia.[8] In the Western Hemisphere, the US and Brazil produce more than two-thirds of the rice. Although some rice is cultivated in upland areas and with no surface or rhizosphere water accumulation (the so-called upland rice), these systems represent only about 10 per cent of all rice production.[9] The following discussion pertains only to the flooded system.

As we have noted repeatedly, one of the main principles of biodiversity conservation in agroecosystems is that the agroecosystem should resemble the natural ecosystems in the region as much as possible. In that sense, it is significant that flooded rice systems have been defined as agronomically managed marshes or a type of freshwater wetland with a cultivated grass. Most flooded rice fields are indeed connected to natural wetlands and share many organisms with these natural ecosystems.[10] It is precisely because of the presence of water in wetland rice that this agroecosystem is an important habitat for a wide variety of aquatic and terrestrial organisms. Rice paddies, along with the adjacent natural wetlands, frequently result in a mosaic of habitats that harbour a rich diversity of flora and fauna. According to ecologists who have studied these biologically rich mosaics, biodiversity is maintained by rapid colonization as well as rapid reproduction and growth of organisms. Although several studies dating back to the 1940s and 1950s have documented the biological richness of the aquatic phase of rice production, few comprehensive studies exist on the ecology and diversity of the rice fields.[11] A survey of an irrigated rice field ecosystem in Sri Lanka documented 494 species of invertebrates, mostly arthropods, belonging to ten phyla, 103 species of vertebrates, 89 species of macrophytes, 39 genera of microphytes, and three species of macrofungi.[12]

Natural wetlands are critical habitats for the survival of many species of both terrestrial and aquatic organisms. However, these ecosystems are suffering a rapid decline in many areas of the world.[13] As a managed wetland, rice agriculture serves as an important habitat refuge for those wetland-dependent species, and as such, is beginning to receive more attention from the conservation community.

The biodiversity of rice fields can be divided into aquatic invertebrates, terrestrial invertebrates, vertebrates, and plants. Here we summarize results from many studies that have examined biodiversity in rice fields.[14]

Aquatic invertebrates inhabiting rice fields include surface water-dwelling insects, zooplankton (protozoans, micro-crustaceans, rotifers, and others), aquatic insects and their larvae, and benthic organisms (bottom-dwelling), such as nematodes, molluscs, and annelid worms. In other words, aquatic organisms in rice fields cover the entire spectrum of freshwater fauna. Of these, the benthic fauna is the most diverse, comprising almost 40 per cent of the species of aquatic invertebrates. Population density of rice field benthic fauna is also high compared to natural and artificial ponds. Freshwater crabs are a common component of the benthic fauna and are important as scavengers and as food for other animals in the ecosystem. Zooplankton is the most studied group in rice fields, where they tend to be as diverse as in ponds, and more diverse than in rivers, streams, and river floodplains. Aquatic insects are also very diverse in the rice fields. Unfortunately, among the most diverse and abundant of the aquatic insects are mosquitoes. Worldwide, 137 species have been listed as breeding in rice fields. Due to human health implications, mosquitoes have received the bulk of the attention from researchers. A few studies have documented other aquatic insects in rice fields, among them a study that listed 117 species of aquatic beetles in rice fields worldwide. Interestingly, almost all of the aquatic insects in the most abundant groups (beetles, bugs, and dragonflies) are predators, including predators of mosquitoes.[15] In China, a survey of natural enemies of rice pests yielded a staggering 1,303 species.[16]

Terrestrial arthropods are also a very diverse component of the rice fauna. One study in Sri Lanka listed 280 species of insects and 60 species of spiders and their relatives for an astonishing total of 340 arthropods in one rice field ecosystem.[17] Several studies have examined the succession of species with the growing cycle, and most of these have examined the trophic interactions, with some constructing food webs of the terrestrial invertebrates. Due to the importance of rice as a food crop, most of the studies of terrestrial arthropod diversity separate the arthropods into herbivores (potential pests) and natural enemies. In this respect, the study from Sri Lanka is illustrative. In this study, the majority of insects belonged to two orders: the Hymenoptera (81 species of wasps and ants), most of which are parasites and predators of other insects, and the Lepidoptera (58 species of moths and butterflies), all of which are herbivores. When combined with the spiders, the natural enemies accounted for at least 41 per cent of the terrestrial arthropod species diversity in these rice field ecosystems. Herbivorous insects represented 38 per cent of the diversity, with many of them (almost 60 per cent) being visitors or spill-overs from weedy and other surrounding vegetation. It has been argued that the long history of rice cultivation in many parts of the world has allowed stable relationships to evolve between rice pests and their natural enemies. Whether this is true or not is hard to tell with short-term studies. Nevertheless, it is evident that in most instances, and in particular when no insecticides are used and there is a high level of landscape heterogeneity, the richness and abundance of natural enemies is greater than that of rice pests.[18]

A wide spectrum of vertebrates belonging to all major groups is found in rice fields. In particular, the rice ecosystem is an important feeding area for aquatic birds. The wetland rice ecosystem acts as a temporary artificial wetland with alternating flooding and dry periods, and usually contains a spatial and temporal heterogeneity that facilitates the establishment of large bird populations. Waterfowl, shore birds, and wading birds utilize rice fields to the greatest extent. The rice ecosystem also harbours species under threat of extinction, like the Bengal florican in Cambodia, an endangered bird species with only two populations remaining.[19] In the last few years, due to the decline in natural wetlands, there has been an increased interest in the potential of rice fields for the conservation of water birds. Indeed, in one California study it appeared that water birds treated flooded rice fields as indistinguishable from natural wetlands.[20] Certification schemes, such as those proposed for cacao and coffee, are less likely to be useful since much of the world's rice crop is grown on small farms for local consumption rather than for international markets where premiums can be obtained.

Fish are another integral component of the rice field vertebrate fauna. Although some authors consider the rice field agroecosystems as essentially an extension of the natural local wetlands that should reflect the diversity of these habitats, a study in the Muda rice irrigation system in Malaysia reported 39 fish species occurring in rice fields, a significant number, yet lower than in natural marshlands.[21] Fish living in the rice fields can also provide an important service to the rice ecosystem. Some fish specialize on mosquito larvae and can be important biological control agents of these vectors of human disease. Likewise, fish that feed on snails may control the vector of schistosomiasis, or may be control agents of the apple snail, a pest of rice.[22]

Rice fields are also important habitats for amphibians and reptiles, many of which are insectivores and implicated in the important ecosystem service of pest control.[23] The study from Sri Lanka mentioned earlier, reported 21 species of reptiles and 18 species of amphibians. In Cambodia, an endangered species of caecilian (a worm-like amphibian, *Ichthyophis bannanicus*) lives in rice fields and is used for medicinal purposes. Some argue that its medicinal value may lead to the controlled cultivation of the species in the rice fields, ultimately saving it from extinction.[24] Amphibians in particular are an important group to study since, with the decline of wetlands, rice fields could be important alternative breeding grounds. However, amphibians are also extremely sensitive to agrochemicals in the water, so increases in pesticide use in rice fields are not likely to be sanguine for this group.[25] To this effect, the introduction of herbicide-resistant (Roundup Ready) rice could have a particularly devastating effect on amphibians since Roundup, the commercial formulation of glyphosate, to which the herbicide-resistant crops are resistant, has been shown to be fatal to amphibians at the recommended application levels.[26]

Many of the mammals that inhabit rice fields are considered pests because they eat rice and also carry human pathogens either directly or via vectors like ticks and fleas. Insectivorous mammals, like shrews and bats, on the other hand, can be beneficial.

Finally, rice fields are an important habitat for a wide variety of plants and other primary producers. This component of the biodiversity of rice fields can be divided into macrophytes, mostly terrestrial and aquatic plants, and microphytes, mostly algae. More than 1,800 weed species are reported in the literature as occurring in rice fields in Asia and Southeast Asia alone, grasses and sedges being the most common. A variety of photosynthetic micro-organisms are present in the aquatic sector of rice fields. Some of these are extremely important for nitrogen fixation and for providing nutrients and organic matter to the system.[27]

The above review is intended to show that rice fields are important habitats for wetland biodiversity, especially in areas where natural wetlands have been destroyed and left in a fragmented state. Likewise, and important for the argument of this book, these artificial wetlands can be important temporary habitats for organisms that are migrating from patches of natural wetlands.[28] In one study in Brazil, for example, it was found that birds in isolated islands of wetlands conformed well to the normal species/area relationship, but that species numbers per area increased when the matrix included wetland rice.[29]

Biodiversity in rice production, and the livelihood of farmers

It comes as no surprise to traditional rice farmers that a rich diversity of aquatic organisms can be found in rice-based agroecosystems. They have nurtured and used this biodiversity for generations. Some rice-based ecosystems contain more than 100 useful species that are important resources for human communities, and can offer a reliable safety net that could protect small-scale farmers in the face of crop failure and other food shortages (Figure 10.1). Studies from Cambodia, China, Laos, and Vietnam documented a total of 232 aquatic species that are collected and used by rural households in these countries. Fishes, crustaceans (shrimps and crabs), amphibians (frogs, toads and salamanders), reptiles (snakes, lizards and turtles), molluscs (snails), plants, and insects supplement the rice diet with important animal protein, fatty acids, and other essential micronutrients that are not found in rice (calcium, iron, zinc, vitamin A, and limiting amino acids).[30] In particular, small fish that are eaten whole provide adequate amounts of calcium and vitamin A. The

Figure 10.1 Girls fishing in a rice field in Madagascar.
Source: Rob Waddington.

calcium comes from the fish bone, and eating only 50 grams of the small fish harvested from the rice fields is enough to get the recommended daily calcium intake for an adult. Indigenous people in Bangladesh recommend eating the small mola fish, which grows in the rice fields, for curing night blindness – a temporary blindness we now understand is caused by vitamin A deficiency. This is a good example of how local indigenous knowledge frequently has a sound scientific basis, since vitamin A is found in the eyes and viscera of fish (which are ingested when the fish is eaten whole).[31]

Farmers who culture fish concurrent with rice have found increases in rice yields, especially on poor soils and unfertilized crops. It has been reported that with reduced cost of pesticides (due to biological control with fish) and additional earnings from fish sales, rice/fish farms report a net income that is 7 to 65 per cent higher than that of rice monoculture farms.[32]

There is no doubt that the aquatic life found in diverse rice ecosystems provides critically important resources for rural livelihoods, and currently contributes to food security, especially for the rural poor in Asia. These resources are also important as a source of income since some are traded in local markets and some are used for medicine, feed for other animals, bait for fishing, decoration and materials for wrapping food. Ironically, most development plans in rice-growing regions focus on increasing rice yields, mostly through the use of HYVs and the use of agrochemicals, which result in a dramatic reduction of these other aquatic resources and likely the predictable subsequent deterioration of the nutritional condition of the rural poor.

Loss of biodiversity and the legacy of the Green Revolution's High-Yielding Varieties (HYVs)

Unfortunately, the high associated biodiversity of the traditional rice eco-systems is declining very rapidly. All major fish farming countries have reported declines in the availability of wild aquatic resources in rice fields. In particular, fish catches from rice fields have diminished considerably. Rice farmers in Xishuangbanna in Yunnan Province, China, claim that they require a full day to collect the aquatic resources they used to collect in one hour a decade ago. In Cambodia, farmers estimate that at current rates of decline, in three to five years there will not be enough fish to make a living. Overall, the contribution of rice-based captures to household consumption has declined from one half, over a decade ago, to less than one-third today.[33] In Japan, from 1968 to 1982 there was an 89 per cent decline in fish in rice fields due to pesticide applications and changes in water management.[34]

Increased fishing pressure on aquatic resources is partially to blame for this decline. However, the transformation of the rice farming system with an emphasis on industrial forms of agriculture is the main culprit for the decline in the abundance and diversity of aquatic resources in the rice ecosystem (Figure 10.2). The history of the rice Green Revolution in Indonesia is illus-trative of the short-sightedness of this approach, suggesting the need for further reflection on the potential benefits of an agroecological approach that can conserve wild and agro-biodiversity and provide a dignified way of life for millions of people in the world.[35]

As discussed in Chapter 4, the perceived success of a high-yield wheat variety developed by Norman Borlaug in Mexico led to the spread of the so-called Green Revolution for maize, rice, and other crops in the rest of the world. In Asia, the International Rice Research Institute (IRRI),[36] together with national research programmes, promoted development of short-duration high-yielding rice varieties, beginning in the 1960s. As a result of this Green Revolution, Indonesia passed from being the largest rice importer in the world to being self-sufficient in rice in the mid-1980s.[37] However, as was the case with the Green Revolution in other countries and for other crops, this achievement in yield increases came at a high environmental and human cost.

In Indonesia, the government contracted with pesticide companies for the aerial spray of phosphamidon, produced by Ciba-Geigy, to control the rice stem borer. This started before the spread of the HYVs in the late 1960s and continued through the 1970s. By 1974, an even more devastating pest, the rice brown planthopper, emerged. This was a classic case of secondary pest outbreak (as discussed in Chapter 4). This pest had not been reported as a pest in rice fields before and emerged as a pest only after massive aerial spraying with broad-spectrum insecticides that eliminated its natural biological control

(a)

(b)

Figure 10.2 (a) Diverse and (b) homogeneous rice landscapes illustrating the loss of biodiversity in Bali, Indonesia.

Source: (a and b) John Vandermeer.

agents. The Indonesian government, rather than reducing pesticide applications to restore the populations of natural enemies in the rice fields, started a programme of insecticide subsidies that made insecticides available to farmers for 20 per cent of the actual cost. By the mid-1980s the cost of the insecticide subsidy plan reached US$120 million per year. In spite of the massive pesticide applications, or rather because of them, damage caused by the rice brown planthopper continued to increase, becoming so severe that, in 1977 alone, the amount of rice lost to the pest was enough to have fed 1.2 million people for a year. That year the IRRI started to multiply and distribute a seed variety from South Asia that was distasteful to the rice brown planthopper. But within three seasons the planthopper had evolved to readily feed on this new variety, and the farmers were back to square one. In the early 1980s, scientists developed new resistant varieties, but a few years later there was a dramatic and sudden breakdown in resistance. Again, the farmers were back to square one. Throughout this time, the Indonesian government had maintained, and even increased, the insecticide subsidy programme at a high economic, human, and environmental cost, and with no positive results in terms of pest control.

Finally, in the mid-1980s, after many researchers were able to demonstrate that the rice planthopper was in fact an insecticide-generated problem, and after strong pressure from civil society, the government took a bold step and banned 57 insecticides for rice, eliminated insecticide subsidies and started a comprehensive Integrated Pest Management (IPM) programme in rice.[37] Since then, many more studies have demonstrated that the rice ecosystem is a complex system that supports high levels of biodiversity, including many natural enemies that act as biological controls of rice pests. Research has also uncovered complex interactions that include links between organic matter and generalist predators through complex trophic cascades, similar to that described earlier for the coffee agroecosystem, and well known in other grain crops. Bacteria and phytoplankton are the base of the aquatic food web in irrigated rice fields. In this case, a rich organic 'soup' originating from crop residues from previous cycles, organic waste brought from villages with irrigation water, and algal growth, increases the abundance of detritivores and plankton feeders, which in turn support the populations of generalist predators, which feed on herbivores. It has also been demonstrated that the rice plant can withstand up to 60 per cent damage without reducing yields, and even higher levels with little yield reductions. Furthermore, it seems that the heterogeneity of the landscape also contributes to pest control. This goes against the idea that synchronous planting over a large area helps control pests because the fallow period (when the fields lie barren and dry for a long period) acts as a bottleneck for the pest population. Comparative studies in Indonesia and Malaysia showed that outbreaks of the rice brown planthopper are more prevalent and severe in synchronously planted areas with continuous rice fields that cover thousands of hectares than in non-synchronously planted areas dominated by smaller rice fields interwoven with natural and home garden vegetation. The reason for this appears to be that in the synchronously planted areas, predators take a much longer time to arrive in the fields than in the non-synchronously planted fields, where predator populations can continue to build up through the years, and even take refuge in the natural and non-rice vegetation that is present throughout the landscape.[39]

In addition to the decline in associated biodiversity, the Green Revolution in rice also resulted in dramatic genetic erosion, that is, a decline in the landraces of rice, many of which were adapted to local environmental conditions. Such a decline increases the vulnerability of rice-producing countries, and threatens food security and food sovereignty. The first high-yielding rice variety was released in 1966 and a decade later a few HYVs had replaced thousands of traditional rice landraces previously cultivated by farmers. Since these HYVs are bred to respond to irrigation, pesticide and fertilizer applications, and mechanization, they spread in those areas with favourable environmental and political conditions (where farmers could afford the technology or, as in the case of Indonesia, where the government provided subsidies to make the technology accessible to farmers). Today, most rice fields in Asia are occupied by a few high-yielding rice varieties. For example, in the Philippines, home of the IRRI, 50 per cent of the rice area is planted in four HYVs; in Cambodia only one variety is cultivated in 90 per cent of the rice fields; and in Pakistan, only four HYVs are planted in 99 per cent of the rice fields.[40] Traditional varieties were collected and stored in national and international seed banks to which farmers do not generally have access. In addition, some traditional varieties are still cultivated in marginal areas, where the HYVs do not perform as well or where farmers can't afford some parts of the technological package.

The story of the rice Green Revolution in Asia illustrates the pitfalls of a narrow technological approach to agriculture. In their efforts to increase yields, researchers and politicians ignored the fact that the rice ecosystems, and their associated landscapes, not only put a bowl of rice on the table of millions of people, but provide for the livelihood of millions of rural people for whom rice is life and culture.

Maize production and the landscape mosaic

The tendency to grow food in large-scale monocultures is not universal, and certainly not very old. Maize is an excellent example of a grain that is frequently produced on a small scale. Even today, traditional maize producers are not clamouring for large extensions of monocultural production. Indeed, in areas of Mexico and Guatemala, where indigenous people still follow traditional agricultural practices, it is commonplace to see maize and beans grown together in the same field, frequently with squash interspersed among the maize and bean plants (Figure 10.3). It is common for the beans, growing as a vine, to use the maize stalk as a trellis, and the squash plants to cover the understorey, choking out undesirable plants. In especially low spots in the

Figure 10.3 A *milpa* system in Guatemala. *Milpa* is a corn-based growing system used throughout Mesoamerica that typically includes corn, bean, and squash. In this system the various components facilitate each other.

Source: Fabian Hanneforth

field, if locally poor drainage is expected, some other crop such as taro or banana is commonly planted instead of the maize/bean mixture. So, a typical maize field, or *milpa*, is frequently a combination of species that take advantage of special features of the environment (the low poorly drained spot) or of each other's special ecological characteristics (the maize using some of the nitrogen the beans fix from the air, the beans using the maize stalk as a trellis) – not at all a true monoculture. Indeed, from our experience of talking with such farmers, their attitude is that it would be foolish not to take advantage of these special ecological conditions when planning the farm.

However, for the arguments presented in this book, it is not so much the micro-management of an individual maize field that matters, but rather the landscape patterns that emerge from individual farmers managing small parcels of land in ways that are dictated by ecological imperatives (Figure 10.4). The environment in which much traditional maize farming is carried out is situated on soils that are characteristically poor at storing nutrients. Thus, a bit of potassium (say) in the soil will rapidly leach out of that soil with the next rain if the crop does not immediately take it up. This is quite different from the

Figure 10.4 A slash and burn agricultural landscape in Belize. In the centre, there is an area that was recently cut and burned and is being prepared for planting. The rest of the landscape consists of areas that are either in cultivation or at various stages of succession.

Source: John Vandermeer.

rich soils of temperate latitudes where the details of soil structure actually hold that potassium in place over many years, like a nutrient bank. This metaphorical bank simply does not exist in some highly weathered tropical soils. Consequently, the nutrients (nitrogen, potassium, phosphorous, calcium, etc.) in tropical ecosystems are contained in the vegetation of the system, more than in the soil. So, when a farmer cuts down and burns the natural vegetation to make a farm field, he or she effectively takes almost all the nutrients in the system and places them in the soil all at once. In the end, if those nutrient elements are not taken up by the crops, they will leach out of the system when the rains come.

The inefficiency of many tropical soils at maintaining nutrients means that crop production dramatically drops after a few years of production. This drop in production requires the farmer to cut down a new piece of vegetation to begin the process again, leaving the previous cropped field in fallow. This cycle of slashing the vegetation, burning it, cropping it, and subsequently leaving it fallow is referred to as 'slash and burn' or swidden agriculture, and was certainly the dominant form of producing grains the world over prior to the imposition of the industrial system,[41] as discussed in Chapter 4. The fallow field would inevitably go through the normal period of ecological succession, with pioneering vegetation establishing in the abandoned field and accompanying any re-sprouting trees and other woody vegetation that remained alive, followed by the establishment of successional species in the march towards a second-growth forest. In ten to 20 years, the vegetation grows back to form a secondary forest with high levels of nutrients stored within the vegetation. At that point, the forest is ready to be cut again in a continuous cycle of cultivation and fallow. Exactly how long it takes for a particular piece of land to grow back to the point when it is ready for the next cultivation depends on the properties of the soil, the amount of rain, and the type of vegetation that grows during the fallow period. Farmers frequently use indicator species to decide when the fallow is ready to be cultivated again. In some places, farmers shorten the fallow period by planting species of trees that accelerate the successional process or increase the amount of nutrients available in the ecosystem.

Among the important consequences of this traditional system of agriculture is the overall landscape that it creates. An individual farming family or collective must have enough land available to be able to set aside relatively large amounts of land to be in fallow between cropping cycles. Leaving 80 or 90 per cent of the land in some sort of recovery stage creates a patchiness of habitats in the landscape reminiscent of the landscape mosaic we described in Chapter 2 (Figure 10.5). Thus, in this traditional mode of maize production, the expected landscape is a mosaic that may provide for a high-quality matrix through which organisms migrate, thus creating the sort of matrix that promotes successful metapopulation structures in the patches of natural vegetation that remain. Indeed, it has been suggested that this form of agriculture actually creates the conditions for higher biodiversity simply because of the habitat heterogeneity it generates.

Nevertheless, recent trends in agriculture have replaced much of the extant traditional maize production (certainly not all), and we must question how future production in traditional maize-producing regions can be organized so as to provide a high-quality matrix for conservation. Leaving large areas in fallow would probably be a good idea but is unfeasible, although the simple notion of a fallow should always remain part of the agricultural planning process (at least in the tropics). However, as has been pointed out by Wes Jackson and others, examining the way the natural world works is a good way to gain ideas for planning agroecosystems. In many ways, the world's ubiquitous rotational systems actually mimic, to some extent, the fallow period (e.g. soybeans are the fallow period from the point of view of maize, in a maize–soybean rotational system). In this way more modern, but still non-industrialized, tropical farmers tend to use complex rotational systems in which maize is a part, but in which beans, squash, vegetables, and some trees, form an equally essential part of the rotational system.

(a)

(b)

(c)

Figure 10.5 (a) A vista of a forested/agriculture area in Bahia, Brazil. (b) A close-up of the area outlined in white in part (a), where the texture of the photograph reveals 'scars' of current and past agriculture. (c) Approximate polygons (in white) around various suspected land use histories (the width of the line is intended to be proportional to the relative time since active cultivation [polygons drawn by eye by the authors]).

Source: (a, b, and c) Google Earth, Digital Globe.

For example, on a traditional farm we recently visited in Chiapas, Mexico, a patchwork of small fields (each approximately a quarter of a hectare) contained maize, beans (sometimes in combination, sometimes separate), broccoli, cabbage, carrots, and cassava, and were bordered by several small fruit trees (especially papaya), with the ubiquitous large mango tree in one corner of the farm. Mixed in with all these fields were several that had not been planted at all and were in various stages of succession (Figure 10.6). While this farm did not look like native vegetation to us, its mosaic nature resulting from small-scale planting decisions created a mosaic that undoubtedly seemed more like some sort of native vegetation to many different kinds of plants, fungi, animals, and other organisms. This was also an organic farm, and the lack of agrochemicals ensured that small insects migrating into one of the fields would have a good chance of reproducing individuals that could migrate to other fields before expiring themselves. In short, this farm created a landscape that was not natural vegetation, nor the sort of high-quality matrix that a rustic coffee farm represents. Nevertheless, it formed a piece of the landscape mosaic that would probably ensure successful migration of many kinds of organisms. It is, in our view, a model that could be emulated, especially in the context of a political system that promotes the idea of food sovereignty.

Figure 10.6 A diverse organic farm near San Cristobal de las Casas, Chiapas, Mexico.
Source: John Vandermeer.

Response from the social movements: where biodiversity conservation meets food sovereignty

Diverse rice and maize farming systems present excellent examples of the type of win–win situation that could ensue when the proper ecological, socio-economic, and cultural conditions are developed. Given the importance of rice for the culture, nutrition, income, and overall livelihood of millions of people in Asia, it is not surprising that the small-scale farmers' organizations and grass-roots social movements in the region are taking a strong stand under the slogan 'Rice is not a commodity: it is life, culture and dignity!' In May 2006, they released a declaration during the Asia Pacific Peoples Conference of Rice and Food Sovereignty (Box 10.1). In this declaration, the participants urged their governments to adopt policies to promote sustainable rice production and family farm-based production. The revival of traditional rice farming practices that rely on high levels of aquatic and terrestrial biodiversity, as well a high diversity of landraces was one of the points within the resolution. Unfortunately, the road towards food sovereignty has been a slow and arduous one in most Asian countries. For example, in Indonesia, after saying that food sovereignty was a goal of his government, President Joko Widodo oversaw an increase in food commodity imports of 3.8 million tons from 2014 to 1017, and in the last ten years Indonesia lost five million farmers.[42] Likewise, in the Philippines, government support for neo-liberal policies and a greater privet sector participation in the market, have led to price spikes of rice for consumers and low buying price for farmers.[43]

Box 10.1: Rice is life, culture and dignity

FINAL DECLARATION OF THE ASIA PACIFIC PEOPLE CONFERENCE ON RICE AND FOOD SOVEREIGNTY

Jakarta – Indonesia, 14–18 May 2006

Rice is Life, Culture and Dignity

We the peasants from Asia and Pacific strongly voice our right to have a better life, to preserve our cultures, and protect the dignity of the people. Rice has been our staple food for centuries, so it is a political issue. Therefore, we demand food sovereignty for the people. Farmers should have the rights to produce food in a sustainable way and be protected from neo-liberal policies. Food sovereignty should prevail over free trade.

Women peasants have protected traditional rural agriculture and sustainable agriculture in their communities for centuries. Therefore, they should be respected and have the same rights as men. Their rights and access to production resources should be protected because women are the main force of production. Women peasants will implement a detailed action plan to protect sustainable agriculture and food sovereignty.

We gathered to formulate the principles of food sovereignty on rice. We urge our governments to adopt policies promoting and supporting sustainable rice production and family-farm based production rather than agro-industry, export-oriented, and high-input rice production. During our discussions, we realized that people's food sovereignty should be implemented at different level: the access and control over natural resources (land, water and seeds), the choice of the mode of production and consumption, and the kind of trade that we want.

1 Land, Water and Seeds

- We demand a genuine agrarian reform that focuses on land distribution to the landless people and provides possibilities for farmers to retain land in their possession.
- Land to the tiller: the land should belong to the small and landless people, and not to landlords and big companies.
- Land and water have to be owned by the local communities with all respect to the community's legal and customary rights to their local and traditional resources.
- It is not enough to have some positive laws. In Asia and Pacific, many countries have agrarian laws, but they are not implemented. Social movements have to be able to pressurize and monitor the implementation of the laws.
- Women should have the equal right in terms of their access to land water and other productive resources.
- We condemn the privatization of water; because water is now being controlled by MNCs [Multinational Corporations]. Governments have to protect their farmers in order to provide them with free irrigation access for production.
- We should be protected from the pollution of the water sources by the industry and chemical farming, especially in rice production systems.
- Seeds are at the heart of agriculture, they are the basis of food sovereignty. We should abolish all patents on seeds, as well as reject any means, system or technology that prevents farmers from saving, improving, and reproducing seeds. We categorically say no to terminator technology on seeds that curtails this freedom.
- Encourage the right to exchange and reproduce seeds by the people and for the people. Seeds can not to be distributed by the MNCs and the governments, as they will make farmers as only the end-user of the chain of seed production.
- Abolish GMOs [Genetically Modified Organisms] and ban its production and trade in rice seed.

2 Rice Production Systems

- We condemn the green revolution because it destroys biodiversity, it foster dependency on chemicals, and leads to environmental degradation and displaces many small farmers from their livelihoods and land.

- We promote sustainable rice production such as organic and natural farming: they use less input and produce better quality outputs.
- Encourage the revival of traditional experience on sustainable rice production systems, for instance the natural farming in India (Karnataka).
- Recognize the importance of food sovereignty in terms of nature and ecology in order to create poverty eradication, protecting ecosystem, land preservation, biodiversity, improving health condition, increasing water quality and food at affordable cost.
- Establish rice quality criteria in favor of people's preferences and needs.
- Strongly push our governments to give support to organizations who promote sustainable agriculture, and to set up formal policies to promote sustainable rice production systems.

3 Post Harvest Activities and Processing

- Develop local rice economies based on local production and processing by farmers.
- Processing activities and local trade should be managed by small family units, with cheap and people adapted technologies.
- The governments should provide services program supporting production and land productivity. They should also facilitate post harvest activities.

4 Trade

- We should ensure adequate remunerative prices and the government should guarantee it through subsidy to cover the costs of production and also to get adequate profit which are related to the cost of living of the farmer.
- Abolish all direct and indirect export subsidies, and ask the government to give subsidy for promoting sustainable rice production, and make sure that the subsidies not to go to the MNCs and large producers.
- The governments must support farmers who produce rice for domestic needs.
- Domestic production should be regulated to prevent surpluses.
- Ban rice imports when countries can produce enough for their own consumption. Most of the time, rice imports are only surpluses dumped on our countries and they kill farmers.
- Promoting family-based rice farming to ensure the domestic rice need and self-sufficiency. We condemn the liberalization of rice trade pushed by the WTO and the Free Trade Agreements. And we demand that WTO out from food and agriculture.

Rice is not only a commodity: it is life, culture and dignity!

Organizations (peasant organizations and civil organizations and NGOs) endorsing this declaration are:

Assembly of the Poor (AOP) of Thailand

PARAGOS of Philippine

UNORKA of Philippine

Viet Nam Farmers Union (VNFU) of Viet Nam

Hasatil of Timor Leste

ANPA of Nepal
KRRS of India
BKU of India
MONLAR of Srilanka
National Family Farmer Coalition (NFFC) of United Sates
Korean Peasant League (KPL) of South Korea
Korean Women Peasant Assosiation (KWPA) of South Korea
Federation of Indonesian Peasant Union (FSPI) of Indonesia
Petani Mandiri of Indonesia
        Source: https://nyeleni.org/spip.php?article30, accessed 13 March 2019

Similarly, maize is an important cultural as well as nutritional element in much of Latin America. Indeed, according to the Mayan creation myth, the god Popol Vuh made people from maize. In Mexico, a coalition of peasants, farm workers, indigenous people, fishers, consumers, and environmental and human rights organizations launched the National Campaign in Defence of Food Sovereignty and the Revitalization of Rural Mexico, under the slogan 'Sin maíz no hay país' (without maize we have no country). According to the national newspaper *La Jornada*, in January 2008 more than 200,000 people marched in Mexico City to demand that NAFTA's agricultural chapter be renegotiated. Among the main points of the campaign are, (1) the renegotiation of NAFTA, removing corn and beans from this pact to protect national grains, (2) the prohibition of transgenic grains in the country, and the protection of native seeds, and (3) the demand for new public policies in defence of the rural areas, peasants, and food sovereignty and security (Box 10.2). Since the formation of the campaign, they have been celebrating the National Day of Maize every 29th of September.

Box 10.2: Ten urgent steps for Mexico

The National Campaign in Defense of Food Sovereignty and the Revitalization of Rural Mexico calls on all Mexicans to join together in support of protecting Mexican maize, defending Mexico's food sovereignty and revitalizing the Mexican countryside. They outline ten urgent steps that need to be taken:

- Remove corn and beans from NAFTA: Install a permanent mechanism for the management of imports and exports of corn and beans (and their derivatives and by-products) by the Mexican Congress.
- Ban the planting of transgenic corn in Mexico: Protect and improve the genetic heritage of Mexican corn, and encourage the production of organic and native corn varieties.
- Approve the Constitutional Right to Food by the Mexican Chamber of Deputies and the Law on Planning for Sovereignty and Food Security and Nutrition by the Mexican Senate.

- Fight the food sector monopolies: Prevent hoarding and speculation, and misleading advertising of junk food.
- Promote Mexican corn and cultural expressions.
- Control prices of the basic food basket: Ensure adequate food supply and create a strategic reserve of food; and promote the consumption of fair trade and small producer-produced food.
- Recognize the rights of peasants and the rights of indigenous peoples to their territories and its natural resources.
- Support higher prices for coffee growers and their access to international markets.
- Promote the conservation of the forests through community-led sustainable management of natural resources.
- Ensure the principle of gender equality in rural policies, as well as the full recognition of the human rights of all citizens including labor, farm workers and migrant workers.

And they call on Mexicans everywhere to:

- Plant corn in homes, sidewalks, camellones and public parks throughout the country.
- Sign and support the ten steps to defend corn and the Mexican countryside.
- Participate in the National Day of Mobilization for the Defense of Food Sovereignty and the Revitalization of Rural Mexico and the 2008 rural budget.
- Educate, organize and act to denounce monopoly agribusiness abuses, and promote the production and consumption of fair, healthy, organic, non-GM foods produced by small farmers.
- Support the efforts of peasant and indigenous organizations' demands for justice, health and sovereignty for Mexico.

Source: https://grassrootsonline.org/in-the-news/
newsarticlesnafta-killing-tradition-corn-mexico/,
accessed 13 March 2019

Agricultural potential in the matrix

The foregoing description of both the history and functioning of various agro-ecosystems (Chapters 4–9) is intended to give the reader the impression that (1) the industrial system has had a peculiar political and social evolutionary pathway, depending on economic and political forces as much as on ecological contingencies, and that (2) the alternative agriculture movement, as well as extant and extensive systems such as coffee and cacao, along with diversified landscape mosaics, suggest that the image of the pesticide-drenched monocultural field is not the only kind of agriculture possible. This point becomes central to our general argument that transforming the agroecosystem into a high-quality matrix is possible given the realities of agriculture, as much as it is essential for conservation. We return to this basic point in our summary in the next chapter.

Notes

1 Aguilar Støen (2009); Méndez *et al.* (2010); Rice (2011); Moguel and Toledo (1999).
2 Rice (2011).
3 FAO (1996).
4 Ricepedia, 'Who grows rice', http://ricepedia.org/rice-as-a-crop/who-grows-rice, accessed 3 November 2018.
5 FAO (2004); Ricepedia, 'The global staple', http://ricepedia.org/rice-as-food/the-global-staple-rice-consumers, accessed 3 November 2018.
6 Paris *et al.* (2000).
7 FAO (2004).
8 FAO (2004); Ricepidia, 'Rice productivity', http://ricepedia.org/rice-as-a-crop/rice-productivity, accessed 3 November 2018.
9 FAO (2004); Halwart *et al.* (2006).
10 Naito *et al.* (2012); Natuhura (2013).
11 Meijen (1940), p. 96; Weerakoon (1957); Fernando (1995, 1996); Odum (1997), p. 330; Bambaradeniya (2000a, 2000b); Bambarabeniya and Amarasinghe (2004); Luo *et al.* (2014); Nayak *et al.* (2018).
12 Bambaradeniya (2000a), p. 525.
13 Luo *et al.* (2014); Katayama *et al.* (2015).
14 Much of the information presented in this section comes from Bambaradeniya and Amarasinghe (2004).
15 Weerakoon and Samarasinghe (1958); Fernando *et al.* (1979); Fernando (1980a, 1980b, 1993); Yano *et al.* (1983); Lacey and Lacey (1990); Rozilah and Ali (1998); Bambaradeniya and Amarasinghe (2004).
16 He (1991), as cite by Luo *et al.* (2014).
17 Bambaradeniya (2000a).
18 Ooi and Shepard (1994); Way and Heong (1994); Settle *et al.* (1996).
19 Smith (2001).
20 Elphick (2000).
21 Amir Sha Ruddin and De Datta (1998).
22 Way and Heong (1994); Amir Sha Ruddin and De Datta (1998); Halwart (2006).
23 Lawler (2001); Naito *et al.* (2012).
24 Halwart *et al.* (2006).
25 Liu *et al.* (2011); Saka *et al.* (2013).
26 Relyea (2005a, 2005b).
27 Moody (1989); Bambaradeniya (2000a); López *et al.* (2011).
28 Naito *et al.* (2012).
29 Guadagnin and Maltchik (2006).
30 FAO (2004); Halwart and Bartley (2005); Halwart *et al.* (2006); Cruz-García and Price (2011).
31 Hansen *et al.* (1998); Elvevoll and James (2000); Roos *et al.* (2002); Halwart (2006).
32 Halwart (1999).
33 Balzer *et al.* (2005); Luo (2005); Halwart *et al.* (2006).
34 Kihira (1983) as reported by Katayama *et al.* (2015).
35 The following section on the history of the rice Green Revolution in Indonesia is based on the account by Settle *et al.* (1996).
36 The IRRI is an international agricultural research and training organization established in 1960 with support from the Ford and the Rockefeller Foundations, in collaboration with the government of the Philippines to promote the Green Revolution in rice in developing countries, especially in Asia. It is part of the Consultative Group on International Agricultural Research.

37 This self-sufficiency was of short duration since they have been importing rice since the mid-1990s. However, in December 2008, the Vice President of Indonesia, Jusuf Kalla, announced that they would not be importing rice in 2009 because the country's rice production target had been achieved.

37 Röling and Van Der Fliert (1994).

38 Zhang (2011); Perfecto *et al.* (2014).

39 Sawada *et al.* (1991); Settle (1996).

40 IRRI (2004).

41 Kleinman *et al.* (1995).

42 VOA (2018).

43 Asia Pacific Network on Food Sovereignty (2018).

11 The new paradigm

Recapping the ecological argument

The theory of biodiversity

The science of ecology has long sought to understand the nature of biodiversity, while the conservation movement has traditionally taken as its core goal the preservation of biodiversity. It would be surprising indeed if the science itself had not, by now, come to some agreement on the nature of its subject matter. We have already described in great detail in Chapter 2 the modern ideas of how biodiversity is maintained. We also alluded to the fact that this is still a contentious and evolving corner of the scientific world, a fact that remains as true today as it was when the first edition of this book was published. However, we also noted that if the cobwebs of that contentiousness are wiped away, a core of ideas emerges, a kind of conventional wisdom among ecologists as to how biodiversity 'works'. It begins with the idea of a metapopulation and focuses on the balance between the perfectly natural tendency for populations to go extinct locally and the tendency for individuals in those populations to migrate to new locations.

The fundamental idea is metaphorically related to the way an infectious disease works. An individual person becomes sick, transmits the sickness to another person, and then gets well. That sort of says it all. Becoming sick is the invasion of a disease organism into the habitat, which is to say, into the body of the host. But the main purpose of the disease organism is to infect someone else, which is to say, migrate to a new habitat. And that migration has to happen before the person's immune system cures the body of the disease, which is to say, causes a local extinction. So, in this metaphor, the human body is the habitat, the organism causing the disease is the population of concern and the transmission rate from person to person is the migration rate from habitat to habitat. The recovery rate, which is a good thing for the person, is a bad thing for the disease organism because it represents a local extinction of that organism. The disease organism can persist over the long run only if its transmission rate is sufficiently large relative to the recovery rate – if its migration rate is sufficiently large relative to its local extinction rate.

It is now fairly well accepted that many populations in nature are subjected to patterns that are similar, formally known as metapopulations.[1] Whether the non-random distribution of trees in a forest or the variable occupation of temporary ponds by amphibians, subpopulations are, to some extent, isolated from one another, and these subpopulations are subjected to many forces that cause a complete elimination of the population from its local environment – a local extinction. However, provided other subpopulations continue to provide propagules, the site originally occupied by a subpopulation that has gone extinct will eventually receive one of these propagules – a regional migrant will arrive. As long as the migration rate balances the extinction rate, the population will survive over the long run as a metapopulation.

To be all-inclusive, this framework must be viewed from three interrelated points of view. First, in a continuous habitat, second, in a heterogenous habitat that is variable such that some sub-habitats are more conducive than others for the survival of a particular species, and third, in the human-dominated landscapes where native habitats (i.e. habitats not under direct management by humans) remain as isolated fragments in a matrix of managed ecosystems, mainly agroecosystems. The basic metapopulational framework operates in varying degrees in all three cases, as we summarize presently.

In continuous habitats it is not easy to establish beyond doubt the existence of metapopulational structures. There is some evidence that the internal dynamics of populations will generate non-random distributions of individuals in space (trees in a tropical forest, for example) and at least one detailed analysis of two different forest sites (in Panama and Ecuador) suggests that most of the tree species indeed have clustered distributions at some spatial scale.[2] Given such clustered distributions, it is likely that mortality effects (individual trees dying) will be related to those clusters since diseases and other natural enemies are likely to encounter these individuals more frequently when they are clustered. Some vertebrate species (e.g. monkeys and wolves) occur in clusters because they are social. Any disease epidemic may act to kill all the individuals in the cluster (family group), being, in effect, a local extinction event. Many other possibilities could be cited in which, even in continuous uniform habitats, populations are likely to have metapopulation dynamics.

In heterogeneous habitats, the existence of metapopulation structures is more evident. Islands and habitat islands are the rule rather than the exception for most species. Many amphibians play out an important phase in their life cycle as larval forms (tadpoles) in habitat islands (ponds and lakes). Sequential sampling of temporary ponds in Michigan, for example, revealed about 30 extinction events in an area that contained only 17 species of amphibians (ponds containing a particular species at one time did not contain that species 20 years later).[3] Furthermore, some heterogeneous habitats result from biological interactions that produce non-random patterns in formerly uniform environments, thus creating a heterogeneous habitat for many other species not involved in those interactions. These later species then are likely to exist in a metapopulational situation.[4]

In fragmented habitats (which are the most important type since they are probably the most common landscapes in the contemporary world), the same situation exists as in heterogeneous habitats. The clusters of individuals that may have existed in the continuous habitats before fragmentation are certainly going to be subject to metapopulation dynamics after fragmentation.

What all these situations have in common is that the probability of extinction at the local level is high, when viewed over the long term (e.g. a family of monkeys subject to an epidemic, a group of trees completely defoliated by a herbivorous insect population attracted to the local abundance of the tree, extinction by genetic drift[5]). The death of all the individuals in a cluster represents a local extinction (the disappearance of that cluster), a fact that is effectively inevitable if taken over a long enough period of time. Local extinctions are thus inevitable, whether caused by humans or not. What stops local extinctions from becoming regional and global (the extinction of the species itself) is the balancing migration that occurs among clusters of individuals or among patches of habitat. Thus, for example, in the case of the 30 extinctions of 17 species of amphibians over a 20-year period, those local extinctions were balanced by almost exactly the same number of migrations. And the basic metapopulation equilibrium of $p = 1 - e/m$ (proportion of habitats occupied is one minus the ratio of extinction to migration) assures us that each of the 17 species of amphibians remains, through the dynamic pattern of extinction and re-migration.

High extinction rates in isolated fragments

If small nature reserves are viewed as fragments in a larger matrix, the concept of the metapopulation immediately comes to mind – a core idea in this book. Each fragment is a subpopulation that need not provide all the necessities to maintain a population in perpetuity, but exists in the context of other fragments that provide propagules for species that have gone locally extinct because a particular fragment is too small to maintain a population over the long term. An important result already cited in Chapter 2 is from long-term experiments on forest fragmentation in Amazonia.[6] While smaller fragments of forest have higher avian extinction rates than larger ones, the actual extinction rates of even the largest patches are surprisingly high. Indeed, some are so high as to suggest that the only acceptable size for a biological preserve is one that is far beyond reasonable expectation under current political circumstances. Similar studies were discussed in Chapter 2. The remarkably high extinction rates for smaller patches contain important lessons for conservation in fragmented landscapes, one of the main messages of this book.

From recent theoretical studies we might expect that for some organisms the spatial extent necessary for their conservation may be unusually large, far beyond what any current or imagined future political arrangement might tolerate, or even what may be available in natural/unmanaged habitat. As an example, an influential analysis postulates that recruitment limitation (or failure to disperse to suitable habitat patches) is one key factor in maintaining tropical tree diversity.[7] Long-distance dispersal events, under this formulation, are important in maintaining species diversity over the long term. Therefore, fragmenting the forest and consequently limiting the rare dispersal event in the original spatially extended forest is likely to cause local extinctions and a concomitant reduction in regional biodiversity. Unfortunately, such expected extinctions are likely to occur far into the future, making the political case for conservation here and now very difficult. However, our point here is that fencing a patch of forest in no way guarantees the protection of biodiversity in perpetuity, since rare migrations (seed dispersal events) from far away are essential.

From these empirical and theoretical studies, we are forced to face the unpleasant conclusion that for some species, protecting a sufficiently large area to avoid major extinctions may not be possible. This conclusion is not meant to imply some sort of surrender, but rather to emphasize a necessary change in focus for conservation planning. The matrix matters, to repeat again the basic theme of this book.

Recapping the agricultural argument

The current world consists mainly of fragments of native or 'natural' habitats. The largest biodiversity preserve in the world is Northeast Greenland National Park (972,000 km^2) and even it appears as a tiny fragment on the globe as a whole. National Parks and biological preserves claim outstanding success in biodiversity conservation the world over, yet there remains an extremely large amount of biodiversity outside of designated natural reserves, scattered in fragmented landscapes from Siberia to the Congo. It is thus critical to view conservation at a landscape level, in which fragments of native habitat almost always exist, but the majority of which are embedded within some sort of agroecological matrix. Our argument is that there is no particular reason to view the fragments or their surrounding matrix as any sort of heaven and hell, and quite certainly not as systems that can or should be treated independently by conservationists or developers. In this context, our approach rejects a focus exclusively on habitat fragments, and proposes that the general dilemma can be better addressed with an increased focus on the matrix and, following current ecological understanding, the interactions it has with remaining fragments of 'natural' habitats.

Increasing agricultural production for conservation

Approximately 90 per cent of the terrestrial surface of the earth is outside of reserves and is used or managed by human beings in one way or another.[8] In the tropics approximately 70 per cent of the land is pastures, agriculture, or a mixture of managed landscapes.[9] In the popular and romantic conceptualization of Nature as a Garden of Eden, many conservationists view agriculture as the defining feature of biodiversity loss. The world is divided into those areas untouched or minimally touched by *Homo sapiens* as contrasted to those areas despoiled by human activity.[10]

Many recent ecological studies have challenged this perception, the two most obvious examples being shade-grown coffee and cacao, as elaborated in Chapters 8 and 9. As both of those examples (and many others) illustrate, when dealing with managed ecosystems it is first necessary to distinguish between two concepts of biodiversity. First is the collection of plants and animals that the manager has decided are part of the managed system – rice in the paddies of Asia, maize and beans in the traditional fields of Native American Mayans, carp in the fish ponds of China, etc. This is the so-called 'planned' biodiversity, to be contrasted with the associated biodiversity, the organisms that live or spend some time in the managed systems but are not intentionally included there by the managers – the fish, aquatic insects, and frogs in the Asian rice paddies, the birds and insects that eat the Mayan's *milpa*, and the crayfish that burrow their way into the sides of the Chinese fish ponds.[11] The pattern of associated biodiversity gain or loss as a function of management intensity is one of the major patterns of biodiversity in the world, as we discussed in Chapter 2. Frequently the managers themselves are determinedly concerned about the planned biodiversity, especially when dealing below the species level (with genetic varieties of crops). However, it is almost certainly the case that the associated biodiversity is the most abundant component of biodiversity in almost all managed ecosystems, and as such, it has received a great deal of attention in recent years. Furthermore, the associated biodiversity may also have important functions in the agroecosystem.[12]

The process of intensification of agriculture provides a conceptual framework for analysing the role of agriculture in biodiversity conservation, especially when concerned with associated biodiversity. Although the term 'agricultural intensification' has a very specific and complex definition in economic history and anthropology, in biodiversity literature the term 'management intensification' or 'agricultural intensification' is taken to be the transition from ecosystems with high planned biodiversity to low planned biodiversity and an increase in the use of agrochemicals and machinery.[13] The ecology of agroecosystems is such that the final stages of intensification usually include the application of agrochemicals to substitute the functions or ecosystem services of some of the biodiversity that is eliminated.

Conservation biologists often fail to fully acknowledge a component of agricultural intensification that can have devastating consequences for biodiversity–pesticide applications. Extinction of the world's small organisms should cause as much concern as the extinction of charismatic megafauna.[14] Insects, mites, nematodes, microbes, and representatives from at least 30 different kingdoms of organisms[15] abound in the soils, leaf litter, and other niches in every environment in the world and are highly susceptible to pesticides. An old cotton field in Nicaragua contains almost no ants, as far as our personal informal faunal surveys could determine. Ants in these fields (and most other insects, except those that have evolved resistance) have gradually disappeared due to the massive spraying of pesticides in a failed 30-year experiment with industrial agriculture that not only did not bring riches to the Nicaraguan people, but also contaminated the land for many years after cotton was abandoned.[16] The 'little things that run the world', as eloquently described by entomologist E. O. Wilson, are disappearing in front of our very eyes and we are not noticing. A recent study documenting the loss of 75 per cent of the insects in less than 30 years in forest fragments in Germany, is forcing us to notice.[17]

Recent studies of amphibians have helped to focus this issue. In a series of studies from the University of California, Atrazine has been found to have complex effects on various species of amphibians.[18] While the popular press has picked up on those studies because of their implications for human health (Atrazine is a chemical that has a similar action to oestrogen), the effects on the amphibians themselves is worrisome regarding their ultimate survivability. In similar studies examining the effect of the herbicide Roundup on the larvae of three species of amphibians, mortality rates of over 90 per cent were reported.[19] These results are surprising given that glyphosate, the active ingredient in Roundup, had been reported to have minimal toxicity on vertebrates.[20] However, the commercial formulation of this herbicide contains a surfactant to assure that the active ingredient adheres to the leaves of plants, and it appears to be the combination of these ingredients that causes the high mortality of amphibians. Regardless of the details about the mortality mechanisms, the main conclusions of these studies are clear and robust: two of the world's most extensively used herbicides do significant damage to amphibians. Given the dynamic turnover pattern of many species of frogs and salamanders, and the fragmented nature of most terrestrial ecosystems, it is essential that these species be able to move through the agricultural landscape to find new breeding ponds.[21] The use of Atrazine or Roundup in the agroecosystem might thus condemn many amphibian species to extinction. In this context, the expansion of Roundup Ready transgenic soybean production in Brazil, Argentina, Paraguay, and Bolivia is of concern.[22]

Intensification, food production, and biodiversity conservation

The increasing demand for food has led some agriculture and development advocates to argue that the best option to meet the challenges of increasing food production and conserving wildlife is to increase yields by intensifying agriculture and spare land for conservation,[23] as we discussed in Chapter 3. To quickly recap this issue, the argument is based on two assumptions. First, that agricultural intensification leads to land sparing, and second, that there is a productivity trade-off: in other words, that the biodiversity value of farmland (including organic and agroecological methods) declines with increasing yield.[24] These two assumptions need to be examined more carefully.

Intensification does not take place in a socio-political vacuum. Frequently, regions that experience agricultural intensification also experience increased economic activity, higher demand for products and services, immigration, road construction, and consequently, in many cases, higher deforestation rates.[25] Furthermore, intensification frequently results in the displacement of small-scale farmers and agricultural workers who then move into nearby marginal land or the agricultural frontier, frequently causing more deforestation elsewhere.[26]

However, the assumption that agricultural intensification leads to land sparing is seductive in its simplicity. Given an arbitrary area, we begin by assuming that there is a target, T, for production of some commodity in that area. Then we can easily construct a simple production function that associates proportion of land in production with total production on that land, in which case we always see an increasing pattern, usually with diminishing returns (see any economics textbook, for example, Simon and Blume[27]). That function will always increase with intensification (by definition). This means that production will always increase with intensification, according to this model framework. To meet the target production of T, the amount of land necessary to put into agriculture is always less with increased intensification, a tautology. This simple approach can be embellished with all sorts of other interesting factors such as discount rates, land rents, input market uncertainties, and many others. We have added, for example, to this basic idea the value of biodiversity itself.[28]

The problem is not with the actual model, and certainly not with the more sophisticated elaborations of the basic idea. The problem is with the basic framing of the problem in the first place.[29] Rarely is it actually of interest to know what a 'target' for production is. Vague notions, such as the world's calorie requirements, have little to do with decisions that are made at the farm, or even regional levels. Indeed, in most parts of the world, if there is a 'target' it is to maximize return on investment, which is only indirectly related to a production target. Farmers frequently fail to engage in particular practices not because they are less productive, but rather because they require a large cost for labour or capital outlays. In the end, the framing of an area with the duo of 'land sparing' (or set-asides, or reserves, or wildlife refuges) and 'agriculture with improved technology' is wrong. Of course, working from within that framing it is difficult to avoid the conclusion that less land will be required for agriculture if it is intensified, thus leaving more land for set-aside. But that is sophistry, even if unintended.

Angelsen and Kaimowitz take a far more sophisticated approach, noting first that there is a fundamental contradiction that is sometimes ignored.[30] First, 'the belief that technological progress in agriculture reduces pressure on forests by allowing farmers to produce the same amount of food in a smaller area, has become almost an article of faith in development and environmental circles.' Second, 'basic economic theory suggests that technological progress makes agriculture more profitable and gives farmers an incentive to expand production on to additional land.' Clearly the belief and the basic theory are at odds. Angelsen and Kaimowitz report on detailed studies that sometimes support one, sometimes the other point of view. We conclude that the 'article of faith' that 'intensification of agriculture reduces pressure on forests' is not supported by data.

In a more extensive work, the same authors edited a series of chapters that include 17 case studies from Latin America, Africa, and Asia.[31] Their conclusion from all these studies is that the issue of intensification of agriculture and its relationship to deforestation is complex and, effectively, that agricultural policy could be modified in such a way as to promote forest-preservation policies rather than policies that, however unintentionally, actually promote more deforestation with 'improved' agricultural technologies. From this analysis, it is difficult to avoid the general conclusion that, for the most part, conventional agricultural intensification causes more deforestation. Granted, the situation is complicated by many factors, and it is certainly sometimes the case that improved agricultural technology has decreased deforestation rates. But examining closely the 17 case studies presented, in 12 of them there was a clear indication that technological change had an effect on deforestation. Of those 12, nine showed *increasing* deforestation as a result of intensification or new agricultural technology (three of the nine suggested it could go either way, depending on circumstances), and only three suggested a necessary decrease in deforestation with intensification. All cases were treated with the complex analysis they deserve, and in our view, negate the simplistic assumption that agricultural intensification leads to land sparing.

The second assumption of the 'intensification–land sparing' argument is that there is a productivity-biodiversity trade-off. In other words, that those agricultural technologies that increase yields decrease biodiversity. The assumption is effectively true in the case of industrial agricultural production systems, especially those following Green Revolution technologies. However, the situation is more complicated when examining other, more complex agroecosystems. For example, reviews comparing organic and conventional agriculture present evidence that, on average, low-input organic farming has the potential to produce as much as industrial farming, but without the negative environmental impacts.[32] Other studies report on specific cases in which organic is less productive[33] and polemics can be constructed on both sides of the issue. A team from the University of Michigan that includes one of the authors (Perfecto) reviewed close to 300 studies that compared 'organic-like' productive activities with conventional ones, and found that, while individual studies could be cherry-picked to support either side of the issue, on average there is no evidence that conventional methods out-perform organic ones in terms of productivity.[34] A problem with all of these studies is that data is compared on a single crop basis when many of the agroecological systems are based on multiple crops. Furthermore, these organic agroecological systems (including organic agriculture, 'natural systems' agriculture, permaculture, etc.) have been shown to be generally more biodiversity-friendly and less environmentally damaging than conventional farming systems.[35] Taken in combination, these studies strongly suggest that a biodiversity–productivity trade-off is not necessarily real. In other words, it is possible for some highly productive farming systems to maintain and promote biodiversity. As we discuss in detail below, it is not only the conversion from a native habitat to agriculture that matters for biodiversity conservation, but also the conversion of agriculture from a biodiversity-friendly type to a biodiversity-unfriendly type that accounts for most biodiversity loss within agricultural landscapes.

Yet, as we have argued throughout this book, there is another important way in which agriculture and biodiversity are related, one that belies the simple formula of the 'intensification–land sparing' argument. Most biodiversity should not be thought of as 'point source' or local. Rather, the collection of those points, the general landscape, must be taken into account. As we have already argued, the evidence regarding local extinctions even within large fragments of natural habitats, and the importance of the agricultural matrix in facilitating or preventing inter-patch migrations, suggest that agricultural landscapes dominated by diverse, ecologically based systems, are frequently our best bet for biodiversity conservation. These ecological farming systems provide a high-quality matrix through which migrations may occur, thus counteracting the extinction rates of populations that invariably exist in a metapopulational context.[36] This suggests that a research or development priority should be to develop highly productive and diversity-friendly agricultural systems, in other words, the agroecological intensification of farming systems.[37] Such systems, from the point of view of biodiversity and, more generally, environmental benefits, stand in stark contrast to the modern industrial agricultural system.

The matrix as repository for biodiversity

The agricultural matrix may contain a substantial amount of biodiversity, as has been shown for traditional 'shade coffee' systems, *cabruca* cacao, and so-called jungle rubber,[38] or many other examples now adequately documented in the literature.[39] While the exact nature of the species encountered in the matrix may not be the same as those encountered in the native habitat, and the matrix may contain many 'fugitive' or 'opportunistic' species, thus far all studies seem to corroborate the fact that certain matrices are 'high-quality' with respect to the biodiversity contained therein.

Nevertheless, the smoke of the underlying prejudices of many conservationists reflects something of an underlying fire. Casual examination of the 'matrix' in areas where industrial agriculture is the norm leaves little appreciation for its contained biodiversity, from wheat fields in the Ukraine, to tulip plantations in the Netherlands, to maize fields in Iowa, to soybean plantations in Brazil. Yet, it is well known that the 'forest edge' habitat is an excellent place to view birds and other biodiversity, and rustic coffee plantations in Mexico contain the same biodiversity in the tree canopy as a natural forest.[40]

Sampling insects on traditional tropical farms can produce biodiversity records that are close to those of nearby forests, while similar sampling in pesticide-drenched, nitrate-poisoned, mechanically tilled sugar plantations yields little more than fire ants.[41] There can be no doubt that some types of agroecosystems contain great amounts of biodiversity while others contain virtually none at all.

The framework for dealing with this issue is the intensification gradient, from the less 'intense' and more traditional systems (such as rustic coffee, jungle rubber, or diversified rice paddies) to the highly 'intense' modern, chemically based agroecosystems, presenting the world with one of the main trends of biodiversity change (as discussed in Chapter 2 – see Figure 2.2). Some authors have objected to the use of the word intensification in this context, since more traditional systems are frequently 'intensely' managed (in terms of the density of agricultural activity in time or space), even though external industrial inputs are minimal. However, as a rough illustration of a qualitative idea, we feel it is appropriate to consider the 'intensification gradient' as one that extends from more traditional diverse systems that use few or no external inputs, to those that are heavily dependent on agrochemicals, tillage, and monocultures. With this general qualitative idea, we ask, 'What are the biodiversity consequences of changing a system along this gradient?'

The literature on this subject is now growing, but, unfortunately, most biodiversity studies in the past have focused on 'natural' habitats. Such biases towards the study of biodiversity in so-called 'pristine' sites preclude any true generalizations. Indeed, from the few studies that have been done, we can say that biodiversity generally is reduced with agricultural intensification, but the exact form of that reduction strongly depends on the taxonomic group in question.[42] For example, as we discussed in Chapter 8, the reduction in ground-foraging ant diversity is most dramatic during later stages of the intensification process in coffee, but the reduction in butterfly diversity is dramatic earlier in the intensification gradient (perhaps because even very diverse coffee plantations remove much of the understorey where the food plants of many specialized butterflies exist).[43]

There is also a growing body of literature examining cases where planned (intentional) biodiversity is compared to associated (non-intentional or wild) biodiversity in both conventional and alternative agricultural systems. It seems to be a general rule that higher-planned biodiversity has a positive effect on associated biodiversity within the agroecological matrix, as well as within local natural forest fragments embedded within the matrix.[44]

Fortunately, the last decade has seen an increased interest in the study of biodiversity in managed ecosystems, including urban settings. If this trend continues, we will soon be able to sort out many of the scientific questions regarding biodiversity in agroecosystems and will be in a better position to develop landscape-level conservation programmes and policies.

Fragment-to-fragment migration

In addition to functioning as direct repositories of biodiversity, we suspect that many fragmented habitats now contain species that function as metapopulations, as stated earlier. If this is so, the equilibrium proportional occupancy of the fragments for any species is expected to be $(1-e/m)$, where e is the extinction rate and m is the migration rate. If migration rate approaches extinction rate, landscape-level extinction occurs. What, in this situation, determines migration? The probability that an organism will travel from one patch to another depends on the nature of the area over which it must travel. And that area, in this context, is the matrix. A high-quality matrix can be thought of as one that permits an easy migration from patch to patch. Thus, even though a species may not be able to persist in the matrix, it will frequently be the case that it must migrate through that matrix to maintain the functioning of the metapopulation dynamics.

This function of the matrix as the medium through which migration or dispersal of organisms must occur among fragments of natural habitat is reminiscent of the debate over corridors.[45] To some extent that debate seems to have died down and most conservation biologists do not feel that corridors, at least in their original formulation, function well in biodiversity conservation. On the other hand, a high-quality matrix may very well function in the way that corridors were expected to function, leading to higher richness in fragments and less effective patch isolation.[46]

Agriculture as a temporary activity

Agriculture is not forever. Much of the Neotropics is currently comprised of a patchwork landscape of different kinds of agriculture, forming a matrix with both active agriculture and areas at different stages of abandonment. A given piece of land, especially in the lowland tropics, is rarely able to produce in perpetuity, which is the principal reason that so many cultures developed some form of shifting (or rotational) agriculture. Furthermore, the dynamics of urban–rural migration are captive to the vicissitudes of socio-economic forces. For example, the North American Free Trade Agreement resulted in a massive exodus from the rural areas in Mexico as a consequence of the imposition of the neo-liberal model and the flooding of local markets with agricultural goods from the US. Similarly, Brazil, in the last 40 years, has gone from a population that was two-thirds rural to one that is three-quarters urban, while conversely, Cuba is engaged in a certain amount of 're-ruralization' due to its agricultural policies, in the period following the collapse of the Soviet Union.[47] Thus, agroecosystems are tremendously dynamic and the subject of post-agricultural succession becomes critical for understanding landscape mosaics. The matrix can be expected to be a mosaic of different forms of agricultural production as well as, and perhaps more importantly, different stages of post-agricultural ecological succession.

Understanding this mix of matrix quality requires an understanding of post-agricultural succession, not only the particulars of what happens to a pasture (say) when it is abandoned, but of the impact the type of agriculture practised has on the details of the successional patterns.[48] Indeed, it has been shown that the post-abandonment successional pattern is dramatically different for slash-and-burn or traditional agroforestry as compared to pasture or modern intensive agriculture.[49] Studies of how various components of biodiversity behave, both in terms of the agroecosystem itself as well as in the post-agricultural successional process, will be crucial to understanding the details of matrix quality. Unfortunately, such studies are thus far very rare.

Recapping the social movement argument

Perhaps it is nothing new. Across the Global South, the legacy of European and US imperialism and neo-colonialism has always encountered resistance and challenges through political protest, whether the revolutionary changes inspired by Mao and Che or the evolutionary changes initiated by Gandhi and Mandela. That resistance continues against the most recent manifestation of these historic political and economic structures: the neo-liberal economic policies. In rural sectors, where much of the protest originates, farmers have always sought to enhance their position through political organizing, sometimes dispersed and informal, at other times concentrated and highly regimented and focused. This tendency of rural social movements to be in constant agitation for political progress is probably a worldwide phenomenon, with deep historical roots. However, the particular conditions faced by these movements have been dynamic and evolving, and characterizing them in any general way is fraught with the pitfalls of oversimplification. Nevertheless, there does seem to be a general structural arrangement that has persisted since the later phases of European–US imperialism and still drives, to some extent, the political ecology of much of the contemporary world, as described in Chapter 6.

As an outgrowth of European–US imperialism, the agricultural systems of the Global South had, and to some extent retain, a dual structure. Vast expanses of land are devoted to export production, that which initially fed the Industrial Revolution with raw materials and an abundance of food to maintain their cheap food policies. However, interspersed in those vast estates are millions of small-scale farmers, *campesinos*, who produce both for their families and the local (and sometimes international) markets. With a few exceptions, like shade-grown coffee and cacao in the Americas, the export sector (and their allies) is now heavily capitalized and frequently labour-intensive. The small farming families constitute a convenient source of the labour so necessary for that export sector. In what has been termed a dual economy, labourers come from the farming sector and retain their flexibility by retaining ties to that farming sector. These ties become important when, invariably, the jobs in the export sector disappear and workers return to the family farm to become farmers again. Increasingly, the ties to the farming sector are cut, forcing fired workers either to transform unused land into new farms, or to move to the cities. This has resulted in a large standing crop of landless poor, some in the cities, some wandering the countryside filling temporary agricultural positions, and some migrating to the Global North. This distancing of workers from the consumption society is extremely efficient for the export sector, unconcerned as it is about local consumer demand, and thus able to drive wages down to minimum levels.[50]

Responses of the subaltern to this arrangement are frequently brutally repressed. Nevertheless, they remain common and sometimes reach national proportions, as in Mexico in the late 1930s, Guatemala in the early 1950s, Cuba in 1959, Vietnam in the 1970s, Nicaragua/El Salvador in the 1980s, and the Zapatistas in Mexico in the mid-1990s.

What is new today, is a growing political consciousness about the origin and maintenance of the condition of the small farming sector in the Global South. Propelled by a rapidly diffused analysis of the neo-liberal agenda, farmers the world over are rejecting the economic model implicit in the conventional agricultural system. Most significant is the development of the idea of food sovereignty by the umbrella organization La Via Campesina[51] (Figure 11.1).

A United Nation goal for a long time, the concept of food security has been a centrepiece of human rights advocates at a global level. In a world of

Figure 11.1 La Via Campesina members holding a food sovereignty banner at a march protesting neoliberal policies that affect peasant livelihoods.

Source: La Via Campesina.

plenty, goes the argument, we ought to be past the time in history when people must go to bed hungry, and socio-political arrangements ought to be organized in such a way that a basic diet is available to all. So-called 'food activists' have been pushing for the acceptance of this idea as a basic human right for many years. Indeed, the right to food is now recognized in Article 11 of the UN's International Covenant on Economic, Social and Cultural Rights, which has been signed by many countries. What is new today is the conjoining of the rights of people to consume food to the rights of people to produce their own food, a synthesis that La Via Campesina refers to as 'food sovereignty'.[52]

Food sovereignty has become a rallying cry for rural social movements across the world and fits snugly with previous political ideas and action associated with land tenure. Those landless people, so the argument goes, are landless because of the underlying political structure that favours the concentration of land in the hands of the wealthy who, frequently, do not even use it 'productively', but rather retain it as a capital investment or speculation. In a variety of political venues, from local courts to land invasions (or as they are sometimes referred to, land rescues), to international protests, this new movement has taken on something of a spontaneous and surprisingly resilient form, from farmers mainly in Asia and Latin America, with participation of small-scale farmers from all over the world, including the developed countries.[53]

In addition to the human rights issues uncovered and publicized by these new rural social movements, a key political conjunction has been between the movements for food sovereignty and those for alternative agriculture. La Via Campesina, for example, has an Agroeocological Commission to promote agroecology and agroecological school for the scaling up of agroecology, which they see one of the pillars of food sovereignty (Figure 11.2). The image that some conservationists in the Global North have of peasants over-populating the land and consequently destroying biodiversity, is supported less each year, as the popular movements make political headway. That most conservation programmes now acknowledge poverty reduction as part of any viable conservation agenda is perhaps at least partly a consequence of the political agitation generated by these new rural social movements. These movements are beginning to exert important influence even in high-level intergovernmental organizations. For example, the International Assessment of Agricultural Knowledge, Science and Technology for Development (IAASTD),[54] a UN- and World Bank-sponsored assessment of the role of agriculture in the reduction of poverty and hunger and towards a more equitable and sustainable development, called for a paradigmatic shift away from conventional large-scale agriculture, emphasized the role of small-scale farmers and local or indigenous knowledge, and (at least in the Latin American and Caribbean regional report) presented options for achieving food sovereignty.[55] After many years of struggle by many peasant organizations under the leadership of La Via Campesina, on 17 December 2018, the United Nations General Assembly adopted the UN Rights of Peasants and Other

Figure 11.2 Discussion at the meeting of the Agroecological Commission of La Via
Campesina, in Malaga, Spain, 2009. Countries represented in this picture
(from top, clockwise) Argentina, Mexico, Puerto Rico, Brazil, Gambia,
Angola, Spain, El Salvador.

Source: John Vandermeer.

People Working in Rural Areas. This declaration is a legal instrument that
can be used by peasant organizations to support regional and national implementation of agroecology and food sovereignty programmes.

Returning to the specific theme of this book, of the two remnants of
European imperialism in the matrix of today's tropical fragmented habitats, it
is perhaps only rhetorical to ask which is likely to create a high-quality
matrix. Large-scale production of bananas, sugarcane, tea, technified coffee
and cacao, soybeans, cotton, pastures, and others are notorious for their
environmental unfriendliness. Small-scale farmers with their usually diverse
farming techniques applying, either by conviction or necessity, organic-like
and environment-friendly techniques, square far more obviously with the
emerging concept of a high-quality matrix.

The vision of a matrix composed of small-scale ecological agriculturalists
may seem utopian to some. But real-world examples do exist, demonstrating
that with proper political will such a vision can dramatically improve the lives
of poor and middle-class people even as it acts to create the high-quality

matrix we seek for conservation purposes. Recall the example of the farming family in Kerala, India, cited in Chapter 10 – farming only one and a half hectares with completely organic methods, producing a diversity of products both for home consumption and sale in local markets. The conditions of that family, especially given India's position in the Global South, are remarkable. A two-storey house that would be regarded as middle class even in the developed world, a car, a motorbike, a satellite dish – all the amenities we normally associate with a middle-class existence, with well-dressed children receiving high-quality education and complete access to health care. The issue is not so much the productivity of the land, but the style of agriculture and, of course, the option for the poor that the state of Kerala had long ago developed as a priority. The vision is not utopian. However, it will require the revitalization of rural areas in ways that will create conditions for a dignified livelihood for rural people. We are not idealizing rural poverty or ignoring the harsh life of the rural poor. On the contrary, we are proposing a change in the social, political, and economic structures that recognizes the contributions of small-scale farmers to food production and biodiversity conservation, and properly rewards those who put food on our tables. Furthermore, we think that it is important to reconstruct rural communities in ways that make it attractive to young people by providing economic, cultural, social, and educational opportunities for them.

An additional fact is worth considering associated with agricultural productivity in the Global South. Contrary to what is frequently an unstated assumption that larger farms tend to be more efficient, an analysis of 15 underdeveloped countries found that production (or productivity – amount produced per unit area) declined as farm size increased,[56] and another study found the same thing in the developed world (in the US).[57] That is, smaller farms tend to be more efficient than larger ones, in terms of producing agricultural products (though not necessarily profit). It is not difficult to see why this is the case. Small farms, especially very small ones, tend to be managed 'ecologically', at least in the sense that the farmer usually knows his or her farm very well, understanding the ecological peculiarities of each and every corner of it. Contrary to the industrial mantra that larger sizes mean more efficiency, when ecology is brought into the equation, the reverse seems to be the case. And, of course, the smaller the farm, in general, the more likely it is to be more ecologically managed.

Putting the three arguments together

New knowledge about the science of ecology; current debates about agriculture (especially in the tropics); and new and powerful rural social movements with a global reach and with strong environmental concerns – these three general tendencies intersect to result in a new paradigm of environmental conservation as applied to biodiversity. Ecological science has converged on the idea that populations exist in a spatially complex context and that at any one point in space any population will experience a local extinction at some time or another, if enough time passes. Key to the maintenance of biodiversity, probably ever since life became complex some 500 million years ago, is the balance of migration to new sites and local extinction. Since there is not much we can do about local extinctions (they have been happening as a matter of course since life became complex), the only potential interventions that make sense must be with the migration rate.

Political realities are such that most biodiversity is located in the tropics, most of which exists in a highly fragmented state, the matrix of which consists mainly of different forms of agriculture. In conjunction with what we know now about how populations behave in nature, it is not within the fragments of remaining natural habitats on which we should focus, but rather on the matrix within which those fragments occur. Inter-fragment migration is key, and a high-quality matrix is key for that. Farming practices that emphasize concepts such as sustainability and ecological planning are most likely to produce that high-quality matrix.

Finally, the new social movements that have arisen throughout the tropics represent a challenge to the large-scale, monocultural, chemical (and now transgenic) basis of agriculture that has developed since World War II. The agenda of these new social movements includes precisely the sorts of agricultural practices that have come to be associated with a high-quality matrix. Furthermore, the extension of the popular human rights initiative of food security to the more inclusive and transformative idea of food sovereignty, incorporates agroecology along with conservation as key political goals. The new paradigm lies at the intersection of these three realities.

To be sure, the paradigm we describe is 'new', and it is likely that many traditional conservationists will recoil against it, and continue promoting agricultural intensification under the banner of land sparing. It is a paradigm that gives preference to more progressive political struggles for land rights over traditional lobbying for set-aside reserves of inviolate nature. In particular, as discussed fully in Chapter 3, the traditional conservation agenda has important foundational elements in a primitive Malthusian formulation, even if not explicitly stated as such. The underlying assumption, frequently left unsaid but strongly implied, is that there is a direct relationship between the amount of land devoted to food production and the size of the population: if there are x people and each requires food from n hectares of land, xn is the number of hectares required for the population. The problem is thus conceived of as a zero-sum game in which every piece of land devoted to agriculture cannot be devoted to biodiversity conservation, with the strong implication that if there were fewer people, or more intensive agriculture producing more food, there would automatically be less land devoted to agriculture, and therefore more biodiversity conservation. With this fundamental land-sparing framing, conservation goals frequently include population control, increasing agricultural production through the intensification of agriculture on land that is already under production, and higher vigilance of areas under 'protected' status.[58] The framework that leads us to the new paradigm is distinctly different, recognizing the well-accepted conclusion that it is not the amount of land in production that determines whether a family has adequate food, but rather the amount of money available to buy food. The irony that the world produces more than enough food and yet hundreds of millions of people go to bed hungry every night is a result of socio-political arrangements, not a simple 'land times population' equation.

Having subtle roots in Malthusian assumptions is certainly a barrier to understanding the complex interplay among land, society, and conservation. However, perhaps the most troubling aspect of the conventional conservation agenda is the failure to realize that not all agriculture has the same effect on biodiversity. The crucial aspects of a high-quality matrix will tend to be eroded with more agricultural intensification, and the landscape effects are likely to be negative overall. From our point of view, this is an especially ironic lacuna in conventional conservation thinking.

Finally, the perceived rationality of building fences and maintaining armed guards, metaphorically speaking, is largely derived from a sense of frustration that most reserve areas, especially in the tropics, are under continuing threat from a variety of elements, from poachers to settlers. That perception is certainly true. However, much as the wall between the US and Mexico is not going to keep immigrants from seeking a better life in the US, fences and armed guards are not going to stop people who are desperate to feed their families and are in search of a better life. This is not to say that corruption and greed are not involved in many cases, but for the most part corrupt officials and greedy entrepreneurs are likely to have as many lawyers and guns as the conservationists and holding back poor people who need to feed their families will only create desperation.

As a counter to the traditional conservationist agenda, the new conservation paradigm avoids many of the usual pitfalls by recognizing what is, we argue, a set of real, on-the-ground contingencies:

1 Advances in ecological research, both theoretical and empirical, show that local extinctions are natural and unavoidable.
2 Conservation of biodiversity must thus concentrate on migration rates rather than extinction rates.
3 The tropics, where most of the world's biodiversity is located, are highly fragmented at the present time, and are likely to be more so in the future.
4 Migration rates among fragments are mainly determined by the kind of agroecosystem in the matrix.
5 Small-scale farmers and agroecological methods are most likely to promote a high-quality matrix.
6 The political struggle for local food sovereignty is part of the struggle to conserve biodiversity.

This new paradigm carries with it normative consequences. It suggests that conservation activities need to interact with the rural masses and their social movements,[59] more than with wealthy donors. Indeed, we suggest that these new rural social movements hold the key to real biodiversity conservation. Joining the worldwide struggle of millions of small-scale farmers clamouring for food sovereignty is more likely to yield long-term biodiversity benefits than buying a patch of so-called 'pristine' forest.

Notes

1 Gilpin and Hanski (1991); Opdam (1991); Eriksson (1996); Harrison and Taylor (1997); Smith and Green (2005).
2 Condit *et al.* (2002).
3 Skelly *et al.* (1999).
4 Pollinators, for example, are attracted to resource patches made by the plants that produce flowers (Heinrich, 1975).
5 Genetic drift is a mechanism of evolution in which the allele frequencies of a population changes over generations due to chance.
6 Ferraz *et al.* (2003).
7 Hubbell (2001).
8 Western and Pearl (1989).
9 McNeely and Scherr (2003).
10 E.g. Terborgh (2004).
11 Vandermeer and Perfecto (2005a).
12 Giller *et al.* (1997); Vandermeer *et al.* (2002).
13 Perfecto *et al.* (1997); Giller *et al.* (1997); Donald (2001, 2004); Vandermeer *et al.* (2002); Perfecto *et al.* (2003); Wickaramasinghe *et al.* (2004); Tscharntke *et al.* (2005); Philpott *et al.* (2006).
14 Wilson (1986).
15 Lecointre and Le Guyader (2006). The number of kingdoms varies with the sources but new phylogenetic studies have dramatically increased the number of kingdoms from the previous five.
16 Thrupp (1988).
17 Hallmann *et al.* (2017).
18 Hayes *et al.* (2005).
19 Relyea (2005b).
20 Smith and Oehme (1992).
21 Werner *et al.* (2007).
22 Kaimowitz and Smith (2001); Trigo and Cap (2003); Pengue (2005).
23 Borlaug (1987, 1997); Waggoner (1994); Waggoner *et al.* (1996); Cassman (1999); Trewavas (2002).
24 Balmford *et al.* (2005); Green *et al.* (2005).
25 Wiersum (1986); Barraclough and Ghimire (1995); Foster *et al.* (1999); Angelsen and Kaimowitz (2001b); Lee and Barrett (2001); Morton *et al.* (2006).
26 Kaimowitz and Smith (2001); Wright and Wolford (2003); Schelhas and Sánchez-Azofeifa (2007); Perfecto and Vandermeer (2010); Phelps *et al.* (2013).
27 Simon and Blume (1994).
28 Perfecto *et al.* (2005).
29 Tscharntke *et al.* (2012); Fischer *et al.* (2014); Kremen (2015).
30 Angelsen and Kaimowitz (2001a).
31 Angelsen and Kaimowitz (2001b).
32 Stanhill (1990); Pretty *et al.* (2003); Halberg *et al.* (2005); Badgley *et al.* (2007).

33 Lockeretz *et al.* (1981); Mäder *et al.* (2002); Trewavas (2004); Seufer *et al.* (2012); Ponisio *et al.* (2015).

34 Badgley *et al.* (2007). Subsequent studies confirmed the analysis of Badgley *et al.* based on the data they analysed. However, more recent data seems to show the conventional production out-yielding organic production (Seufert *et al.*, 2012; Ponisio *et al.*, 2015). In the years after 2007 the basic relationship that existed before seems to have changed, causing some to question the quality of the data in the first place. It would seem that either experiments before 2007 were biased in favour of organic or experiments after 2007 were biased in favour of conventional.

35 For a meta-analysis, see Bengtsson *et al.* (2005).

36 Jonsen *et al.* (2001); White *et al.* (2002); Bender and Fahrig (2005); Donald and Evans (2006); Vaughan *et al.* (2007).

37 Matson and Vitousek (2006); Badgley *et al.* (2007).

38 Gouyon *et al.* (1993); Perfecto *et al.* (1996); Greenberg *et al.* (1997); Moguel and Toledo (1999); Perfecto and Armbrecht (2003); Gouyon *et al.* (1993); Schroth and Harvey (2007).

39 E.g. Andow (1991); Altieri (1994); Collins and Qualset (1998); Griffith (2000); Schroth *et al.* (2000).

40 Greenberg *et al.* (1997).

41 Perfecto *et al.* (1997).

42 Roth *et al.* (1994); Swift *et al.* (1995, 1996); Perfecto and Snelling (1995); Perfecto *et al.* (1997); Burel *et al.* (1998); Daily *et al.* (2001, 2003); Steffan-Dewenter (2002).

43 Perfecto and Snelling (1995); Perfecto *et al.* (2003).

44 Vandermeer and Perfecto (2005b, 2007a); Perfecto and Vandermeer (2008a).

45 Simberloff and Cox (1987); Simberloff *et al.* (1992); Mann and Plummer (1993); Haddad and Tewksbury (2005); Kirchner *et al.* (2003).

46 Vandermeer and Carvajal (2001); Ricketts (2001); Perfecto and Vandermeer (2002); Steffan-Dewenter (2002); Weibull *et al.* (2003); Tscharntke *et al.* (2002); Armbrecht and Perfecto (2003); Fahrig (2003); Perfecto and Vandermeer (2008a).

47 Wright (2008).

48 Fischer and Vasseur (2000); Boucher *et al.* (2001).

49 Ferguson *et al.* (2003).

50 De Janvry (1981).

51 Desmarais (2002, 2007).

52 Menezes (2001).

53 The spontaneous nature of these movements represents something rather new, according to Hardt and Negri (2005).

54 IAASTD (2009b).

55 IAASTD (2009a).

56 Cornia (1985).

57 Rosset (1999).

58 Oates (1999); Terborgh (2004); Wilson (2016).

59 Vandermeer and Perfecto (2007b).

References

Adams, J. S. and McShane, T. O. (1997) *The Myth of Wild Africa*, University of California Press, Berkeley

Adams, W. B. (2013) *Against Extinction: The Story of Conservation*, Routledge, Abingdon, UK

Adams, W. M. and Hutton, J. (2007) 'People, parks and poverty: political ecology and biodiversity conservation', *Conservation and Society*, vol 5, no 2, pp. 147–183

Agarwal, B. (2014) 'Food sovereignty, food security and democratic choice: Critical contradictions, difficult conciliations', *Journal of Peasant Studies*, vol 41, no 6, pp. 1247–1268

Agrawal, A. and Redford, K. (2006) 'Poverty, development, and biodiversity conservation: Shooting in the dark?' World Conservation Society Working Paper ISSN 1530–4426, Ann Arbor MI

Aguilar-Støen, C. (2009) 'Gardens in the forest: Peasants, coffee and biodiversity in Canderalia Loxicha, Mexico', PhD Dissertation, Norwegian University of Life Sciences, Aas, Norway

Aide, T. M. and Grau, H. R. (2004) 'Globalization, migration and Latin American ecosystems', *Science*, vol 305, no 5692, pp. 915–916

Allan, E., Manning, P., Alt, F., Binkenstein, J., Blaser, S., Blüthgen, N., Böhm, S., Grassein, F., Hölzel, N., Klaus, V. H., Kleinebecker, T., Morris, E. K., Oelmann, Y., Prati, D., Renner, S. C., Rillig, M. C., Schaefer, M., Schloter, M., Schmitt, B., Schöning, I., Schrumpf, M., Solly, E., Sorkau, E., Steckel, J., Steffen-Dewenter, I., Stempfhuber, B., Tschapka, M., Weiner, C. N., Weisser, W. W., Werner, M., Westphal, C., Wilcke, W., and Fischer, M. (2015) 'Land use intensification alters ecosystem multifunctionality via loss of biodiversity and changes to functional composition', *Ecology Letters*, vol 18, no 8, pp. 834–843

Altieri, M. A. (1987) *Agroecology: The Scientific Bases of Alternative Agriculture*, Westview Press, Boulder, CO

Altieri, M. A. (1990) 'Why study traditional agriculture?', in C. R. Carroll, J. H. Vandermeer, and P. Rosset (eds) *Agroecology*, McGraw-Hill, New York, pp. 551–564

Altieri, M. A. (1994) *Biodiversity and Pest Management in Agroecosystems*, Haworth Press, New York

Altieri, M. A. and Nicholls, C. I. (2017) 'The adaptation and mitigation potential of traditional agriculture in a changing climate', *Climatic Change*, vol 140, no 1, pp. 33–45

Altieri, M. A. and Toledo, V. M. (2011) 'The agroecological revolution in Latin America: rescuing nature, ensuring food sovereignty and empowering peasants', *Journal of Peasant Studies*, vol 38, no 3, pp. 587–612

Alves, M. C. (1990) 'The role of cocoa plantations in the conservation of the Atlantic Forests of Southern Bahia, Brazil', MS thesis, University of Florida, Gainesville, FL

Amir Shah Ruddin, M. S. and De Datta, S. K. (1998) 'Fish distribution in the irrigation and drainage canals of the Muda area', in B. N. Nashriya, N. K. Ho, B. S. Ismail, B. A. Ahyaudin, and K. Y. Lum (eds) *Rice Agroecosystem of the Muda Irrigation Scheme, Malaysia*, Malaysian Institute of Nuclear Technology Research, and Muda Agricultural Development Authority

Anderson, D. and Grove, R. (1987) *Conservation in Africa: People, Policies, and Practices*, Cambridge University Press, Cambridge

Andow, D. A. (1991) 'Vegetational diversity and arthropod population response', *Annual Review of Entomology*, vol 36, pp. 561–586

Angelsen, A. and Kaimowitz, D. (2001a) 'When does technological change in agriculture promote deforestation?', in D. R. Lee and C. B. Barrett (eds) *Tradeoffs or Synergies? Agricultural Intensification, Economic Development and the Environment*, CABI Publishing, Wallingford, pp. 89–114

Angelsen, A. and Kaimowitz, D. (eds) (2001b) *Agricultural Technologies and Tropical Deforestation*, CABI Publishing, Wallingford

Armbrecht, I. and Gallego, M. C. (2007) 'Testing ant predation on the coffee berry borer in shaded and sun coffee plantations in Colombia', *Entomologia Experimentalis et Applicata*, vol 124, pp. 261–267

Armbrecht, I. and Perfecto, I. (2003) 'Litter ant's diversity and predation potential in two Mexican coffee matrices and forest fragments', *Agriculture, Ecosystems and Environment*, vol 97, pp. 107–115

Armbrecht, I., Perfecto, I., and Vandermeer, J. (2004) 'Enigmatic biodiversity correlations: ant diversity responds to diverse resources', *Science*, vol 304, pp. 284–286

Asia Pacific Network on Food Sovereignty (2018) 'Call for government strong support in rice trade to curb price spikes', Asia Pacific Network on Food Sovereignty, 27 February, available at www.apnfs.info/2018/02, accessed 26 December 2018

Bacon, C., Mendez, E., and Brown, M. (2005) 'Participatory action research and support for community development and conservation: Examples from shade coffee landscapes in Nicaragua and El Salvador', Center for Agroecology & Sustainable Food Systems, Research Briefs, University of California, Santa Cruz, CA

Badgley, C., Moghtader, J., Quintero, E., Zakem, E., Chappell, M. J., Aviles-Vazquez, K., Samulon, A., and Perfecto, I. (2007) 'Organic agriculture and the global food supply', *Renewable Agriculture and Food Systems*, vol 22, pp. 86–108

Balehegn, M. (2015) 'Unintended consequences: The ecological repercussions of land grabbing in Sub-Saharan Africa', *Environment*, vol 57, no 2, pp. 4–21

Balfour, E. B. (1950) *The Living Soil: Evidence of the Importance to Human Health of Soil Vitality, with Special Reference to National Planning*, Faber and Faber, London

Balmford, A., Green, E., and Scharlemann, J. P. W. (2005) 'Sparing land for nature: exploring the potential impact of changes in agricultural yield on the area needed for crop production', *Global Change Biology*, vol 11, pp. 1594–1605

Balzer, T., Balzer, P., and Pon, S. (2005) 'Traditional use and availability of aquatic biodiversity in rice-based ecosystems, I. Kampong Thom Province, Kingdom of Cambodia', in M. Halwart and D. Bartley (eds) *Aquatic biodiversity in rice-based ecosystems. Studies and reports from Cambodia, China, Lao People's Democratic Republic and Viet Nam*, [CD-ROM], FAO, Rome, available at www.fao.org/tempref/FI/CDrom/Awarness AgrBiodiv/Kampong%20thom%20reportA.pdf, accessed 16 March 2019

Bambaradeniya, C. N. B. (2000a) 'Ecology and biodiversity in an irrigated rice field ecosystem in Sri Lanka', PhD thesis, University of Peradeniya, Sri Lanka

Bambaradeniya, C. N. B. (2000b) 'Rice fields: an important man-made habitat of amphibians', *Lyriocephalus*, vol 4, nos 1 and 2, pp. 57–63

Bambaradeniya, C. N. B. and Amarasinghe, F. P. (2004) 'Biodiversity associated with rice field agro-ecosystem in Asian countries: a brief review', International Water Management Institute, Working Paper 63

Barraclough, S. L. and Ghimire, K. B. (2000) *Agricultural Expansion and Tropical Deforestation: Poverty, International Trade and Land Use*, Earthscan, London

Barrett, C. B. and Arcese, P. (1995) 'Are integrated conservation-development projects (ICDPs) sustainable? On the conservation of large mammals in sub-Saharan Africa', *World Development*, vol 23, no 7, pp. 1073–1084

Bawa, K. S., Joseph, G., and Setty, S. (2007) 'Poverty, biodiversity and institutions in forest-agriculture ecotones in Western Ghats and Eastern Hilalaya ranges of India', *Agriculture Ecosystems Environment*, vol 121, pp. 287–295

BBC News (2018) 'Brazil's Bolsonaro plans threaten Amazon, experts say', BBC News 31 October

Beketov, M. A., Kefford, B. J., Schäfer, R. B., and Liess, M. (2013), 'Pesticides reduce regional biodiversity of stream invertebrates', *Proceedings of the National Academy of Sciences, USA*, vol 110, no 27, pp. 11039–11043

Benchimol, M. and Peres, C. A. (2014) 'Predicting primate local extinctions within "real-world" forest fragments: a pan-neotropical analysis', *American Journal of Primatology*, vol 76, no 3, pp. 289–302

Bender, D. J. and Fahrig, L. (2005) 'Matrix structure obscures the relationship between interpatch movement and patch size and Isolation', *Ecology*, vol 86, pp. 1023–1033

Bengtsson, J., Ahnström, J., and Weibull, A. C. (2005) 'The effects of organic agriculture on biodiversity and abundance: a meta-analysis', *Journal of Applied Ecology*, vol 42, pp. 261–269

Benjaminsen, T. A. and Bryceson, I. (2012) 'Conservation, green/blue grabbing and accumulation by dispossession in Tanzania', *Journal of Peasant Studies*, vol 39, no 2, pp. 335–355

Benton, T. G., Vickery, J. A., and Wilson, J. D. (2003) 'Farm biodiversity: Is habitat heterogeneity the key?', *Trends in Ecology and Evolution*, vol 18, no 4, pp. 182–188

Beuchelt, T. D. and Zeller, M. (2011) 'Profits and poverty: Certification's troubled link for Nicaragua's organic and fairtrade coffee producers', *Ecological Economics*, vol 70, no 7, pp. 1316–1324

Bird Life International (2018) *State of the World's Birds: Taking the Pulse of the Planet*, Cambridge, UK, Bird Life International, available at www.birdlife.org/sites/default/files/attachments/BL_ReportENG_V11_spreads.pdf, accessed 22 December 2018

Black, H. I. J. and Okwakol, M. J. N. (1997) 'Agricultural intensification, soil biodiversity and ecosystem function in the tropics: The role of termites', *Applied Soil Ecology*, vol 6, no 1, pp. 37–53

Blackman, A., and Naranjo, M. A. (2012) 'Does eco-certification have environmental benefits? Organic coffee in Costa Rica', *Ecological Economics*, vol 83, pp. 58–66

Boillat, S., Gerber, J. F., and Funes-Monzote, F. R. (2012) 'What economic democracy for degrowth? Some comments on the contribution of socialist models and Cuban agroecology', *Futures*, vol 44, no 6, pp. 600–607

Bonner, R. (1993) *At the Hand of Man: Peril and Hope for Africa's Wildlife*, Knopf, New York

Borlaug, N. E. (1987) *Making Institutions Work: A Scientist's Viewpoint*, Texas A & M University Press, College Station, TX

Borlaug. N. E. (1997) 'Feeding a world of 10 billion people: the miracle ahead', *Plant Tissue Culture and Biotechnology*, vol 3, pp. 119–127

Borras Jr, S. M., Hall, R., Scoones, I., White, B., and Wolford, W. (2011) 'Towards a better understanding of global land grabbing: an editorial introduction', *Journal of Peasant Studies*, vol 38, no 2, pp. 209–216

Boserup, E. (2005) *The Conditions of Agricultural Growth: The Economics of Agrarian Change Under Population Pressure*, Aldine Transactions, Piscataway, New Jersey

Boucher, D., Roquemore, S., and Fitzhugh, E. (2013) 'Brazil's success in reducing deforestation', *Tropical Conservation Science*, vol 6, no 3, pp. 426–445

Boucher, D., Vandermeer, J. H., Granzow de la Cerda, I., Mallona, M. A., Perfecto, I., and Zamora, N. (2001) 'Post-agriculture versus post-hurricane succession in southeastern Nicaraguan rain forest', *Plant Ecology*, vol 156, no 2, pp. 131–137

Bourguignon, F. (2017) *The Globalization of Inequality*, Princeton University Press, Princeton, NJ

Brechin, S. R., Wilshusen, P. R., Forthwangler, C. L., and West, P. C. (2002) 'Beyond the square wheel: Toward a more comprehensive understanding of biodiversity conservation as social and political process', *Society and Natural Resources*, vol 15, pp. 41–64

Bretveld, R. W., Thomas, C. M., Scheepers, P. T., Zielhuis, G. A., and Roeleveld, N. (2006) 'Pesticide exposure: the hormonal function of the female reproductive system disrupted?', *Reproductive Biology and Endocrinology*, vol 4, no 1, article no 30

Brinck, K., Fischer, R., Groeneveld, J., Lehmann, S., De Paula, M. D., Pütz, S., Sexton, J. O., Song, D., and Huth, A. (2017) 'High resolution analysis of tropical forest fragmentation and its impact on the global carbon cycle', *Nature Communications*, vol 8, article no 14855

Brittain, C. A., Vighi, M., Bommarco, R., Settele, J. and Potts, S. G. (2010) 'Impacts of a pesticide on pollinator species richness at different spatial scales', *Basic and Applied Ecology*, vol 11, no 2, pp. 106–115

Brock, W. H. (2002) *Justus Von Liebig: The Chemical Gatekeeper*, Cambridge University Press, Cambridge

Brockington, D. (2001) *Fortress Conservation: The Preservation of the Mkomazi Game Reserve*, James Currey, Oxford

Brockington, D., Duffy, R., and Igoe, J. (2008) *Nature Unbound: The Past, Present and Future of Protected Areas*, Earthscan, London

Brooks, J., Waylen, K. A., and Mulder, M. B. (2013) 'Assessing community-based conservation projects: a systematic review and multilevel analysis of attitudinal, behavioral, ecological, and economic outcomes', *Environmental Evidence*, vol 2, no 1, article no 2

Brooks, T. M., Pimm, S. L., and Oyugi, J. O. (1999) 'Time lag between deforestation and bird extinction in tropical forest fragments', *Conservation Biology*, vol 13, pp. 1140–1150

Browder, J. O. and Godfrey, B. J. (1997) *Rainforest Cities: Urbanization, Development, and Globalization of the Brazilian Amazon*, Columbia University Press, New York

Brown, D., Brown, J. C., and Desposato, S. (2007) 'Promoting and preventing political change through international-funded NGO activity', *Latin American Research Review*, vol 42, no 1, pp. 126–138

Brown, J. C. (2006a) 'Placing local environmental protest within global environmental networks: Colonist farmers and sustainable development in the Brazilian Amazon', in C. Mauch, N. Stoltzfus, and D. Weiner (eds) *Shades Of Green*, Lanham Md, Rowman and Littlefield, pp. 197–218

Brown, J. C. (2006b) 'Productive conservation and its representation: The case of Beekeeping in the Brazilian Amazon', in K. Zimmerer (ed.) *Globalization and the new Geographies Of Conservation*, University of Chicago Press, Chicago, pp. 92–115

Brown, J. C. and Purcell, M. (2005) 'There's nothing inherent about scale: political ecology, the local trap, and the politics of development in the Brazilian Amazon', *Geoforum*, vol 36, no 5, pp. 607–624

Brown, J. H. and Kodric-Brown, A. (1977) 'Turnover rates in insular biogeography: effect of immigration and extinction', *Ecology*, vol 58, pp. 445–449

Brush, S. B. (2007) 'Farmers' rights and protection of traditional agricultural knowledge', *World Development*, vol 35, no 9, pp. 1499–1514

Bunker, S. (1985) *Underdeveloping the Amazon: Extraction, Unequal Exchange, and the Failure of the Modern State*, University of Chicago Press, Chicago

Burel, F., Baudry, J., Butet, A., Clergeau, P., Delettre, Y., Le Coeur, D., Dubs, F., Morvan, N., Paillat, G., Petit, S., Thenail, C., Brunel, E., and Lefeuvre, J. C. (1998) 'Comparative biodiversity along a gradient of agricultural landscapes', *Acta Oecologia*, vol 19, pp. 47–60

Büscher, B., Dressler, W., and Fletcher, R. (2014) *Nature, Inc: Environmental Conservation in the Neoliberal Age*, University of Arizona Press: Tucson

Büscher, B., Sullivan, S., Neves, K., Igoe, J., and Brockington, D. (2012) 'Towards a synthesized critique of neoliberal biodiversity conservation', *Capitalism Nature Socialism*, vol 23, no 2, pp. 4–30

Buttel, F. H. (1990) 'Social relations and the growth of modern agriculture', in C. R. Carroll, J. Vandermeer, and P. M. Rosset (eds) *Agroecology*, McGraw-Hill, New York

California Water Boards (2013) *Communities that Relay on a Contaminated Water Source for Drinking Water*, State Water Resources Control Board Report to the Legislature

Campos, M. T. and Nepstad, D. C. (2006) 'Smallholders, the Amazon's new conservationists', *Conservation Biology*, vol 20, pp. 1553–1556

Cardinale, B. J., Duffy, J. E., Gonzalez, A., Hooper, D. U., Perrings, C., Venail, P., Narwani, A., Mace, G. M., Tilman, D., Wardle, D. A., Kinzig, A. P., Daily, G. C., Loreau, M., Grace, J. B., Larigauderie, A., Srivastava, D., and Naeem, S. (2012) 'Biodiversity loss and its impact on humanity', *Nature*, vol 486, no 7401, pp. 59–67

Carrington, D. (2016) 'The Anthropocene epoch: scientists declare dawn of human-influenced age', *Guardian*, 29 August

Carson, R. (1962) *Silent Spring*, Houghton Mifflin, Boston

Carvalho, G., Nepstad, D., McGrath, D., Vera Diaz, M. C., Santilli, M., and Barros, A. N. (2002) 'Frontier expansion in the Amazon: balancing development and sustainability', *Environment*, vol 44, no 3, pp. 34–42

Cassano, C. R., Schroth, G., Faria, D., Delabie, J. H., and Bede, L. (2009) 'Landscape and farm scale management to enhance biodiversity conservation in the cocoa producing region of southern Bahia, Brazil', *Biodiversity and Conservation*, vol 18, no 3, pp. 577–603

Cassman, K. G. (1999) 'Ecological intensification of cereal production systems: yield potential, soil quality, and precision agriculture', *Proceedings of the National Academy of Sciences, USA*, vol 96, pp. 5952–5959

Castro, E. V. de (2016) *The Relative Native: Essays on Indigenous Conceptual Worlds*, Hau Press, Chicago, IL

Ceballos, G., Ehrlich, P. R., and Dirzo, R. (2017) 'Biological annihilation via the ongoing sixth mass extinction signaled by vertebrate population losses and declines', *Proceedings of the National Academy of Sciences, USA*, vol 114, no 30, pp. E6089–E6096

Chacon, J. C. and Gliessman, S. R. (1982) 'Use of the "non-weed" concept in traditional tropical agroecosystems of southeastern Mexico', *Agroecosystems*, vol 8, pp. 1–11

Chandler, R. B., King, D. I., Raudales, R., Trubey, R., Chandler, C., and Arce Chávez, V. J. (2013) 'A small-scale land-sparing approach to conserving biological diversity in tropical agricultural landscapes', *Conservation Biology*, vol 27, no 4, pp. 785–795

Chaplin-Kramer, R., O'Rourke, M. E., Blitzer, E. J., and Kremen, C. (2011) 'A meta-analysis of crop pest and natural enemy response to landscape complexity', *Ecology Letters*, vol 14, no 9, pp. 922–932

Chappell, M. J., Wittman, H., Bacon, C., Ferguson, B. G., Garcia Barrios, L., Garcia Barrios, R., Jaffee, D., Lima, J., Mendez, V. E., Morales, H., Soto-Pinto, L., Vandermeer, J., and Perfecto, I. (2013) 'Food sovereignty: an alternative paradigm for poverty reduction and biodiversity conservation in Latin America', *F1000Research* 2013, vol 2, p. e235, available at doi: 10.12688/f1000research.2-235.v1, accessed on 13 March 2019

Chazdon, R. L., Brancalion, P. H., Laestadius, L., Bennett-Curry, A., Buckingham, K., Kumar, C., Moll-Rocek, J., Vieira, I. C., and Wilson, S. J. (2016) 'When is a forest a forest? Forest concepts and definitions in the era of forest and landscape restoration', *Ambio*, vol 45, no 5, pp. 538–550

Chomsky, N., Bourchier, D., Ham, K. P. P., Kammen, D., Van Klinken, G., and McDonald, H. (2006) *Masters of Terror: Indonesia's Military and Violence in East Timor*, Rowman & Littlefield Publishers, Lenham, MD

Cobera, E. and Schroeder, H (2017) 'REDD+ crossroads post review: politics, lessons and interplays', *Forests*, vol 8, no 12, pp. 508

Colborn, T. Dumanoski, D., and Peterson Myers, J. (1997) *Our Stolen Future: Are We Threatening Our Fertility, Intelligence and Survival?* Plum, New York

Collins, W. W. and Qualset, C. O. (1998) *Biodiversity in Agroecosystems*, CRC Press, Boca Raton, FL

Condit, R., Chisholm, R. A., and Hubbell, S. P. (2012) 'Thirty years of forest census at Barro Colorado and the importance of immigration in maintaining diversity', *PloS One*, vol 7, no 11, p. e49826

Condit, R., Hubbell, S. P., Lafrankie, J. V., Sukumar, R., Manokaran, N., Foster, R. B., and Ashton, P. S. (1996) 'Species-area and species-individual relationships for tropical trees: a comparison of three 50-ha plots', *Journal of Ecology*, vol 84, no 4, pp. 549–562

Condit, R., Pitman, B. N. C. A., Leigh, E. G. Jr., Chave, J., Terborgh, W. J., Foster, R. B., Nunez Vargas, P., Aguilar, S., Valencia, R., Villa Munoz, G., Muller-Landau, H. C., Losos, E. C., and Hubbell, S. P. (2002) 'Beta-diversity in tropical forest trees', *Science*, vol 295, pp. 666–669

Conford, P. (2001) *The Origins of the Organic Movement*, Floris Books, Edinburgh

Conservation International (2007) 'Biodiversity hot spots: Atlantic forest', available at www.biodiversityhotspots.org/xp/hotspots/atlantic_forest/Pages/default.aspx, accessed 26 February 2009

Coria, J. and Calfucura, E. (2012) 'Ecotourism and the development of indigenous communities: The good, the bad, and the ugly', *Ecological Economics*, vol 73, pp. 47–55

Cornia, G. A. (1985) 'Farm size, land yields and the agricultural production function: An analysis for fifteen developing countries', *World Development*, vol 13, pp. 513–534

Cox, T. S., Van Tassel, D. L., Cox, C. M., and DeHaan, L. R. (2010) 'Progress in breeding perennial grains', *Crop and Pasture Science*, vol 61, no 7, pp. 513–521

Cruz-Garcia, G. S. and Price, L. L. (2011) 'Ethnobotanical investigation of wild food plants used by rice farmers in Kalasin, Northeast Thailand', *Journal of Ethnobiology and Ethnomedicine*, vol 7, no 1, article no 33

Cullen, L. Jr., Alger, K., and Rambaldi, D. M. (2005) 'Land reform and biodiversity conservation in Brazil in the 1990s: Conflict and the articulation of mutual interests', *Conservation Biology*, vol 19, pp. 747–755

Culman, S. W., Young-Mathews, A., Hollander, A. D., Ferris, H., Sánchez-Moreno, S., O'Geen, A. T., and Jackson, L. E. (2010) 'Biodiversity is associated with indicators of soil ecosystem functions over a landscape gradient of agricultural intensification', *Landscape Ecology*, vol 25, no 9, pp. 1333–1348

Curry, J. P. (1994) *Grassland Invertebrates: Ecology, Influence on Soil Fertility and Effects on Plant Growth*, Springer, New York

D'Arcy, W. G. and Hammel, B. (1985) 'The plants of "Ocoquili" Island, San Plas coast, Panama', *Annals of the Missouri Botanical Garden*, vol 72, no 2, pp. 264–267

Daily, G. C., Ceballos, G., Pacheco, J., Suzan, G., and Sanchez-Azofeifa, G. A. (2003) 'Countryside biogeography of neotropical mammals: conservation opportunities in agricultural landscapes of Costa Rica', *Conservation Biology*, vol 17, pp. 1814–1826

Daily, G. C., Ehrlich, P. R., and Sanchez-Azofeifa, G. A. (2001) 'Countryside biogeography: use of human-dominated habitats by the avifauna of southern Costa Rica', *Ecological Applications*, vol 11, pp. 1–13

Danielsen, F., Beukema, H., Burgess, N. D., Parish, F., Brühl, C. A., Donald, P. F., Murdiyarso, D., Phalan, B., Reijnders, L., Struebig, M., and Fitzherbert, E. B. (2009) 'Biofuel plantations on forested lands: double jeopardy for biodiversity and climate', *Conservation Biology*, vol 23, no 2, pp. 348–358

Darwin, C. (1881) *The Formation of Vegetable Mould through the Action of Worms, with Observations of their Habits*, Murray, London

Davis, K. F., D'Odorico, P., and Rulli, M. C. (2014) 'Land grabbing: a preliminary quantification of economic impacts on rural livelihoods', *Population and Environment*, vol 36, no 2, pp. 180–192

Davis, M. (2001) *Late Victorian Holocausts: El Niño Famines and the Making of the Third World*, Verso, London

Davis, M. (2006) *Planet of Slums*, Penguin Random House, New York

De Beenhouwer, M., Aerts, R., and Honnay, O. (2013) 'A global meta-analysis of the biodiversity and ecosystem service benefits of coffee and cacao agroforestry', *Agriculture, Ecosystems & Environment*, vol 175, pp. 1–7

De Gisi, S., Petta, L., and Wendland, C. (2014) 'History and technology of Terra Preta sanitation', *Sustainability*, vol 6, no 3, pp. 1328–1345

De Janvry, A. (1981) *The Agrarian Question and Reformism in Latin America*, Baltimore, MD

De Janvry, A., Gonzalez-Navarro, M., and Sadoulet, E. (2014) 'Are land reforms granting complete property rights politically risky? Electoral outcomes of Mexico's certification program', *Journal of Development Economics*, vol 110, pp. 216–225.

De Ponti, T., Rijk, B., and Van Ittersum, M. K. (2012) 'The crop yield gap between organic and conventional agriculture', *Agricultural Systems*, vol 108, pp. 1–9

deLaplante, K., Brown, B., and Peacock, K. A. (2011) 'The biodiversity-ecosystem function debate in ecology', *Philosophy of Ecology*, vol 11, pp. 169–200

Delgado, A. and Rommetveit, K. (2012) '"Our strength is diversity": imaginaries of nature and community in a Brazilian social movement', *International Journal of Sustainable Development*, vol 15, no 4, pp. 353–373

Descola, P. (2013) *Beyond Nature and Culture*, University of Chicago Press, Chicago, IL

Desmarais, A. A. (2002) 'Peasants speak – The Via Campesina: Consolidating an international peasant and farm movement', *Journal of Peasant Studies*, vol 29, pp. 91–124

Desmarais, A. A. (2007) *La Via Campesina: Globalization and the Power of Peasants*, Pluto Press, Cambridge

Diamond, J. M. (1975) 'The island dilemma: lessons of modern biogeographic studies for the design of natural reserves', *Biological Conservation*, vol 7, no 2, pp. 129–146

Diamond, J. M. and May, R. M. (1976) 'Island biogeography and the design of natural reserves', in R. May (ed.) *Theoretical Ecology: Principles and Applications*, Blackwell, Oxford, pp. 163–186

Diaz, R. J. and Rosenberg, R. (2008) 'Spreading dead zones and consequences for marine ecosystems', *Science*, vol 321, pp. 926–929

Doane, M. (2012) *Stealing Shining Rivers: Agrarian Conflict, Market Logic, and Conservation in a Mexican Forest*, University of Arizona Press: Tucson

Donahue, B., Burke, J. Anderson, M, Beal, A., Kelly, T., Lapping, M, Ramer, H., Libby, R., and Berlin, L. (2014) *A New England Food Vision*, Food Solutions New England, University of New Hampshire, Durham, NH

Donald, P. F. (2001) 'Agricultural intensification and the collapse of Europe's farmland bird populations', *Proceedings of the Royal Society B*, vol 268, no 1462, pp. 25–29

Donald, P. F. (2004) 'Biodiversity impacts of some agricultural commodity production systems', *Conservation Biology*, vol 18, no 1, pp. 17–38

Donald, P. F. and Evans, A. D. (2006) 'Habitat connectivity and matrix restoration: the wider implications of agri-environment schemes', *Journal of Applied Ecology*, vol 43, pp. 209–218

Dreher, A. and Gaston, N. (2008) 'Has globalization increased inequality?', *Review of International Economics*, vol 16, no 3, pp. 516–536

Drinkwater, L. E. and Snapp, S. S. (2007) 'Nutrients in agroecosystems: rethinking the management paradigm', *Advances in Agronomy*, vol 92, pp. 163–186

Durham, W. (1979) *Scarcity and Survival in Central America*, Stanford University Press, Palo Alto

Ehlers Smith, D. A., Si, X., Ehlers Smith, Y. C., Kalle, R., Ramesh, T., and Downs, C. T. (2018) 'Patterns of avian diversity across a decreasing patch-size gradient in a critically endangered subtropical forest system', *Journal of Biogeography*, vol 45, no 9, pp. 2118–2132

Elphick, C. S. (2000). 'Functional equivalency between rice fields and seminatural wetland habitats', *Conservation Biology*, 14(1), pp. 181–191

Elvevoll, E. O. and James, D. G. (2000) 'Potential benefits of fish for maternal, fetal and neonatal nutrition: a review of the literature', *Food Nutrition and Agriculture (FAO)*, vol 27, pp. 28–39

Ennis, K. K. and Philpott, S. M. (2017) 'Strong influences of a dominant, ground-nesting ant on recruitment, and establishment of ant colonies and communities', *Biotropica*, vol 49, no 4, pp. 521–530

Erickson, C. L. (2006) 'Domesticated landscapes of the Bolivian Amazon', in W. Balée and C. Erickson (eds) *Time and Complexity in Historical Ecology*, Columbia University Press, NY, pp. 187–233

Eriksson, O. (1996) 'Regional dynamics of plants: a review of evidence for remnant, source-sink and metapopulations' *Oikos*, vol 77, no 2. pp. 248–258

Erwin, T. L. (1982) 'Tropical forests: their richness in Coleoptera and other arthropod species', *Coleopterists Bulletin*, vol 36, no 1, pp. 34–35

Escobar, A. (1995) *Encountering Development: The Making and Unmaking of the Third World*, Princeton University Press, Princeton

Escobar, A. (1998) 'Whose knowledge, whose nature? Biodiversity, conservation, and the political ecology of social movements', *Journal of Political Ecology*, vol 5, no 1, pp. 53–82

ETC Group (2017) 'Who will feed us? The peasant food web versus the industrial food chain', ETC Group, available at www.etcgroup.org/sites/www.etcgroup.org/files/files/whowillfeedcover.png, accessed 9 December 2018

Ewel, J. J. (1999) 'Natural systems as models for the design of sustainable systems of land use', *Agroforestry Systems*, vol 45, no 1–3, pp. 1–21

Fagan, M. E., DeFries, R. S., Sesnie, S. E., Arroyo, J. P., Walker, W., Soto, C., Chazdon, R. L., and Sanchun, A. (2013) 'Land cover dynamics following a deforestation ban in northern Costa Rica', *Environmental Research Letters*, vol 8, no 3, p. e034017

Fahrig, L. (2003) 'Effects of habitat fragmentation on biodiversity', *Annual Review of Ecology, Evolution, and Systematic*, vol 34, pp. 487–515

FAO (1996) 'Report on the State of the World's Plant Genetic Resources for Food and Agriculture', prepared for the International Technical Conference on Plant Genetic Resources, Leipzig, Germany, 17–23 June 1996

FAO (2002) *The State of FoodIinsecurity in the World 2002*, FAO, Rome

FAO (2004) *International Year of Rice, 2004. Rice is Life: Aquatic Biodiversity in Rice Fields*, FAO, Rome

FAO (2014) 'The state of food and agriculture: innovation in family farming', FAO, Rome, available at www.fao.org/3/a-i4040e.pdf, accessed 9 December 2018

FAO (2017) 'The future of food and agriculture: trends and challenges', available at www.fao.org/3/a-i6583e.pdf, accessed September 2018

FAO (2018) 'FAO's work on agroecology: A pathway to achieving the SGDs', available at www.fao.org/3/i9021en/I9021EN.pdf, accessed 9 December 2018

Faria, D., Laps, R. R., Baumgarten, J., and Cetra, M. (2006) 'Bat and bird assemblages from forests and shade cacao plantations in two contrasting landscapes in the Atlantic Forest of southern Bahia, Brazil', *Biodiversity and Conservation*, vol 15, no 2, pp. 587–612

Faria, D., Paciencia, M. L. B., Dixo, M., Laps, R. R., and Baumgarten, J. (2007) 'Ferns, frogs, lizards, birds and bats in forest fragments and shade cacao plantations in two contrasting landscapes in the Atlantic forest, Brazil', *Biodiversity and Conservation*, vol 16, no 8, pp. 2335–2357

Ferguson, B. G. and Morales, H. (2010) 'Latin American agroecologists build a powerful scientific and social movement', *Journal of Sustainable Agriculture*, vol 34(4), pp. 339–341

Ferguson, B. G., Vandermeer, J., Morales, H., and Griffith, D. (2003) 'Post agricultural succession in El Peten, Guatemala', *Conservation Biology*, vol 17, pp. 818–828

Fernandes Nogueira, R., Roitman, I., Alvim Carvalho, F., Taboada Soldati, G., and Jacobson, T. K. B. (2018) 'Challenges for agroecological and organic management of Caburuca cocoa agroecosystems in three rural settlements in south Bahia, Brazil: perceptions from local actors', Agroforest Systems, available at https://doi-org.proxy.lib.umich.edu/10.1007/s10457-018-0303-x, accessed 13 March 2019

Fernandes, B. M. (2013) 'Re-peasantization, resistance and subordination: the struggle for land and Agrarian reform in Brazil', *Agrarian South: Journal of Political Economy*, vol 2, no 3, pp. 269–289

Fernández, M. W., Figueroa, G., Graddy-Lovelace, G., Machado, M., Vázquez, L., Pérez N., Casimiro, L., Romero, G., and Funes-Aguilar, F. (2018) 'New opportunities, new challenges: Harnessing Cuba's advances in agroecology and sustainable agriculture in the context of changing relations with the United States', *Elementa Science of the Anthropocene*, vol 6, p. e76, available at doi: https://doi.org/10.1525/elementa.337, accessed 14 March 2019

Fernando C. H. (1980a) 'The freshwater invertebrate fauna of Sri Lanka', *Spolia Zeylanica*, vol 35, pp. 5–39

Fernando C. H. (1980b) 'The freshwater zooplankton of Sri Lanka, with a discussion of tropical freshwater zooplankton composition', *Internationale Revue Der Gesamten Hydrobiologie*, vol 65, pp. 85–125

Fernando C. H. (1993) 'Rice field ecology and fish culture – an overview', *Hydrobiologia*, vol 259, pp. 91–113

Fernando C. H. (1995) 'Rice fields are aquatic, semi-aquatic, terrestrial and agricultural: a complex and questionable limnology', *Tropical Limnology*, vol 1, pp. 121–148

Fernando C. H. (1996) 'Ecology of rice fields and its bearing on fisheries and fish culture', in S. S. de Silva (ed.) *Perspectives in Asian Fisheries*, Asian Fisheries Society, Manila, Philippines, pp. 217–237

Fernando C. H., Furtado J. I., and Lim, R. P. (1979) 'Aquatic fauna of the world's rice fields', *Wallaceana Supplement* (Kuala Lumpur), vol 2, pp. 1–105

Fernside, P. M. (1993) 'Deforestation in Brazilian Amazonia: The effect of population and land tenure', *Ambio*, vol 22, pp. 537–545

Ferraro, P. J., Hanauer, M. M., and Sims, K. R. (2011) 'Conditions associated with protected area success in conservation and poverty reduction', *Proceedings of the National Academy of Sciences, USA*, vol 108, no 34, pp. 13913–13918

Ferraz, G., Russell, G. J. Stouffer, P. C. Bierregaard, R. O., Jr., Pimm, S. L., and Lovejoy, T. E. (2003) 'Rates of species loss from Amazonian forest fragments', *Proceedings of the National Academy of Sciences, USA*, vol 100, pp. 14069–14073

Finegan, B. and Nasi, R. (2004) 'The biodiversity and conservation potential of shifting cultivation landscapes', in G. Schroth, G. A. B. da Fonseca, C. A. Harvey, C. Gascon, H. L. Vasconcelos, and A. M. N. Izac (eds) *Agroforestry and Biodiversity Conservation in Tropical Landscapes*, Island Press, Washington, DC, pp. 153–197

Fischer, A. and Vasseur, L. (2000) 'The crisis in shifting cultivation practices and the promise of agroforestry: a review of the Panamanian experience', *Biodiversity and Conservation*, vol 9, pp. 739–756

Fischer, J., Abson, D. J., Butsic, V., Chappell, M. J., Ekroos, J., Hanspach, J., Kuemmerle, T., Smith, H. G., and von Wehrden, H. (2014) 'Land sparing versus land sharing: moving forward', *Conservation Letters*, vol 7, no 3, pp. 149–157

Fischer, J., Batáry, P., Bawa, K. S., Brussaard, L., Chappell, M. J., Clough, Y., Daily, G. C., Dorrough, J., Hartel, T., Jackson, L. E., and Klein, A. M. (2011) 'Conservation: limits of land sparing', *Science*, vol 334, no 6056, pp. 593–593

Fish, S. K. (2000) 'Hohokam impacts on Sonoran Desert environment', in D. L. Lentz (ed.) *Imperfect Balance: Landscape Transformations in the Precolumbian Americas*, Columbia University Press, New York, pp. 251–280

Flynn, D. F., Gogol-Prokurat, M., Nogeire, T., Molinari, N., Richers, B. T., Lin, B. B., Simpson, N., Mayfield, M. M., and DeClerck, F. (2009) 'Loss of functional diversity under land use intensification across multiple taxa', *Ecology Letters*, vol 12, no 1, pp. 22–33

Foresta, R. (1991) *Amazon Conservation in the Age of Development: The Limits of Providence*, University of Florida Press, Gainesville, FL

Foster, A. D., Behrma, J. R., and Barber, W. (1999) 'Population growth, technical change and forest degradation', Department of Economics Working Paper, Brown University, Providence, RI

Foster, D. R. (2017) *Wildlands and Woodlands: Farmlands and Communities*, Harvard University Forest, Harvard University Press, Cambridge, MA

Foster, D. R., Aber, J., Cogbill, C., Hart, C., Colburn, E., D'Amato, A., Donahue, B., Driscoll, C., Ellison, A., Fahey, T., and Hall, B. (2010) *Wildlands and Woodlands: A Vision for the New England Landscape*, Harvard Forest; Harvard University Press, Cambridge, MA

Foufopoulos, J. and Ives, A. R. (1999) 'Reptile extinctions on land-bridge islands: life-history attributes and vulnerability to extinction', *American Naturalist*, vol 153, pp. 1–25

Franke, R. and Chasin, B. (1980) *Seeds of Famine: Ecological Destruction and the Development Dilemma in the West African Sahel*, Allen-Held, Osman, NJ

Funes, F., García, L., Bourque, M., Pérez, N., and Rosset, P. (2002) *Sustainable Agriculture and Resistance: Transforming Food Production in Cuba*, Food First Books, Oakland, CA

Galaty, J. G. (2013) 'Land grabbing in the Eastern African rangelands', in A. Catley, I. Scoones, and J. Lind (eds) *Pastoralism and Development in Africa: Dynamic Change at the Margins*, Routledge, London and New York, pp. 143–153

Garcia, C. A., Bhagwat, S. A., Ghazoul, J., Nath, C. D., Nanaya, K. M., Kushalappa, C. G., Raghuramulu, Y., Nasi, R., and Vaast, P. (2010) 'Biodiversity conservation in agricultural landscapes: challenges and opportunities of coffee agroforests in the Western Ghats, India', *Conservation Biology*, vol 24, no 2, pp. 479–488

García-Barrios, L., Galván-Miyoshi, Y. M., Valsieso-Pérez, I. A., Masera, O. R., Bocco, G., and Vandermeer, J. (2009) 'Neotropical forest conservation, agricultural intensification, and rural out-migration: the Mexican experience', *Bioscience*, vol 59, no 10, pp. 863–873

García-Barrios, R., García Barrios, L., and Alvarez Buylla, E. (1991) *Lagunas, Deterioro Ambiental y Tecnológico en el Campo Semipreletarizado*, El Colegio de México, Mexico City

Gateau-Rey, L., Tanner, E. V., Rapidel, B., Marelli, J. P., and Royaert, S. (2018) 'Climate change could threaten cocoa production: Effects of 2015–16 El Niño-related drought on cocoa agroforests in Bahia, Brazil', *PloS One*, vol 13, no 7, p. e0200454

Geiger, F., Bengtsson, J., Berendse, F., Weisser, W. W., Emmerson, M., Morales, M. B., Ceryngier, P., Liira, J., Tscharntke, T., Winqvist, C., Eggers, S., Bommarco,

R., Pärt, T., Bretagnolle, V., Plantegenest, M., Clement, L. W., Dennis, C., Palmer, C., Oñate, J. J., Guerrero, I., Hawro, V., Aavik, T., Thies, C., Flohre, A., Hänke, S., Fischer, C., Goedhart, P. W., and Inchausti, P. (2010) 'Persistent negative effects of pesticides on biodiversity and biological control potential on European farmland', *Basic and Applied Ecology*, vol 11, no 2, pp. 97–105

Geisseler, D. and Scow, K. M. (2014) 'Long-term effects of mineral fertilizers on soil microorganisms – A review', *Soil Biology and Biochemistry*, vol 75, pp. 54–63

Gibbons, M. A. and Harcourt, A. H. (2009) 'Biological correlates of extinction and persistence of primates in small forest fragments: a global analysis', *Tropical Conservation Science*, vol 2, no 4, pp. 388–403

Gibson, L., Lynam, A. J., Bradshaw, C. J., He, F., Bickford, D. P., Woodruff, D. S., Bumrungsri, S., and Laurance, W. F. (2013) 'Near-complete extinction of native small mammal fauna 25 years after forest fragmentation', *Science*, vol 341, no 6153, pp. 1508–1510

Giller, K. E., Beare, M. J., Lavelle, P., Izac, A. M. N., and Swift, M. J. (1997) 'Agricultural intensification, soil biodiversity and agroecosystem function', *Applied Soil Ecology*, vol 6, no 1, pp. 3–16

Gilpin, M. and Hanski, I. (ed.) (1991) *Metapopulation Dynamics: Empirical and Theoretical Investigations*, Academic Press, London

Gliessman, S. (2013) 'Agroecology: Growing the roots of resistance', *Agroecology and Sustainable Food Systems*, vol 37, no 1, pp. 19–31

Gliessman, S. R. (2006) *Agroecology: The Ecology of Sustainable Food Systems*, CRC Press, Boca Raton, FL

Gomes, C. V. A., Vadjunec, J. M., and Perz, S. G. (2012) 'Rubber tapper identities: Political-economic dynamics, livelihood shifts, and environmental implications in a changing Amazon', *Geoforum*, vol 43, no 2, pp. 260–271

Gonthier, D. J., Ennis, K. K., Philpott, S. M., Vandermeer, J., and Perfecto, I. (2013) 'Ants defend coffee from berry borer colonization', *BioControl*, vol 58, no 6, pp. 815–820

González González, C., López Martínez, R., Hernández López, S., and Benítez, M. (2016) 'A dynamical model to study the effect of landscape agricultural management on the conservation of native ecological networks', *Agroecology and Sustainable Food Systems*, vol 40, no 9, pp. 922–940

Goodman, D., Sorj, B., and Wilkinson, J. (1987) *From Farming to Biotechnology: A Theory of Agro-Industrial Development*, Basil Blackwell, Oxford

Goulart, F. F., Carvalho-Ribeiro, S., and Soares-Filho, B. (2016) 'Farming-biodiversity segregation or integration? Revisiting land sparing versus land sharing debate', *Journal of Environmental Protection*, vol 7, no 7, pp. 1016–1032

Goulart, F. F., Vandermeer, J., Perfecto, I., and Matta-Machado, R. (2011) 'Frugivory by five bird species in agro-forest homegardens of Pontal do Paranepanema, Brazil', *Agroforestry Systems*, 82 pp. 239–246

Goulart, F. F., Galán, Á. L., Nelson, E., and Soares-Filho, B. (2018) 'Conservation lessons from Cuba: Connecting science and policy', *Biological Conservation*, vol 217, pp. 280–288

Gould, S. J. (1990) *Wonderful Life: The Burgess Shale and the Nature of History*, W. W. Norton, London

Gouyon, A., de Foresta, H., and Levang, P. (1993) 'Does "jungle rubber" deserve its name? An analysis of rubber agroforestry systems in southeast Sumatra', *Agroforestry Systems*, vol 22, pp. 181–206

Goyes, D. R. and South, N. (2016) 'Land-grabs, biopiracy and the inversion of justice in Colombia', *British Journal of Criminology*, vol 56, no 3, pp. 558–577

Graeub, B. E., Chappell, M. J., Wittman, H., Ledermann, S., Kerr, R. B., and Gemmill-Herren, B. (2016) 'The state of family farms in the world', *World Development*, vol 87, pp. 1–15

Grau, H. R., Hernández, M. E., Gutierrez, J., Gasparri, N. I., Casavecchia, M. C., Flores-Ivaldi, E. E., and Paolini, L. (2008) 'A peri-urban neotropical forest transition and its consequences for environmental services', *Ecology and Society*, vol 13, no 1, article no 35.

Green, R. E., Cornell, S. J., Scharlemann, J. P. W., and Balmford, A. (2005) 'Farming and the fate of wild nature', *Science*, vol 307, pp. 550–555

Greenberg, R., Bichier, P., and Sterling, J. (1997) 'Bird populations in rustic and planted shade coffee plantations of Eastern Chiapas, México', *Biotropica*, vol 29, pp. 501–514

Griffith, D. M. (2000) 'Agroforestry: a refuge for tropical biodiversity after fire', *Conservation Biology*, vol 14, pp. 325–326

Guadagnin, D. L. and Maltchik, L. (2006) 'Habitat and landscape factors associated with neotropical waterbird occurrence and richness in wetland fragments', *Vertebrate Conservation and Biodiversity*, pp. 405–418, Springer, Dordrecht

Guardian (2018) 'Brazil's Jair Bolsonaro threatens purge of leftwing activists', theguardian.com, 12 October

Guha, R. and Martinez-Alier, J. (1997) *Varieties of Environmentalism: Essays North and South*, Earthscan, London

Gunapala, N. and Scow, K. M. (1998), 'Dynamics of soil microbial biomass and activity in conventional and organic farming systems', *Soil Biology and Biochemistry*, vol 30, no 6, pp. 805–816.

Gutiérrez-Vélez, V. H., DeFries, R., Pinedo-Vásquez, M., Uriarte, M., Padoch, C., Baethgen, W., Fernandes, K., and Lim, Y. (2011) 'High-yield oil palm expansion spares land at the expense of forests in the Peruvian Amazon', *Environmental Research Letters*, vol 6, no 4, p. e044029

Haddad, N. M. and Tewksbury, J. J. (2005) 'Low-quality habitat corridors as movement conduits for two butterfly species', *Ecological Applications*, vol 15, pp. 250–257

Halberg, N., Alrøe, H. F., Knudsen, M. T., and Kristensen, E. S. (eds) (2005) *Global Development of Organic Agriculture: Challenges and Promises*, CAB International, Wallingford

Hall, A. (ed.) (2000) *Amazonia at the Crossroads: The Challenge of Sustainable Development*, Institute of Latin American Studies, London

Hallmann, C. A., Sorg, M., Jongejans, E., Siepel, H., Hofland, N., Schwan, H., Stenmans, W., Müller, A., Sumser, H., Hörren, T., and Goulson, D. (2017) 'More than 75 percent decline over 27 years in total flying insect biomass in protected areas', *PloS One*, vol 12, no 10, p. e0185809

Halwart, M. (1999) 'Fish in rice-based farming systems – Trends and prospects', in D. van Tran (ed.) *International Rice Commission – Assessment and Orientation Towards the 21st Century*, Proceedings of the 19th Session of the International Rice Commission, Cairo, Egypt, 7–9 September 1998, p. 260

Halwart, M. (2006) 'Biodiversity and nutrition in rice-based aquatic ecosystems', *Journal of Food Composition and Analysis*, vol 19, pp. 747–751

Halwart, M. and Bartley, D. (eds) (2005) *Aquatic biodiversity in rice-based ecosystems. Studies and reports from Cambodia, China, Lao PDR and Vietnam,* [CD-ROM] Rome, FAO, available at www.fao.org/3/a-i3841e.pdf, accessed 14 March 2019

Halwart, M., Bartley, D., Burtlingame, B., Funge-Smith, S., and James, D. (2006) 'FAO Regional Technical Expert Workshop on aquatic biodiversity, its nutritional composition, and human consumption in rice-based systems', *Journal of Food Composition and Analysis*, vol 19, pp. 752–755

Hansen, M. C., Potapov, P. V., Moore, R., Hancher, M., Turubanova, S. A. A., Tyukavina, A., Thau, D., Stehman, S. V., Goetz, S. J., Loveland, T. R., and Kommareddy, A. (2013) 'High-resolution global maps of 21st-century forest cover change', *Science*, vol 342, no 6160, pp. 850–853

Hansen, M., Thilsted, S. H., Sandstrom, B., Kongsbak, K., Larsen, T., Jensen, M., and Sorensen, S. S. (1998) 'Calcium absorption from small softboned fish', *Journal of Trace Elements in Medicine and Biology*, vol 12, pp. 148–154

Hanski, I. (1999) *Metapopulation in Ecology (Oxford Series in Ecology and Evolution)*, Oxford University Press, Oxford

Hanski, I. and Gilpin, M. (1991) 'Metapopulation dynamics: brief history and conceptual domain', *Biological Journal of the Linnean Society*, vol 42, pp. 3–16

Hardin, R. and Bahuchet, S. (2011) 'Concessionary politics: property, patronage, and political rivalry in central African forest management', *Current Anthropology*, vol 52, no S3, S113–S125

Hardt, M. and Negri, A. (2005) *Multitude: War and Democracy in the Age of Empire*, Penguin, London

Harrison, S. and Taylor, A. D. (1997) 'Empirical evidence for metapopulation dynamics', in, I. Hanski and M. E. Gilpin (eds) *Metapopulation Biology: Ecology, Genetics and Evolution*, Academic Press, Cambridge, pp. 27–42

Hartmann, M., Frey, B., Mayer, J., Mäder, P., and Widmer, F. (2015) 'Distinct soil microbial diversity under long-term organic and conventional farming', *ISME Journal*, vol 9, no 5, pp. 1177–1194

Harvey, C. A., Dickson, B., and Kormos, C. (2010) 'Opportunities for achieving biodiversity conservation through REDD', *Conservation Letters*, vol 3, no 1, pp. 53–61

Harvey, C. A., Komar, O., Chazdon, R., Ferguson, B. G., Fynegan, B., Griffith, D. M., Martinez-Ramos, M., Morales, H., Nigh, R., Soto-Pinto, L., van Breugel, M., and Wishniw, M. (2008) 'Integrating agricultural landscapes with biodiversity conservation in the MesoAmerican hotspot', *Conservation Biology*, vol 22, no 1, pp. 8–15

Harvey, D. (2005) *A Brief History of Neoliberalism*, Oxford University Press, Oxford

Hayes, R. C., Newell, M. T., DeHaan, L. R., Murphy, K. M., Crane, S., Norton, M. R., and Cox, T. S. (2012) 'Perennial cereal crops: An initial evaluation of wheat derivatives', *Field Crops Research*, vol 133, pp. 68–89

Hayes, T. B., Case, P., Chui, S., Chung, D., Haefele, C., Haston, K., Lee, M., Mai, V. P., Marjuoa, Y., Parker, J. and Tsui, M. (2005) 'Pesticide mixtures, endocrine disruption, and amphibian declines: Are we underestimating the impact?', *Environmental Health Perspectives*, vol 114 (Suppl. 1), pp. 40–50

Hecht, S. B. (2010) 'The new rurality: Globalization, peasants and the paradoxes of landscapes', *Land Use Policy*, vol 27, no 2, pp. 161–169

Hecht, S. B. (2012) 'From eco-catastrophe to zero deforestation? Interdisciplinarities, politics, environmentalisms and reduced clearing in Amazonia', *Environmental Conservation*, vol 39, no 1, pp. 4–19

Hecht, S. B. (2014) 'Forests lost and found in tropical Latin America: the woodland "green revolution"', *Journal of Peasant Studies*, vol 41, no 5, pp. 877–909

Hecht, S. B. and Cockburn, A. (1989) The Fate of the Forest: Developers, Destroyers, and Defenders of the Amazon, Verso, London

Hecht, S. B., and Saatchi, S. S. (2007) 'Globalization and forest resurgence: changes in forest cover in El Salvador'. BioScience, vol 57, no 8, pp. 663–672

Hecht, S. B., Morrison, K., and Padoch, C. (eds) (2014) The Social Lives of Forests: Past, Present, and Future of Woodland Resurgence, University of Chicago Press: Chicago.

Heckenberger, M. J., Russell, J. C., Fausto, C., Toney, J. R., Schmidt, M. J., Pereira, E., Franchetto, B., and Kuikuro, A. (2008) 'Pre-Columbian urbanism, anthropogenic landscapes, and the future of the Amazon', Science, vol 321(5893), pp. 1214–1217

Heckman, J. (2006) 'A history of organic farming: Transition from Sir Albert Howard's war in the soil to USDA National Organic Program', Renewable Agriculture and Food Systems, vol 21, pp. 143–150

Heinrich. B. (1975) 'Energetics of pollination', Annual Review of Ecology and Systematics, vol 6, pp. 139–170

Heller, M. C. and Keoleian, G. A. (2003) 'Assessing the sustainability of the US food system: a life cycle perspective', Agricultural Systems, vol 76, pp. 1007–1041

Hemming, J. (2004) Red Gold: The Conquest of the Brazilian Indians, Pan Macmillan, New York

Himmelberg, R. F. (2001) The Great Depression and the New Deal, Greenwood Press, Santa Barbara, CA

Hochstetler, K. and Keck, M. E. (2007) Greening Brazil: Environmental Activism in State and Society, Duke University Press, Durham, NC

Hoffmann, E. and Fox, M. (2017) 'Food sovereignty, one occupation at a time: An interview with Ester Hoffman of Brazil's Landless Workers' Movement (MST) on sustainability, agro-ecology, and the future of the hemisphere's largest social movement', NACLA Report on the Americas, vol 49, no 4, pp. 451–456

Holmes, G. (2014) 'What is a land grab? Exploring green grabs, conservation, and private protected areas in southern Chile', Journal of Peasant Studies, vol 41, no 4, pp. 547–567

Holt-Giménez, E. and Altieri, M. A. (2013) 'Agroecology, food sovereignty, and the new green revolution', Agroecology and Sustainable Food Systems, vol 37, no 1, pp. 90–102

Holyoak, M. and Lawler, S. P. (1996) 'Persistence of an extinction-prone predator-prey interaction through metapopulation dynamics', Ecology, vol 77, no 6, pp. 1867–1879

Howard, A. (1940) An Agricultural Testament, Oxford University Press, Oxford

Howard, P. H. (2009) 'Consolidation in the North American organic food processing sector, 1997 to 2007', International Journal of Sociology of Agriculture & Food, vol 16, no 1, pp. 13–30

Hsieh, H. Y., Liere, H., Soto, E. J., and Perfecto, I. (2012) 'Cascading trait-mediated interactions induced by ant pheromones', Ecology and Evolution, vol 2, no 9, pp. 2181–2191

Hubbell, S. P. (2001) The Unified Neutral Theory of Biodiversity and Biogeography, Princeton University Press, Princeton

Hufbauer, R. A., Szücs, M., Kasyon, E., Youngberg, C., Koontz, M. J., Richards, C., Tuff, T., and Melbourne, B. A. (2015) 'Three types of rescue can avert extinction in a changing environment', Proceedings of the National Academy of Sciences, USA, vol 112, no 33, pp. 10557–10562

Hughes, J. D. (2001) An Environmental History of the World: Humankind's Role in the Community of Life, Routledge, London

Hughes, R. and Flintan, F. (2001) *Integrating conservation and development experience: a review and bibliography of the ICDP literature*, International Institute for Environment and Development, London

Hulme, M. (2009) *Why We Disagree About Climate Change: Understanding Controversy, Inaction and Opportunity*, Cambridge University Press, Cambridge, MA

Hulme, M. F., Vickery, J. A., Green, R. E., Phalan, B., Chamberlain, D. E., Pomeroy, D. E., Nalwanga, D., Mushabe, D., Katebaka, R., Bolwig, S., and Atkinson, P. W. (2013) 'Conserving the birds of Uganda's banana-coffee arc: land sparing and land sharing compared', *PloS One*, vol 8, no 2, p. e54597

Hutchinson, G. E. (1961) 'The paradox of the plankton', *American Naturalist*, vol 95, no 882, pp. 137–145

Hyvönen, T., Ketoja, E., Salonen, J., Jalli, H., and Tiainen, J. (2003) 'Weed species diversity and community composition in organic and conventional cropping of spring cereals', *Agriculture, Ecosystems and Environment*, vol 97, no 1–3, pp. 131–149

IAASTD (2009a) *International Assessment of Agricultural Knowledge, Science and Technology for Development: Summary for Decision Makers: Latin America and Caribbean Report*, Island Press, Washington, DC

IAASTD (2009b) *International Assessment of Agricultural Knowledge, Science and Technology for Development: The Synthesis Report*, B. D. McIntyre, H. R. Herren, J. Wakhungu, and R. T. Watson (eds), Island Press, Washington, DC

Ibañez, M., and Blackman, A. (2016) 'Is eco-certification a win–win for developing country agriculture? Organic coffee certification in Colombia', *World Development*, vol 82, pp. 14–27

Ingvarsson, P. K. (2002) 'Conservation biology: lone wolf to the rescue', *Nature*, vol 420, no 6915, p. 472

IRRI (2004) *World Rice Statistics*, available at www.irri.org/science/ricestat/index.asp, accessed 22 March 2009

Jackson, W. (1980) *New Roots for Agriculture (Farming and Ranching)*, University of Nebraska Press, Lincoln, NE

Jackson, W. (1985) *New Roots for Agriculture*, University of Nebraska Press, Lincoln, NE

Jackson, W. (1996) *Becoming Native to this Place*, Counterpoint, New York

Jacobsen, T. and Adams, R. M. (1958) 'Salt and silt in ancient Mesopotamian agriculture', *Science*, vol 128, pp. 1251–1258

Jaikumar, N. S., Snapp, S. S., Murphy, K., and Jones, S. S. (2012) 'Agronomic assessment of perennial wheat and perennial rye as cereal crops', *Agronomy Journal*, vol 104, no v6, pp. 1716–1726

Jha, S. and Dick, C. W. (2008) 'Shade coffee farms promote genetic diversity of native trees', *Current Biology*, vol 18, no 24, pp. R1126–R1128

Jha, S. and Dick, C. W. (2010) 'Native bees mediate long-distance pollen dispersal in a shade coffee landscape mosaic', *Proceedings of the National Academy of Sciences, USA*, vol 107, no 31, pp. 13760–13764

Jha, S., Allen, D., Liere, H., Perfecto, I., and Vandermeer, J. (2012) 'Mutualisms and population regulation: mechanism matters', *PLoS One*, vol 7, no 8, p. e43510

Jha, S., Bacon, C. M., Philpott, S. M., Ernesto Mendez, V., Läderach, P., and Rice, R. A. (2014) 'Shade coffee: update on a disappearing refuge for biodiversity', *BioScience*, vol 64, no 5, pp. 416–428

Jha, S., Vandermeer, J. H., and Perfecto, I. (2009) 'Population dynamics of Coccus viridis, a ubiquitous ant-tended agricultural pest, assessed by a new photographic method', *Bulletin of Insectology*, vol 62, no 2, pp. 183–189

Johns, N. D. (1999) 'Conservation in Brazil's chocolate forest: the unlikely persistence of the traditional cocoa agroecosystem', *Environmental Management*, vol 23, no 1, pp. 23–47

Johnson, C. (2004) *Blowback: The Costs and Consequences of America's Empire*, Holt, New York

Johnston, B. R. (1994) *Who Pays the Price: The Sociocultural Context of Environmental Crisis*, Island Press, Washington, DC

Jones, K. (2004) *Who's Afraid of the WTO?*, Oxford University Press, Oxford

Jonsen, I. D., Bourchier, S. R., and Roland, J. (2001) 'The influence of matrix habitat on *Aphthona* flea beetle immigration to leafy spurge patches', *Oecologia*, vol 127, pp. 287–294

Joppa, L. N., Loarie, S. R., and Pimm, S. L. (2008) 'On the protection of "protected areas"', *Proceedings of the National Academy of Sciences, USA*, vol 105, no 18, pp. 6673–6678

Kaimowitz, D. (1996) *Livestock and Deforestation in Central America in the 1980s and 1990s: A Policy Perspective*, CIFOR Special Publication, Center for International Forestry Research, Jakarta

Kaimowitz, D. and Smith, J. (2001) 'Soybean technology and the loss of natural vegetation in Brazil and Bolivia', in A. Angelsen and D. Kaimowitz (eds) *Agricultural Technologies and Tropical Deforestation*, CAB International, Wallingford, pp. 195–211

Karp, D. S. and Daily, G. C. (2014) 'Cascading effects of insectivorous birds and bats in tropical coffee plantations', *Ecology*, vol 95, no 4, pp. 1065–1074

Karstensen, J., Fiedler, B., Schütte, F., Brandt, P., Körtzinger, A., Fischer, G., Zantopp, R., Hahn, J., Visbeck, M., and Wallace, D. W. (2015) 'Open ocean dead zones in the tropical North Atlantic Ocean', *Biogeosciences (BG)*, vol 12, pp. 2597–2605

Katayama, N., Baba, Y. G., Kusumoto, Y., and Tanaka, K. (2015) 'A review of postwar changes in rice farming and biodiversity in Japan', *Agricultural Systems*, vol 132, pp. 73–84

Keay, J. (1991) *The Honourable Company: A History of the English East India Company*, MacMillan, New York

Khadse, A., Rosset, P. M., Morales, H., and Ferguson, B. G. (2018) 'Taking agroecology to scale: The zero budget natural farming peasant movement in Karnataka, India', *Journal of Peasant Studies*, vol 45, no 1, pp. 192–219

Kirchner, F., Ferdy, J. B., Andalo, C., Colas, B., and Moret, J. (2003) 'Role of corridors in plant dispersal: and example with the endangered *Ranunculus nodif lorus*', *Conservation Biology*, vol 17, pp. 401–410

Klein, N. (2008) *The Shock Doctrine: The Rise of Disaster Capitalism*, Metropolitan Books, Macmillan, New York

Kleinman, P. J. A., Pimentel, D., and Bryant, R. B. (1995) 'The ecological sustainability of slash-and-burn agriculture', *Agriculture, Ecosystems and Environment*, vol 52, no 2–3, pp. 235–249

Knapp, R. A. (1996) 'Non-native trout in Natural Lakes of the Sierra Nevada: an analysis of their distribution and impacts on native biota', *Sierra Nevada Ecosystem Project: Final Report to Congress, V.III*, University of California, Davis, Center for Water and Wildland Resources

Koh, L. P., Miettinen, J., Liew, S. C., and Ghazoul, J. (2011) 'Remotely sensed evidence of tropical peatland conversion to oil palm', *Proceedings of the National Academy of Sciences, USA*, p. e201018776.

Kolbert, E. (2014) *The Sixth Extinction: An Unnatural History*, A & C Black, Edinburgh

Koohafkan, P. and Altieri, M. A. (2011) *Globally Important Agricultural Heritage Systems: A Legacy for the Future*, Rome: Food and Agriculture Organization of the United Nations

Krauss, J., Gallenberger, I., and Steffan-Dewenter, I. (2011) 'Decreased functional diversity and biological pest control in conventional compared to organic crop fields', *Plos One*, vol 6, no 5, p. e19502

Kreft, H., Jetz, W., Mutke, J., Kier, G., and Barthlott, W. (2008) 'Global diversity of island floras from a macroecological perspective', *Ecology Letters*, vol 11 no 2, pp. 116–127

Kremen, C. (2015) 'Reframing the land-sparing/land-sharing debate for biodiversity conservation', *Annals of the New York Academy of Sciences*, vol 1355, no 1, pp. 52–76

Kremen, C., Merenlender, A. M., and Murphy, D. D. (1994) 'Ecological monitoring: a vital need for integrated conservation and development programs in the tropics', *Conservation Biology*, vol 8, no 2, pp. 388–397

Kremen, C., Williams, N. M., and Thorp, R. W. (2002) 'Crop pollination from native bees at risk from agricultural intensification', *Proceedings of the National Academy of Sciences, USA*, vol 99, no 26, pp. 16812–16816

Kuussaari, M., Nieminen, M., and Hanski, I. (1996) 'An experimental study of migration in the Glanville fritillary butterfly *Melitaea cinxia*', *Journal of Animal Ecology*, vol 65, pp. 791–801

Lacey, L. A. and Lacey, C. M. (1990) 'The medical importance of land rice mosquitoes and their control using alternatives to chemical insecticides', *Journal of the American Mosquito Control Association*, vol 6 (Supplement), pp. 1–93

Laliberte, E., Wells, J. A., DeClerck, F., Metcalfe, D. J., Catterall, C. P., Queiroz, C., Aubin, I., Bonser, S. P., Ding, Y., Fraterrigo, J. M., and McNamara, S. (2010) 'Land-use intensification reduces functional redundancy and response diversity in plant communities', *Ecology Letters*, vol 13, no 1, pp. 76–86

Landell Mills, N. (1992) 'Organic farming in seven European countries', Report for the ECPA (European Crop Protection Association), Brussels

Lappé, F. M., Collins, J., Rosset, P., and Esparza, L. (1998) *World Hunger: 12 Myths*. Grove Press, New York, N.Y.

Latour, B. (2004) *Politics of Nature*, Harvard University Press, Cambridge, MA

Latour, B. (2012) *We Have Never Been Modern*, Harvard University Press, Cambridge, MA

Latta, A. and Wittman, H. (eds) (2012) *Environment and Citizenship in Latin America: Natures, Subjects and Struggles*, CEDLA Series, Berghahn Books: Oxford, New York

Laurance, W. F., Camargo, J. L., Luizão, R. C., Laurance, S. G., Pimm, S. L., Bruna, E. M., Stouffer, P. C., Williamson, G. B., Benítez-Malvido, J., Vasconcelos, H. L., Van Houtan. S., Zartman, C. E., Boyle, S. A., Didham, R. K., Andrade, A., Lovejoy, T. E. (2011) 'The fate of Amazonian forest fragments: a 32-year investigation', *Biological Conservation*, vol 144, no 1, pp. 56–67

Lawler, S. P. (2001) 'Rice fields as temporary wetlands: a review', *Israel Journal of Zoology*, vol 47, no 4, pp. 513–528

Le-Pelley, R. H. (1973) 'Coffee insects', *Annual Review of Entomology*, vol 18, pp. 121–142

Lecointre, G. and Le Guyader, H. (2006) *The Tree of Life: A Phylogenetic Classification*, Belknap Press of Harvard University Press, Cambridge, MA

Lee, D. R. and Barrett, B. (eds) (2001) *Tradeoffs or Synergies? Agricultural Intensification, Economic Development and the Environment*, CAB International, Wallingford

Leopold, A. 1949. *A Sand County Almanac*. Oxford University Press, Oxford

Leslie, J. (2010) 'How gross is my valley', *The New Republic*, 30 June 2010: available at https://newrepublic.com/article/75946/how-gross-my-valley, accessed September 2018

Letourneau, D. K., Jedlicka, J. A., Bothwell, S. G., and Moreno, C. R. (2009) 'Effects of natural enemy biodiversity on the suppression of arthropod herbivores in terrestrial ecosystems', *Annual Review of Ecology, Evolution, and Systematics*, vol 40, pp. 573–592

Levins, R. (1969) 'Some demographic and genetic consequences of environmental heterogeneity for biological control', *Bulletin of the Entomological Society of America*, vol 15, pp. 237–240

Levins, R. (1990) 'The struggle for ecological agriculture in Cuba', *Capitalism, Nature and Socialism*, vol 16, no 3, pp. 7–25

Levis, C., de Souza, P. F., Schietti, J., Emilio, T., Pinto, J., Clement, C. R., and Costa, F. R. C. (2012) 'Historical human footprint on modern tree species composition in the Purus-Madeira interfluve, Central Amazonia', *Plos One*, vol 7, no 11, p. e48559, available at doi: https://doi.org/10.1371/journal.pone.0048559, accessed 14 March 2019

Lewontin, R. (1982) 'Agricultural research and the penetration of capital', *Science for the People*, vol 14, no 1, January–February, pp. 12–17

Lewontin, R. (1998) 'The maturing of capitalist agriculture: farmer as proletarian', *Monthly Review*, vol 50, no 3, pp. 72–85

Lewontin, R. and Berlan, J. P. (1986) 'Technology research and the penetration of capital: The case of US agriculture', *Monthly Review*, vol 38, pp. 21–35

Liere, H., and Larsen, A. (2010) 'Cascading trait-mediation: disruption of a trait-mediated mutualism by parasite-induced behavioral modification', *Oikos*, vol 119, no 9, pp. 1394–1400

Liere, H. and Perfecto, I. (2008) 'Cheating in a mutualism: Indirect benefits of ant attendance to a coccidophagous coccinelid', *Ecological Entomology*, vol 37, pp. 143–149

Lin, B. B., Chappell, M. J., Vandermeer, J., Smith, G. R., Quintero, E., Bezner-Kerr, R., Griffith, D., Ketchum, S., Latta, S., McMichael, P., McGuire, K., Nigh, R., Rocheleau, D., Soluri, J., and Perfecto, I. (2011) 'Effects of industrial agriculture on global warming and the potential of small-scale agroecological farming to mitigate those effects', *CAB Reviews: Perspectives in Agriculture, Veterinary Science, Nutrition and Natural Resources*, vol 6, no 20, pp. 1–18

Liu, W. Y., Wang, C. Y., Wang, T. S., Fellers, G. M., Lai, B. C., and Kam, Y. C. (2011) 'Impacts of the herbicide butachlor on the larvae of a paddy field breeding frog (*Fejervarya limnocharis*) in subtropical Taiwan', *Ecotoxicology*, vol 20, no 2, pp. 377–384

Lockeretz, W., Shearer, G., and Kohl, D. H. (1981) 'Organic farming in the corn belt', *Science*, vol 211, pp. 540–547

Loehle, E. and Eschenback, W. (2012) 'Historical bird and terrestrial mammal extinction rates and causes', *Diversity and Distributions*, vol 18, no 1, pp. 84–91

López-Carr, D. and Burgdorfer, J. (2013) 'Deforestation drivers: population, migration, and tropical land use', *Environment: Science and Policy for Sustainable Development*, vol 55, no 1, pp. 3–11

Luo, A. (2005) 'Traditional use and availability of aquatic biodiversity in rice-based ecosystems. II. Xishuangbanna, Yunnan province, P.R. China', J. Margraf (guest ed.) *Aquatic Biodiversity in Rice-Based Ecosystems, Studies and Reports from Cambodia, China, Lao People's Democratic Republic and Vietnam* [CD-ROM], FAO, Rome, available at www.researchgate.net/publication/265106098_Traditional_use_and_availability_of_aquatic_biodiversity_in_rice-based_ecosystems, accessed 13 March 2019

Luo, Y., Fu, H., and Traore, S. (2014) 'Biodiversity conservation in rice paddies in China: toward ecological sustainability', *Sustainability*, vol 6, no 9, pp. 6107–6124

Maas, B., Karp, D. S., Bumrungsri, S., Darras, K., Gonthier, D., Huang, J. C. C., Lindell, C. A., Maine J. J., Mestre, L., Michel, N. L., Morrison, E. B., Perfecto, I., Philpott S. M. , Şekercioğlu Ç. H., Silva R. M., Taylor, P. J., Tscharntke, T., Van Bael, S. A., Whelan C. J., Williams-Guillén, K., and Morrison, E. B. (2016) 'Bird and bat predation services in tropical forests and agroforestry landscapes', *Biological Reviews*, vol 91, no 4, pp. 1081–1101

Maathai, W. (2006) *Unbowed: A Memoir*, Knopf, New York

MacArthur, R. H. and Wilson, E. O. (2001) *The Theory of Island Biogeography*, Princeton University Press, Princeton, NJ

MacCleery, D. (2008) 'Re-inventing the USFS: Evolution from Custodial Management to Production Forestry to Ecosystem Management', in *Re-inventing Forestry Agencies: Experiences of institutional restructuring in Asia and the Pacific.*, United Nations Food and Agriculture Organization, RAP Publication 2008/5: 2008

MacPhee, R. D. and Sues, H. D. (eds) (2013) *Extinctions in Near Time: Causes, Contexts, and Consequences* (vol 2), Springer Science, Berlin

Mäder, P., Fliebach, A., Dubois, D., Gunst, L., Fried, P., and Niggli, U. (2002) 'Soil fertility and biodiversity in organic farming', *Science*, vol 296, pp. 1694–1697

Maezumi, S. Y., Alves, D., Robinson, M., de Souza, J. G., Levis, C., Barnett, R. L., de Oliveira, E. A., Urrego, D., Schaan, D., and Iriarte, J. (2018) 'The legacy of 4,500 years of polyculture agroforestry in the eastern Amazon', *Nature Plants*, vol 4, no 8, pp. 540–547

Mann, C. C. (2005) *1491: New Revelations of the Americas before Columbus*, Knopf, New York

Mann, C. C. (2008) 'Ancient earthmovers of the Amazon', *Science*, vol 321, no 5893, pp. 1448–1452

Mann, C. C. (2011). *1493: Uncovering the New World Columbus Created*, Vintage, New York

Mann, C. C. and Plummer, M. L. (1993) 'The high cost of biodiversity', *Science*, vol 260, pp. 1868–1871

Marquardt, K., Vico, G., Glynn, C., Weih, M., Eksvärd, K., Dalin, P., and Björkman, C. (2016) 'Farmer perspectives on introducing perennial cereal in Swedish farming systems: a sustainability analysis of plant traits, farm management, and ecological implications', *Agroecology and Sustainable Food Systems*, vol 40, no 5, pp. 432–450

Martinez-Reyes, J. E. (2016) *Moral Economy of a Forest: The Nature Industry and Maya Post-Conservation*, University of Arizona Press: Tucson

Martínez-Torres, M. E. (2006) *Organic Coffee: Sustainable Development by Mayan Farmers*, Ohio University Press, Colombus, OH

Martínez-Torres, M. E. (2008) 'The benefits and sustainability of organic farming by peasant coffee farmers in Chiapas, Mexico', in J. A. Fox (ed.) *Confronting the Coffee crisis: Fair Trade, Sustainable Livelihoods, and Ecosystems in Mexico and Central America* MIT Press, Cambridge, MA, pp. 99–126.

Martínez-Torres, M. E. and Rosset, P. M. (2014) 'Diálogo de saberes in La Vía Campesina: food sovereignty and agroecology', *Journal of Peasant Studies*, vol 41, no 6, pp. 979–997

Mather, A. S. (1992) 'The forest transition', *Area*, vol 24, no 4, pp. 367–379

Mather, A. S. and Needle, C. L. (1998) 'The forest transition: a theoretical basis', *Area*, vol 30, no 2, pp. 117–124

Matson, P. A. and Vitousek, P. M. (2006) 'Agricultural intensification: Will land spared from farming be spared for nature?', *Conservation Biology*, vol 20, pp. 709–710

Matson, P. A., Parton, W. J., Power, A. G., and Swift, M. J. (1997) 'Agricultural intensification and ecosystem properties', *Science*, vol 277, pp. 504–509

Matteucci, S. D., Totino, M., and Arístide, P. (2016) 'Ecological and social consequences of the Forest Transition Theory as applied to the Argentinean Great Chaco', *Land Use Policy*, vol 51, pp. 8–17

May, R. M. (1988) 'How many species are there on Earth?', *Science*, vol 241, pp. 1441–1449

McNeely, J. A. and Scherr, S. J. (2003) *Ecoagriculture: Strategies to Feed the World and Save Biodiversity*, Island Press, Washington, DC

McNeill, W. (1977) *Plagues and Peoples*, Anchor, New York

McShane, T. O., and Wells, M. P. (eds) (2004) *Getting Biodiversity Projects to Work: Towards more Effective Conservation and Development*, Columbia University Press

Meek, D. (2014) 'Agroecology and radical grassroots movements' evolving moral economies', *Environment and Society*, vol 5, no 1, pp. 47–65

Meek, D. (2016) 'The cultural politics of the agroecological transition', *Agriculture and Human Values*, vol 33, no 2, pp. 275–290

Meek, D. and Simonian, L. T. (2017) 'Transforming space and society? The political ecology of education in the Brazilian Landless Workers' Movement's Jornada de Agro-ecologia', *Environment and Planning D: Society and Space*, vol 35, no 3, pp. 513–532

Meijen V. A. (1940) *Fish Culture in Rice Fields*, Food Industry Publishers, Moscow, (Russian). English translation: Fernando, C. H. (ed.) (1993), SUNY, Geneseo, New York

Melo, F. P., Arroyo-Rodríguez, V., Fahrig, L., Martínez-Ramos, M., and Tabarelli, M. (2013) 'On the hope for biodiversity-friendly tropical landscapes', *Trends in Ecology & Evolution*, vol 28, no 8, pp. 462–468.

Melville, E. G. (1994) *A Plague of Sheep: Environmental Consequences of the Conquest of Mexico*, Cambridge University Press, Cambridge

Méndez, V. E., Bacon, C. M., Cohen, R., and Gliessman, S. R. (2015) *Agroecology: A Transdisciplinary, Participatory and Action-oriented Approach*, CRC Press, Boca Ratón, FL

Méndez, V. E., Bacon, C. M., Olson, M., Morris, K. S., and Shattuck, A. (2010) 'Agrobiodiversity and shade coffee smallholder livelihoods: a review and synthesis of ten years of research in Central America', *The Professional Geographer*, vol 62, no 3, pp. 357–376

Menezes, F. (2001) 'Food Sovereignty: A vital requirement for food security in the context of globalization', *Development*, vol 44, pp. 29–30

Menon, S. (2006) 'Fearful symmetry. Essay (Endpaper)', *Natural History*, December 2006/January 2007, p. 72.

Meyfroidt, P. and Lambin, E. F. (2009) 'Forest transition in Vietnam and displacement of deforestation abroad', *Proceedings of the National Academy of Sciences, USA*, vol 106, no 38, pp. 16139–16144

Miller, S. W. (2007) *An Environmental History of Latin America*, Cambridge University Press, Cambridge

Mittermeier, R. A., Myers, N., Thomsen, J. E., and Olivieri, S. (1998) 'Biodiversity hotspots and major wilderness areas: Approaches to setting conservation priorities', *Conservation Biology*, vol 12, no 3, pp. 516–520

Moguel, P. and Toledo, V. M. (1999) 'Biodiversity conservation in traditional coffee systems of Mexico', *Conservation Biology*, vol 12, pp. 1–11

Mongabay (2018) 'Amazon deforestation at high levels in ten years, says Brazil', Mongabay.com, 24 November, available at https://news.mongabay.com/2018/11/amazon-deforestation-at-highest-level-in-10-years-says-brazil/, accessed 18 December 2018

Moody K. (1989) *Weeds Reported in Rice in South and Southeast Asia*, IRRI, Los Baños, Philippines

Morales, H. and Perfecto, I. (2000) 'Traditional knowledge and pest managements in the Guatemalan highlands', *Agriculture and Human Values*, vol 17, no 1, pp. 49–63

Moran, E. (1996) 'Deforestation in the Brazilian Amazon', in L. E. Sponsel, T. N. Headland, and R. C. Bailey (eds) *Tropical Deforestation: The Human Dimension*, Columbia University Press, New York, pp. 149–164

Moran, E., Brondizio, E. S., Tucker, J., da Silva-Forsberg, M. C., Falesi, I. C., and McCracken, S. (2000) 'Strategies for Amazonian forest restoration: Evidence for afforestation in five regions of the Brazilian Amazon', in A. Hall (ed.) *Amazonia at the Crossroads: The Challenge of Sustainable Development*, Institute of Latin American Studies, London

Morris, J. R. and Perfecto, I. (2016) 'Testing the potential for ant predation of immature coffee berry borer (*Hypothenemus hampei*) life stages', *Agriculture, Ecosystems & Environment*, vol 233, pp. 224–228

Morris, J. R., Jimenez-Soto, E., Philpott, S. M., and Perfecto, I. (2018) 'Ant-mediated (Hymenoptera: Formicidae) biological control of the coffee berry borer: diversity, ecological complexity, and conservation biocontrol', *Myrmecological News*, vol 26, pp. 1–17

Morris, J. R., Vandermeer, J., and Perfecto, I. (2015) 'A keystone ant species provides robust biological control of the coffee berry borer under varying pest densities', *PloS One*, vol 10, no 11, p. e0142850

Morton, D. C., DeFries, R. S., Shimabukuro, Y. E., Anderson, L. O., Arai, E., del Bon Espirito-Santo, F., Freitas, R., and Morisette, J. (2006) 'Cropland expansion changes deforestation dynamics in the southern Brazilian Amazon', *Proceedings of the National Academy of Sciences, USA*, vol 103, pp. 14637–14641

Naito, R., Yamasaki, M., Imanishi, A., Natuhara, Y., and Morimoto, Y. (2012) 'Effects of water management, connectivity, and surrounding land use on habitat use by frogs in rice paddies in Japan', *Zoological Science*, vol 29, no 9, pp. 577–584

Nassauer, J. I. and Opdam, P. (2008) 'Design in science: Extending the landscape ecology paradigm', *Landscape Ecology*, vol 23, pp. 633–644

National Academy of Sciences (2018) *Progress Toward Restoring the Everglades: Seventh Biennial Review*, National Academy Press: Washington, DC

National Oceanic and Atmospheric Administration (2017) 'Gulf of Mexico "dead zone" is the largest ever measured', 2 August, available at https://coastalscience.noaa.gov/news/gulf-mexico-dead-zone-largest-ever-measured/, accessed 13 March 2019

National Research Council (1989) *Alternative Agriculture*, National Academy Press, Washington, DC

Natuhara, Y. (2013) 'Ecosystem services by paddy fields as substitutes of natural wetlands in Japan', *Ecological Engineering*, vol 56, pp. 97–106

Nayak, P. K., Nayak, A. K., Panda, B. B., Lal, B., Gautam, P., Poonam, A., Shahid, M., Tripathi, R., Kumar, U., Mohapatra, S. D., and Jambhulkar, N. N. (2018) 'Ecological mechanism and diversity in rice based integrated farming system', *Ecological Indicators*, vol 91, pp. 359–375

Neher, D. A. (1999) 'Soil community composition and ecosystem processes: Comparing agricultural ecosystems with natural ecosystems', *Agroforestry Systems*, vol 41, no 1–3, pp. 159–185

Nepstad, D., McGrath, D., Alencar, A., Barros, A. C., Carvalho, G., Santilli, M., and Vera Diaz, M. C. (2002) 'Frontier governance in Amazonia', *Science*, vol 295, pp. 630–631

Neumann, R. P. (1998) *Imposing Wilderness: Struggles over Livelihood and Nature Preservation in Africa*, University of California Press, Berkeley, CA

New York Times (2018) 'With 10 Million Acres in Patagonia, a National Park System is Born', *New York Times*, 19 February

Newmark, W. D. (1995) 'Extinction of mammal populations in Western North American national parks', *Conservation Biology*, vol 9, pp. 512–526

Niggli, U. and Lockeretz, W. (1996) 'Development of research in organic agriculture', in T. Oestergaard (ed.) *Fundamentals of Organic Agriculture*, IFOAM, Tholey-Theley, Germany, pp. 9–23

Oates, J. F. (1999) *Myth and Reality in the Rain Forest: How Conservation Strategies are Failing in West Africa*, University of California Press, Berkeley, CA

Oberlack, C., Tejada, L., Messerli, P., Rist, S., and Giger, M. (2016) 'Sustainable livelihoods in the global land rush? Archetypes of livelihood vulnerability and sustainability', *Global Environmental Change*, vol 41, no 2016, pp. 153–171

Odum, E. P. (1997) *Ecology: A Bridge Between Science and Society*, Sinauer Associates, Sunderland, MA

Ojeda, D. (2012) 'Green pretexts: Ecotourism, neoliberal conservation and land grabbing in Tayrona National Natural Park, Colombia', *Journal of Peasant Studies*, vol 39, no 2, pp. 357–375

Oliveira, G. and Hecht, S. (2016) 'Sacred groves and sacrifice zones', *Journal of Peasant Studies*, vol 43, no 2, pp. 251–285

Oliveira, G. and Hecht S. (2018) *Soy, Globalization and Environmental Politics in Latin America*, Routledge, New York

Oliveira, L. C., Hankerson, S. J., Dietz, J. M., and Raboy, B. E. (2010) 'Key tree species for the golden-headed lion tamarin and implications for shade-cocoa management in southern Bahia, Brazil', *Animal Conservation*, vol 13, no 1, pp. 60–70

Ondetti, G. (2008) *Land, Protest, and Politics: The Landless Movement and the Struggle for Agrarian Reform in Brazil*, University of Pennsylvania Press, Philadelphia, PA

Ooi, P. A. C. and Shepard, B. M. (1994) 'Predators and parasitoids of rice insects', in E. A. Heinrichs (ed.) *Biology and Management of Rice Insects*, Wiley Eastern Ltd., India and IRRI, Manila, Philippines, pp. 613–656

Opdam, P. (1991) 'Metapopulation theory and habitat fragmentation: a review of holarctic breeding bird studies', *Landscape Ecology*, vol 5, no 2, pp. 93–106

Ozorio de Almeida, A. L. (1992) *The Colonization of the Amazon*, University of Texas Press, Austin

Pacheco, P. (2009) 'Agrarian reform in the Brazilian Amazon: its implications for land distribution and deforestation', *World Development*, vol 37, no 8, pp. 1337–1347

Padoch, C. and Pinedo-Vasquez, M. (2010) 'Saving slash-and-burn to save biodiversity', *Biotropica*, vol 42, no 5, pp. 550–552

Padoch, C., Brondizio, E., Costa, S., Pinedo-Vasquez, M., Sears, R. R. and Siqueira, A. (2008) 'Urban forest and rural cities: multi-sited households, consumption patterns, and forest resources in Amazonia', *Ecology and Society*, vol 13, no 2

Pahnke, A. (2015) 'Institutionalizing economies of opposition: explaining and evaluating the success of the MST's cooperatives and agroecological repeasantization', *Journal of Peasant Studies*, vol 42, no 6, pp. 1087–1107

Palma, I. P., Toral, J. N., Vázquez, M. R. P., Fuentes, N. F., and Hernández, F. G. (2015) 'Historical changes in the process of agricultural development in Cuba', *Journal of Cleaner Production*, vol 96, pp. 77–84

Palmeirim, A. F., Benchimol, M., Vieira, M. V., and Peres, C. A. (2018) 'Small mammal responses to Amazonian forest islands are modulated by their forest dependence', *Oecologia*, vol 187, no 1, pp. 191–204

Palmer, M. W. (1994) 'Variation in species richness: Toward unification of hypotheses', *Folia Geobotanica*, vol 29, no 4, pp. 511–530

Paoletti, M. G., Pimentel, D., Stinner, B. R., and Stinnerc, D. (1992) 'Agroecosystem biodiversity: matching production and conservation biology', in M. G. Paoletti and D. Pimentel (eds) *Biotic Diversity in Agroecosystems*, Elsevier, Berlin

Parés-Ramos, I. K., Gould, W. A., and Aide, T. M. 'Agricultural abandonment, suburban growth, and forest expansion in Puerto Rico between 1991 and 2000', *Ecology and Society* vol 13, no 2

Paris, T., Singh, A., Hossain, M., and Luis, J. (2000) 'Using gender analysis in characterizing and understanding farm-household systems in rainfed lowland rice environments', in T. P. Toung, S. P. Kam, L. Wade, S. Pandey, B. A. M. Bouman, and P. Hardy (eds) *Characterizing and Understanding Rainfed Environments*, International Rice Research Institute, Los Baños, Philippines

Patel, R. (2009) 'Food sovereignty', *Journal of Peasant Studies*, vol 36, no 3, pp. 663–706

Patel, R. (2012) *Stuffed and Starved: The Hidden Battle for the World Food System*, Melville House Publications, Brooklyn, NY

Peet, J. R. (1969) 'The spatial expansion of commercial agriculture in the Nineteenth century: A VonThunen interpretation', *Economic Geography*, vol 45, pp. 283–301

Pellegrini, P. (2009) 'Knowledge, identity and ideology in stances on GMOs: The case of the Movimento Sem Terra in Brazil', *Science & Technology Studies*, available at https://sciencetechnologystudies.journal.fi/article/view/55245, accessed 13 March 2019

Pengue, W. A. (2005) 'Transgenic crops in Argentina: the ecological and social debt', *Bulletin of Science, Technology and Society*, vol 25, pp. 314–322

Perfecto, I. (1992) 'Pesticide exposure to farm workers and the international connection', in B. Bryant and P. Mohai (eds) *Race and the Incidence of Environmental Hazards*, Westview Press, Boulder, CO, pp. 177–203

Perfecto, I. and Armbrecht, I. (2003) 'The coffee agroecosystem in the Neotropics: Combining ecological and economic goals', in J. Vandermeer (ed.) *Tropical Agroecosystems*, CRC Press, Boca Ratón, FL, pp. 159–194

Perfecto, I. and Snelling, R. (1995) 'Biodiversity and tropical ecosystem transformation: ant diversity in the coffee agroecosystem in Costa Rica', *Ecological Applications*, vol 5, pp. 1084–1097

Perfecto, I. and Vandermeer, J. (2002) 'The quality of the agroecological matrix in a tropical montane landscape: ants in coffee plantations in southern Mexico', *Conservation Biology*, vol 16, pp. 174–182

Perfecto, I. and Vandermeer, J. (2006) 'The effect of an ant–hemipteran mutualism on the coffee berry borer (*Hypothenemus hampei*) in southern Mexico', *Agriculture, Ecosystems and Environment*, vol 117, pp. 218–221

Perfecto, I. and Vandermeer, J. (2008a) 'Biodiversity conservation in tropical agroecosystems: A new paradigm', *Annals of the New York Academy of Science, (The Year in Ecology and Conservation Biology 2008)*, vol 1134, pp. 173–200

Perfecto, I. and Vandermeer, J. (2008b) 'Spatial pattern and ecological process in the coffee agroecosystem', *Ecology*, vol 89, (Special Feature) pp. 915–920

Perfecto, I. and Vandermeer, J. (2010) 'The agroecological matrix as alternative to the land-sparing/agriculture intensification model', *Proceedings of the National Academy of Sciences, USA*, vol 107, pp. 5786–5791

Perfecto, I. and Vandermeer, J. (2015) *Coffee Agroecology: A New Approach to Understanding Agricultural Biodiversity, Ecosystem Services and Sustainable Development*, Routledge, Abingdon, UK

Perfecto, I. and Vandermeer, J. (2017) 'A landscape approach to integrating food production and nature conservation', in I. J. Gordon, H. T. Prins, and G. R. Squire, *Food Production and Nature Conservation: Conflicts and Solutions*, Routledge, New York, pp. 131–151

Perfecto, I., Armbrecht, I., Philpott, S. M., Soto Pinto, L., and Dietsch, T. V. (2007) 'Shade coffee and the stability of forest margins in Northern Latin America', in T. Tscharntke, M. Zeller, and C. Leuschner (eds) *The Stability of Tropical Rainforest Margins: Linking Ecological, Economic and Social Constraints*, Springer, Berlin

Perfecto, I., Hansen, P., Vandermeer, J., and Cartín, V. (1997) 'Arthropod biodiversity loss and the transformation of a tropical agro-ecosystem', *Biodiversity and Conservation*, vol 6, pp. 935–945

Perfecto, I., Mas, A., Dietsch, T. V., and Vandermeer, J. (2003) 'Species richness along an agricultural intensification gradient: A tri-taxa comparison in shade coffee in southern Mexico', *Biodiversity and Conservation*, vol 12, pp. 1239–1252

Perfecto, I., Rice, R., Greenberg, R., and Van der Voolt, M. (1996) 'Shade coffee as refuge of biodiversity', *BioScience*, vol 46, pp. 698–608

Perfecto, I., Vandermeer, J., López, G., Ibarra-Nuñez, G., Greenberg, R., Bichier, P., and Langridge, S. (2004) 'Greater predation of insect pests in a diverse agroecosystem: The role of resident Neotropical birds in shade coffee farms', *Ecology*, vol 85, pp. 2677–2681

Perfecto, I., Vandermeer, J., and Philpott, S. M. (2014) 'Complex ecological interactions in the coffee agroecosystem', *Annual Review of Ecology, Evolution, and Systematics*, vol 45, pp. 137–158

Perfecto, I., Vandermeer, J., Mas, A., and Soto Pinto, L. (2005) 'Biodiversity, yield and shade coffee certification', *Ecological Economics*, vol 54, pp. 435–446

Phalan, B., Balmford, A., Green, R. E., and Scharlemann, J. P. (2011) 'Minimising the harm to biodiversity of producing more food globally', *Food Policy*, vol 36, pp. S62–S71

Phelps, J., Carrasco, L. R., Webb, E. L., Koh, L. P., and Pascual, U. (2013) 'Agricultural intensification escalates future conservation costs', *Proceedings of the National Academy of Sciences, USA*, vol 110, no 19, pp. 7601–7606

Phillips, P. and Swanson, B. J. (2018) 'A genetic analysis of dragonfly population structure', *Ecology and Evolution*, vol 8, no 14, pp. 7206–7215

Philpott, S. M., Bichier, P., Rice, R., and Greenberg, R. (2007) 'Field-testing ecological and economic benefits of coffee certification programs', *Conservation Biology*, vol 21, no 4, pp. 975–985.

Philpott, S. M., Perfecto, I., and Vandermeer, J. (2006) 'Effect of management intensity and season on arboreal ant diversity and abundance in coffee agroecosystems', *Biodiversity and Conservation*, vol 15, pp. 139–155

Picketty, T. (translated by A. Goldhammer) (2014) *Capital in the 21st Century*, Harvard University Press, Cambridge, MA.

Pimentel, D. (1996) 'Green revolution agriculture and chemical hazards', *The Science of the Total Environment*, vol 188, Suppl. 1, pp. S86–S98

Pimentel, D. and Lehman, H. (1993) *The Pesticide Question: Environment, Economics, and Ethics*, Springer, Berlin

Pimentel, D., Greiner, A., and Bashore, T. (2003) 'Economic and environmental costs of pesticide use', *Environmental Toxicology: Current Developments*, vol 7, pp. 121–152

Pimentel, D., Harvey, C., Resosudarmo, P., Sinclair, K., Kurz, D., McNair, M., Crist, S., Shpritz, L., Fitton, L., Saffouri, R., and Blair, R. (1995) 'Environmental and economic cost of soil erosion and conservation benefits', *Science*, vol 267, pp. 1117–1123

Pleasants, J. M. and Oberhauser, K. S. (2013) 'Milkweed loss in agricultural fields because of herbicide use: effect on the monarch butterfly population', *Insect Conservation and Diversity*, vol 6, no 2, pp. 135–144

Polanyi, K. (2001) *The Great Transformation: The Political and Economic Origins of Our Time*, Beacon Press, Boston, MA

Ponisio, L. C., M'Gonigle, L. K., Mace, K. C., Palomino, J., de Valpine, P., and Kremen, C. (2015) 'Diversification practices reduce organic to conventional yield gap', *Proceeding of the Royal Society of London B*, vol 282, p. 20141396.

Power, A. G. (2010) 'Ecosystem services and agriculture: tradeoffs and synergies', *Philosophical Transactions of the Royal Society of London B: Biological Sciences*, vol 365, no 1554, pp. 2959–2971

Prakash, O. (1998) *European Commercial Enterprise in Pre-Colonial India*, Cambridge University Press, Cambridge

Pretty, J. and Smith, D. (2004) 'Social capital in biodiversity conservation', *Conservation Biology*, vol 18, pp. 631–638

Pretty, J. N., Morison, J. I. L., and Hine, R. E. (2003) 'Reducing food poverty by increasing agricultural sustainability in developing countries', *Agriculture, Ecosystems and Environment*, vol 95, pp. 217–234

Prime, R. (2002) *Vedic Ecology: Practical Wisdom for Surviving the 21st Century*, Springer Science & Business, Berlin.

Pyne, S. J. (1982) *Fire in America: A Cultural History of Wildland and Rural Fire*, Princeton University Press, Princeton, NJ

Rabinowitz, A. (1999) 'Nature's last bastions: Sustainable use of our tropical forests may be wishful thinking', *Natural History*, vol 108, pp. 70–72

Raynolds, L. T. (2002) 'Consumer/producer links in fair trade coffee networks', *Sociologia Ruralis*, vol 42, no 4, pp. 404–424

Raynolds, L. T., Murray, D., and Heller, A. (2007) 'Regulating sustainability in the coffee sector: A comparative analysis of third-party environmental and social certification initiatives', *Agriculture and Human Values*, vol 24, no 2, pp. 147–163

Reardon, T. Timmer, C. P. Berrett, C. B., and Berdegue, J. (2003) 'The rise of supermarkets in Africa, Asia and Latin America', *American Journal of Agricultural Economics*, vol 85, pp. 1140–1146

Redford, K. H., Padoch, C., and Sunderland, T. (2013) 'Fads, funding, and forgetting in three decades of conservation', *Conservation Biology*, vol 27, no 3, pp. 437–438

Redo, D. J., Grau, H. R., Aide, T. M., and Clark, M. L. (2012) 'Asymmetric forest transition driven by the interaction of socioeconomic development and environmental heterogeneity in Central America', *Proceedings of the National Academy of Sciences, USA*, vol 109, no 23, pp. 8839–8844

Relinger, R. (2010) 'NAFTA and US corn subsidies: Explaining the displacement of Mexico's corn farmers', *Prospect: Journal of international affairs at UCSD*. available at https://prospectjournal.org/2010/04/19/nafta-and-u-s-corn-subsidies-explaining-the-displacement-of-mexicos-corn-farmers/comment-page-1/, accessed 18 December 2018

Relyea, R. A. (2005a) 'The impact of insecticides and herbicides on the biodiversity and productivity of aquatic communities', *Ecological Applications*, vol 15, pp. 618–627

Relyea, R. A. (2005b) 'The lethal impact of Roundup on aquatic and terrestrial amphibians', *Ecological Applications*, vol 15, pp. 1118–1124

Revkin, A. (2004) *The Burning Season: The Murder of Chico Mendes and the Fight for the Amazon Forest*, Island Press, Washington, DC

Reyes-García, V., Aceituno-Mata, L., Calvet-Mir, L., Garnatje, T., Gómez-Baggethun, E., Lastra, J. J., Ontillera, R., Parada, M., Rigat, M., Vallès, J., and Vila, S. (2014) 'Resilience of traditional knowledge systems: the case of agricultural knowledge in home gardens of the Iberian Peninsula', *Global Environmental Change*, vol 24, pp. 223–231

Ricciardi, V., Ramankutty, N., Mehrabi, Z., Jarvis, L., and Chookolingo, B. (2018) 'How much of the world's food do smallholders produce?', *Global Food Security*, vol 17, pp. 64–72.

Rice, R. A. (1999) 'A place unbecoming: The coffee farm of northern Latin America', *Geographical Review*, vol 89

Rice, R. A. (2011) 'Fruits from shade trees in coffee: how important are they?', *Agroforestry Systems*, vol 83, no 1, pp. 41–49.

Rice, R. A. and Greenberg, R. (2000) 'Cacao cultivation and the conservation of biological diversity', *Ambio*, vol 29, no 3, pp. 167–173

Rich, B. (1994) *Mortgaging the Earth: The World Bank, Environmental Impoverishment and the Crisis of Development*, Beacon, Boston, MA

Richards, J. F. (2003) *The Unending Frontier: An Environmental History of the Early Modern World*, University of California Press, Berkeley, CA

Ricketts, T. H. (2001) 'The matrix matters: Effective isolation in fragmented landscapes', *American Naturalist*, vol 158, pp. 87–99

Ricketts, T. H., Soares-Filho, B. S., da Fonseca, G. A. B.. Nepstad, D., Pfaff, A., Petsonk, A., Anderson, A., Boucher, D., Cattaneo, A., Conte, M., Creighton, K., Linden, L., Maretti, C., Moutinho, P., Ullman, R., and Victurine, R. (2010) 'Indigenous lands, protected areas, and slowing climate change', *Plos Biology*, vol 8, no 3, p. e1000331

Roberts, P. (2008) *The End of Food*, Houghton Mifflin, Boston, MA

Rocheleau, D. E. (2015) 'Networked, rooted and territorial: Green grabbing and resistance in Chiapas', *Journal of Peasant Studies*, vol 42, no 3–4, pp. 695–723

Rojas Rabiela, T. (1983) *Agricultura chinampera: Compilación histórica*, Universidad Autónoma Chapingo, Chapingo, Mexico

Rojas Rabiela, T. (ed.) (1994) *Agricultura Indígena: pasado y presente*, CIESA, Mexico DF

Röling, N. and Van De Fliert, E. (1994) 'Transforming extension for sustainable agriculture: the case of integrated pest management in rice in Indonesia', *Agriculture and Human Values*, vol 11, no 2–3, pp. 96–108

Roos, N., Leth, T., Jakobsen, J., and Thilsted, S. H. (2002) 'High vitamin A content in some small indigenous fish species in Bangladesh: perspectives for food-based strategies to reduce vitamin A deficiency', *International Journal of Food Sciences and Nutrition*, vol 53, pp. 425–437

Rosset, P. M. (1999) 'The multiple functions and benefits of small farm agriculture', *Food First Policy Brief* 4

Rosset, P. M. (2006) *Food Is Different: Why We Must Get the WTO Out of Agriculture*, Zed Books, London

Rosset, P. M. and Altieri, M. A. (2017) *Agroecology: Science and Politics*, Practical Action Publishing, Rugby, Warwickshire, UK

Rosset, P. M. and Martínez-Torres, M. E. (2012) 'Rural social movements and agroecology: context, theory, and process', *Ecology and Society*, vol 17, no 3

Rosset, P. M., Machín Sosa, B., Roque Jaime, A. M., and Ávila Lozano, D. R. (2011) 'The Campesino-to-Campesino agroecology movement of ANAP in Cuba: social process methodology in the construction of sustainable peasant agriculture and food sovereignty', *Journal of Peasant Studies*, vol 38(1), pp. 161–191

Roth, D. S., Perfecto, I., and Rathke, B. (1994) 'The effects of management systems on ground-foraging ant diversity in Costa Rica', *Ecological Applications*, vol 4, pp. 423–436

Rothenberg, D. and Ulvaeus, M. (eds) (2001) *The World and the Wild: Expanding Wilderness beyond its American Roots*, University of Arizona Press, Tucson, AZ

Rozilah, I. and Ali, A. B. (1998) 'Aquatic insect populations in the Muda rice agroecosystem', in B. N. Nashriya, N. K. Ho, B. S. Ismail, B. A. Ahyaudin, and K. Y. Lum (eds) *Rice Agroecosystem of the Muda Irrigation Scheme, Malaysia*, Malaysian Institute of Nuclear Technology Research and Muda Agricultural Development Authority

Rudel, T. K., Coomes, O. T., Moran, E., Achard, F., Angelsen, A., Xu, J., and Lambin, E. (2005) 'Forest transitions: towards a global understanding of land use change', *Global Environmental Change*, vol 15, no 1, pp. 23–31

Rudel, T. K., Defries, R., Asner, G. P., and Laurance, W. F. (2009) 'Changing drivers of deforestation and new opportunities for conservation', *Conservation Biology*, vol 23, no 6, pp. 1396–1405

Rudel, T. K., Perez-Lugo, M., and Zichal, H. (2000) 'When fields revert to forest: development and spontaneous reforestation in post-war Puerto Rico', *The Professional Geographer*, vol 52, no 3, pp. 386–397

Rueda, X. and Lambin, E. F. (2013) 'Responding to globalization: impacts of certification on Colombian small-scale coffee growers', *Ecology and Society*, vol 18, no 3, p. e21

Rueda, X., Thomas, N. E., and Lambin, E. F. (2015) 'Eco-certification and coffee cultivation enhance tree cover and forest connectivity in the Colombian coffee landscapes', *Regional Environmental Change*, vol 15, no 1, pp. 25–33

Rulli, M. C., Saviori, A., and D'Odorico, P. (2013) 'Global land and water grabbing', *Proceedings of the National Academy of Sciences, USA*, vol 110, no 3, pp. 892–897

Runnels, C. N. (1995) 'Environmental degradation in Ancient Greece', *Scientific American*, vol March, pp. 72–75

Russell, E. (2001) *War and Nature: Fighting Humans and Insects with Chemicals from World War I to Silent Spring*, Cambridge University Press, Cambridge, MA

Saka, M., Tada, N., and Kamata, Y. (2013) 'Application of an amphibian (*Silurana tropicalis*) metamorphosis assay to the testing of the chronic toxicity of three rice paddy herbicides: Simetryn, mefenacet, and thiobencarb', *Ecotoxicology and Environmental Safety*, vol 92, pp. 135–143

Sambuichi, R. H., Vidal, D. B., Piasentin, F. B., Jardim, J. G., Viana, T. G., Menezes, A. A., Mello, D. L., Ahnert, D., and Baligar, V. C. (2012) 'Cabruca agroforests in southern Bahia, Brazil: tree component, management practices and tree species conservation', *Biodiversity and Conservation*, vol 21, no 4, pp. 1055–1077

Sauer, C. O. (1955) 'The agency of Man on the Earth', in W. L. Thomas (ed.) *Man's Role in Changing the Face of the Earth*, University of Chicago Press, Chicago, IL, pp. 49–69

Sauer, S., Almeida, W. (2011) *Territorios na Amazônia, Demandas, Desafios e Perspectivas*, Editorial Universitaria de Brasilia: Brasilia

Sawada, H., Subroto, S. W. G., Mustaghfirin, and Wijaya, E. S. (1991) 'Immigration, population development and outbreaks of the brown planthopper, *Nilaparvata lugens* (Stal), under different rice cultivation patterns in Central Java', *Indonesia Technical Bulletin*, no 130, Asian Pacific Food & Fertilizer Technology Center, Taipei, Republic of China

Sax, D. F. and Gaines, S. D. (2008) 'Species invasions and extinction: the future of native biodiversity on islands', *Proceedings of the National Academy of Sciences, USA*, vol 105, Suppl. 1, pp. 11490–11497

Scarano, F. R. and Ceotto, P. (2015) 'Brazilian Atlantic forest: impact, vulnerability, and adaptation to climate change', *Biodiversity and Conservation*, vol 24, no 9, pp. 2319–2331

Schelhas, J. and Sánchez-Azofeifa, G. A. (2007) 'Post-frontier change adjacent to Braulio Carrillo National Park, Costa Rica', *Human Ecology*, vol 34, pp. 407–431

Schroth, G. and Harvey, C. A. (2007) 'Biodiversity conservation in cocoa production landscapes: an overview', *Biodiversity and Conservation*, vol 16, pp. 2237–2244

Schroth, G., Krauss, U., Gasparotto, L., Duarte Aguilar, J. A., and Vohland, K. (2000) 'Pests and diseases in agroforestry systems of the humid tropics', *Agroforestry Systems*, vol 50, pp. 199–241

Schwendler, S. F. and Thompson, L. A. (2017) 'An education in gender and agroecology in Brazil's Landless Rural Workers' Movement', *Gender and Education*, vol 29, no 1, pp. 100–114

Settle, W. H., Ariawan, H., Tri Astuti, E., Cahyana, W., Luckman Hakim, A., Hindayana, D., Sri Lestari, A., and Sarnato, P. (1996) 'Managing tropical rice pests through conservation of generalist natural enemies and alternative prey', *Ecology*, vol 77, no 7, pp. 1975–1988

Seufert, V., Ramankutty, N., and Foley, J. A. (2012) 'Comparing the yields of organic and conventional agriculture', *Nature*, vol 485, no 7397, p. 229

Shanley, P., Da Silva, F. C., and MacDonald, T. (2011) 'Brazil's social movement, women and forests: a case study from the National Council of Rubber Tappers', *International Forestry Review*, vol 13, no 2, pp. 233–244

Simberloff, D. and Abele, L. G. (1982) 'Refuge design and island biogeographic theory: effects of fragmentation', *American Naturalist*, vol 120, no 1, pp. 41–50

Simberloff, D. and Cox, J. (1987) 'Consequences and costs of conservation corridors', *Conservation Biology*, vol 1, pp. 63–71

Simberloff, D., Farr, J. A., Cox, J., and Mehlman, D. W. (1992) 'Movement corridors: Conservation bargains or poor investments?', *Conservation Biology*, vol 6, pp. 493–504

Simon, C. P. and Blume, L. (1994) *Mathematics for Economists*, W. W. Norton & Company, Inc., New York

Skelly, D. K., Werner, E. E., and Cortwright, S. A. (1999) 'Long-term distributional dynamics of a Michigan amphibian assemblage', *Ecology*, vol 80, pp. 2326–2337

Sloan, S. (2015) 'The development-driven forest transition and its utility for REDD+', *Ecological Economics*, vol 116, pp. 1–11

Smil, V. (2004) *Enriching the Earth: Fritz Haber, Carl Bosch, and the Transformation of World Food Production*, MIT Press, Cambridge, MA

Smith P., Bustamante, M., Ahammad, H., Clark, H., Dong, H., Elsiddig, E. A., Haberl, H., Harper, R., House, J., Jafari, M., Masera, O., Mbow, C., Ravindranath, N. H., Rice, C. W., Robledo Abad, C., Romanovskaya, A., Sperling, F., and Tubiello, F. (2014) 'Agriculture, Forestry and Other Land Use (AFOLU)', in O. Edenhofer, R. Pichs-Madruga, Y. Sokona, E. Farahani, S. Kadner, K. Seyboth, A. Adler, I. Baum, S. Brunner, P. Eickemeier, B. Kriemann, J. Savolainen, S. Schlömer, C. von Stechow, T. Zwickel, and J. C. Minx (eds) *Climate Change 2014: Mitigation of Climate Change*, Contribution of Working Group III to the Fifth Assessment Report of the Intergovernmental Panel on Climate Change, Cambridge University Press, Cambridge, United Kingdom and New York, NY

Smith, E. A. and Oehme, E. W. (1992) 'The biological activity of glyphosate to plants and animals: A literature review', *Veterinary and Human Toxicology*, vol 34, pp. 531–543

Smith, J. D. (ed.) (2001) *Biodiversity, the Life of Cambodia: Cambodian Biodiversity Status Report*, Cambodia Biodiversity Enabling Activity, Phnom Penh, Cambodia

Smith, M. A. and Green, D. M. (2005) 'Dispersal and the metapopulation paradigm in amphibian ecology and conservation: Are all amphibian populations metapopulations?', *Ecography*, vol 28, no 1, pp. 110–128

Soares-Filho, B. S., Moutinho, P., Nepstad, D., Anderson, A., Rodrigues, H., Garcia, R., Dietzsch, L., Merry, F., Bowman, M., Hissa, L., Silvestrini R., and Maretti, C. (2010) 'Role of Brazilian Amazon protected areas in climate change mitigation', *Proceedings of the National Academy of Sciences, USA*, vol 107, no 24, pp. 10821–10826

Soares-Filho, B. S., Nepstad, D. C., Curran, L. M., Cerqueira, G. C., Garcia, R. A., Ramos, C. A., Voll, E., McDonald, A., Lefebvre, P., and Schlesinger, P. (2006) 'Modelling conservation in the Amazon basin', *Nature*, vol 440, no 7083, pp. 520–523

Soule, J. D. and Piper, J. K. (1991) *Farming in Nature's Image: An Ecological Approach to Agriculture*, Island Press, Washington, DC

Spence, M. D. (1999) *Dispossessing the Wilderness: Indian Removal and the Making of the National Parks*, Oxford University Press, Oxford

Speth, J. G. (2009) *The Bridge at the End of the World: Capitalism, Environment and Crossing from Crisis to Sustainability*, Yale University Press, New Haven, CT

Sponsel, L. E, Headland, T. N., and Bailey, R. C. (eds) (1996) *Tropical Deforestation: The Human Dimension*, Columbia University Press, New York

St. Barber Baker, R. (1985) *My Life My Trees*, Findhorn Press, Forres, Scotland

Stahler-Sholk, R. (2010) 'The Zapatista social movement: innovation and sustainability', *Alternatives*, vol 35, no 3, pp. 269–290

Stanhill, G. (1990) 'The comparative productivity of organic agriculture', *Agriculture, Ecosystems and Environment*, vol 3, pp. 1–26

Steffan-Dewenter, I. (2002) 'Landscape context affects trap-nesting bees, wasps, and their natural enemies', *Ecological Entomology*, vol 27, pp. 631–637

Steingraber. S. (ed.) (2018) *Rachel Carson: Silent Spring and Other Writings on the Environment* (LOA #307), Library of America Stiglitz, J. (2003) *Globalization and Its Discontents*, Norton, New York

Stiglitz, J. (2005) *A Chance for the World Bank*, Anthem Press, London

Stiglitz, J. (2007) *Making Globalization Work*, Norton, New York

Stork, N. E., McBroom, J., Gely, C., and Hamilton, A. J. (2015) 'New approaches narrow global species estimates for beetles, insects, and terrestrial arthropods', *Proceedings of the National Academy of Sciences, USA*, vol 112, no 24, pp. 7519–7523

Swift, M. J., Izac, A. M. N., and van Noordwijk, M. (2004) 'Biodiversity and eco-system services in agricultural landscapes – are we asking the right questions?' *Agriculture, Ecosystems and Environment*, vol 104, no 1, pp. 113–134

Swift, M., Vandermeer, J., Ramakrishan, R., Anderson, J., and Ong, C. (1995) 'Agroecosystems', in V. H. Heywood and R. T. Watson (eds) *Global Biodiversity Assessment*, Cambridge University Press, Cambridge, pp. 443–446

Swift, M., Vandermeer, J., Ramakrishan, R., Anderson, J., Ong, C., and Hawkins, B. (1996) 'Biodiversity and agroecosystem function', in H. A. Mooney, T. Cushman, E. Medina, O. E. Sala, and E. D. Schulze (eds) *Functional Roles of Biodiversity: A Global Perspective*, SCOPE, John Wiley and Sons, New York

Takahashi, R. and Todo, Y. (2017) 'Coffee certification and forest quality: evidence from a wild coffee forest in Ethiopia', *World Development*, vol 92, pp. 158–166

Tapia-Armijos, M. F., Homeier, J., Espinosa, C. I., Leuschner, C., and de la Cruz, M. (2015) 'Deforestation and forest fragmentation in South Ecuador since the 1970s–losing a hotspot of biodiversity', *PloS One*, vol 10, no 9, p. e0133701

Tejeda-Cruz, C., Silva-Rivera, E., Barton, J. R., and Sutherland, W. J. (2010) 'Why shade coffee does not guarantee biodiversity conservation', *Ecology and Society*, vol 15, no 1, p. e13

Terborgh, J. (2004) *Requiem for Nature*, Island Press, Washington, DC

Terui, A., Ishiyama, N., Urabe, H., Ono, S., Finlay, J. C., and Nakamura, F. (2018) 'Metapopulation stability in branching river networks', *Proceedings of the National Academy of Sciences, USA*, 26 June 2018, 115 (26) pp. E5963–E5969

Thrupp, A. (1988) 'Pesticides and policies: Approaches to pest-control dilemmas in Nicaragua and Costa Rica', *Latin American Perspectives*, vol 15, pp. 37–70

Tilman, D. (1999) 'Global environmental impacts of agricultural expansion: the need for sustainable and efficient practices', *Proceedings of the National Academy of Sciences, USA*, vol 96, pp. 5995–6000

Tinker, P. B., Ingram, J. S. I., and Struw, S. (1996) 'Effects of slash-and-burn agriculture and deforestation on climate change', *Agriculture, Ecosystems and Environment*, vol 58, pp. 13–22

Trewavas, A. (2002) 'Malthus foiled again and again', *Nature*, vol 418, pp. 668–670

Trewavas, A. (2004) 'A critical assessment of organic farming-and-food assertions with particular respect to the UK and the potential environmental benefits of no-till agriculture', *Crop Protection*, vol 23, pp. 757–781

Trigo, E. J. and Cap, E. J. (2003) 'The impact of the introduction of transgenic crops in Argentinean agriculture', *AgBio Forum*, vol 6, no 3, pp. 87–94

Tscharntke, T., Clough, Y., Jackson, J., Motzke, I. Perfecto, I., Vandermeer, J., Wanger, T. C., and Whitbread, A. (2012) 'Global food security, biodiversity conservation and the future of agricultural intensification', *Biological Conservation*, vol 51, no 1, pp. 53–59

Tscharntke, T., Klein, A. M., Cruess, A., Steffan-Dewenter, I., and Thies, C. (2005) 'Landscape perspective on agricultural intensification and biodiversity: ecosystem service management', *Ecology Letters*, vol 8, no 8, pp. 857–874

Tscharntke, T., Milder, J. C., Schroth, G., Clough, Y., DeClerck, F., Waldron, A., Rice, R., and Ghazoul, J. (2015) 'Conserving biodiversity through certification of tropical agroforestry crops at local and landscape scales', *Conservation Letters*, vol 8, no 1, pp. 14–23

Tscharntke, T., Steffan-Dewenter, I., Kruess, A., and Thies, C. (2002) 'Contribution of small habitat fragments to conservation of insect communities of grassland-cropland landscapes', *Ecological Applications*, vol 12, pp. 354–363

Tucker, R. P. (2000) *Insatiable Appetite: The United States and Ecological Degradation of the World*, University of California Press, Berkeley, CA

Tucker, R. P. (2018) 'War and the Environment', in D. Kieran, E. A. Martini, S. Conway-Lanz, and S. Aune (eds) *At War: The Military and American Culture in the Twentieth Century and Beyond*, Rudgers University Press, New Brunswick, NJ

Turcotte, D. L. and Malamud, B. D. (2004) 'Landslides, forest fires, and earthquakes: examples of self-organized critical behavior', *Physica A: Statistical Mechanics and its Applications*, vol 340, pp. 580–589

United States Global Change Research Program (USGCRP) (2018) *Impacts, Risks, and Adaptation in the United States: Fourth National Climate Assessment, Volume II*, D. R. Reidmiller, C. W. Avery, D. R. Easterling, K. E. Kunkel, K. L. M. Lewis, T. K. Maycock, and B. C. Stewart (eds), US Global Change Research Program, Washington, DC

USA Today (2017) 'Mega-mergers in agribusiness raise concerns about food costs, biodiversity', *USA Today*, usatoday.com, 29 August

Van den Bosch, R. (1978) *The Pesticide Conspiracy*, Doubleday, New York

van der Ploeg, J. D. (2009). *The New Peasantries: Struggles for Autonomy and Sustainability in an Era of Empire and Globalization*, Routledge, Abingdon, UK

van der Ploeg, J. D. (2013). *Peasants and the Art of Farming: A Chayanovian Manifesto* (no 2). Fernwood, Winnipeg

van Mansvelt, J. D. and Mulder, J. A. (1993) 'European features for sustainable development: a contribution to the dialogue', *Landscape and Urban Planning*, vol 27, no 2–4, pp. 67–90

Vandermeer, J. (2010) *The Ecology of Agroecosystems*, Barlett and Jones, Boston

Vandermeer, J. and Carvajal, R. (2001) 'Metapopulation dynamics and the quality of the matrix', *American Naturalist*, vol 158, pp. 211–220

Vandermeer, J. and Perfecto, I. (2005a) *Breakfast of Biodiversity: The Political Ecology of Deforestation (second edition)*, Institute for Food and Development Policy, Oakland, CA

Vandermeer, J. and Perfecto, I. (2005b) 'The future of farming and conservation', *Science*, vol 308, pp. 1257–1258

Vandermeer, J. and Perfecto, I. (2007a) 'The agricultural matrix and a future paradigm for conservation', *Conservation Biology*, vol 21, pp. 274–277

Vandermeer, J. and Perfecto, I. (2007b) 'Tropical conservation and grass roots social movements: Ecological theory and social justice', *Bulletin of the Ecological Society of America*, vol 88, pp. 171–175

Vandermeer, J. and Perfecto, I. (2013) 'Complex traditions: Intersecting theoretical frameworks in agroecological research', *Agroecology and Sustainable Food Systems*, vol 37, no 1, pp. 76–89

Vandermeer, J. and Perfecto, I. (2017a) *Ecological Complexity and Agroecology*, Routledge, Abingdon, UK

Vandermeer, J. and Perfecto, I. (2017b) 'Ecological complexity and agroecosystems: seven themes from theory', *Agroecology and Sustainable Food Systems*, vol 41, no 7, pp. 697–722

Vandermeer, J., Aga, A., Allgeier, J., Badgley, C., Baucom, R., Blesh, J., Fink-Shapiro, L., Hoey, L., Jain, M., Jones, A. D., Perfecto, I., and Wilson, M. L. (2018) 'Feeding Prometheus: An interdisciplinary approach for solving the global food crisis', *Frontiers in Sustainable Food Systems*, vol 2, p. e39, available at doi: 10.3389/fsufs.2018.00039, accessed 14 March 2019

Vandermeer, J., Lawrence, D., Symstad, A., and Hobbie, S. (2002) 'Effect of biodiversity on ecosystem function in managed ecosystems', in M. Laureau, S. Naeem, and P. Inchausti (eds) *Biodiversity and Ecosystem Functioning*, Oxford University Press, Oxford

Vandermeer, J., Perfecto, I., and Philpott, S. M. (2008) 'Clusters of ant colonies and robust criticality in a tropical agroecosystem', *Nature*, vol 451, pp. 457–459

Vandermeer, J., Perfecto, I., and Philpott, S. (2010a) 'Ecological complexity and pest control in organic coffee production: uncovering an autonomous ecosystem service', *BioScience*, vol 60, no 7, pp. 527–537

Vandermeer, J., Perfecto, I., and Schellhorn, N. (2010b) 'Propagating sinks, ephemeral sources and percolating mosaics: conservation in landscapes', *Landscape Ecology*, vol 25, no 4, pp. 509–518

Vara-Sánchez, I. and Padilla, M. C. (2013) 'Biodiversidad cultivada: una cuestión de coevolución y transdisciplinariedad', *Revista Ecosistemas*, vol 22, no 1, pp. 5–9

Vaughan, C., Ramirez, O., Herrera, G., and Guries, R. (2007) 'Spatial ecology and conservation of two sloth species in a cacao landscape in Limón, Costa Rica', *Biodiversity and Conservation*, vol 16, pp. 2293–2310

Velázquez Runk, J. (2017) *Craftin Wounan Landscapes: Identity, Art, and Environmental Governance in Panama's Darien*, University of Arizona Press: Tucson

Vermeulen, S. J., Campbell, B. M., and Ingram, J. S. I. (2012) 'Climate change and food systems', *Annual Review of Environmental Resources*, vol 37, pp. 195–222

VOA (2018) 'Indonesia's goal of food sovereignty remains a work in progress', VOA, 14 November, available at https://learningenglish.voanews.com/a/indonesia-s-goal-of-food-sovereignty-remains-a-work-in-progress/4652007.html, accessed 26 December 2018

Vorley, B. (2003) *Food, Inc.: Corporate Concentration from Farm to Consumer*, UK Food Group, London

Voss, S. R., Woodcock, M. R., and Zambrano, L. (2015) 'A tale of two axolotls', *BioScience*, vol 65, no 12, pp. 1134–1140

Waggoner, P. E. (1994) 'How much land can ten billion people spare for nature: Task Force Report', Council for Agricultural Science and Technology No. 121, Report 0194–4088, Council for Agricultural Science and Technology, Ames, IA

Waggoner, P. E., Ausuel, J. H., and Wernick, I. K. (1996) 'Lightening the thread of population on the land: American examples', *Population and Development Review*, vol 22, pp. 531–545

Wakeford, T. (2001) *Liaisons of Life: From Hornworts to Hippos, How the Unassuming Microbe has Driven Evolution*, Wiley, New Jersey

Walker, R., Arima, E., Messina, J., Soares, B., Perz, S., Vergara, D., Sales, M., Pereira, R., and Castro, W. (2013) 'Modeling spatial decisions with graph theory:

logging roads and forest fragmentation in the Brazilian Amazon', *Ecological Applications*, vol 23, no 1, pp. 239–254

Wallerstein, I. (2004) *World Systems Analysis: An Introduction*, Duke University Press, Durham

Way M. J. and Heong K. L. (1994) 'The role of biodiversity in the dynamics and management of insect pests of tropical irrigated rice – a review', *Bulletin of Entomological Research*, vol 84, pp. 567–587

Weerakoon, A. C. J. (1957) 'Some animals of the paddy field', *Loris*, vol 7, pp. 335–343

Weerakoon, A. C. J. and Samarasinghe, E. L. (1958) 'Mesofauna of the soil of a paddy field in Ceylon – a preliminary survey', *Ceylon Journal of Science (Biological Sciences)*, vol 1, no 2, pp. 155–170

Weibull, A. C., Östman, Ö., and Granqvist, A. (2003) 'Species richness in agroecosystems: the effect of landscape, habitat and farm management', *Biodiversity and Conservation*, vol 12, pp. 1335–1355

Werner, E. E., Yurewicz, K. L., Skelly, D. K., and Relyea, R. A. (2007) 'Turnover in an amphibian metacommunity: the role of local and regional factors', *Oikos*, vol 116, pp. 1713–1725

Western, D. and Pearl, M. C. (1989) *Conservation for the Twenty-First Century*, Oxford University Press, New York

Wezel, A., Bellon, S., Doré, T., Francis, C., Vallod, D., and David, C. (2009) 'Agroecology as a science, a movement and a practice. A review', *Agronomy for Sustainable Development*, vol 29, no 4, pp. 503–515

Wheeler, K., Wenger, S. J., Walsh, S. J., Martin, Z. P., Jelks, H. L., and Freeman, M. C. (2018) 'Stream fish colonization but not persistence varies regionally across a large North American river basin', *Biological Conservation*, vol. 223, pp. 1–10

White, G. M., Boshier, D. H. and Powell, W. (2002) 'Increased pollen flow contracts fragmentation in a tropical dry forest: An example from *Swietenia humilis* Zuccarini', *Proceedings of the National Acadademy of Sciences. USA*, vol 99, pp. 2038–2042

Whitesell, T. (1993) 'Changing courses: the Jurua River, its people and Amazonian extractive reserves', PhD dissertation, University of California, Berkeley, CA

Whitesell, T. (1996) 'Local struggles over rain-forest conservation in Alaska and Amazonia', *Geographical Review*, vol 86, no 3, pp. 414–436

Wickaramasinghe, L. P., Harris, S., Jones, G., and Jennings, N. V. (2004) 'Abundance and species richness of nocturnal insects on organic and conventional farms: effects of agricultural intensification on bat foraging', *Conservation Biology*, vol 18, pp. 1283–1292

Wiersum, K. F. (1986) 'The effect of intensification of shifting cultivation in Africa on stabilizing land-use and forest conservation', *Netherlands Journal of Agricultural Science*, vol 34, pp. 485–488

Willer, H. and Lernoud, J. (2017) *Organic Agriculture Worldwide 2017: Current Statistics*, Research Institute of Organic Agriculture (FiBL), Biofash 2017, 15 February, Frick, Switzerland

Williams-Guillen, K., Perfecto, I., and Vandermeer, J. (2008) 'Bats limit insects in a neotropical agroforestry system', *Science*, vol 320, p. 70

Williams, G. W. (1994) *References on the American Indian Use of Fire in Eco-Systems*, available at www.nrcs.usda.gov/Internet/FSE_DOCUMENTS/nrcs144p2_051334.pdf, accessed 14 March 2019

Williams, M. (2003) *Deforesting the Earth: From Prehistory to Global Crisis*, University of Chicago Press: Chicago

Wilshusen, P. R., Brechin, S. R., Fortwangler, C. L., and West, P. C. (2000) 'Reinventing a square wheel: Critique of a resurgent "protection paradigm" in international biodiversity conservation', *Society and Natural Resources*, vol 27, pp. 300–311

Wilson, E. O. (2016) *Half-Earth: Our Planet's Fight for Life*, W. W. Norton & Company, New York

Wilson, E. O. (ed.) (1986) *Biodiversity*, National Academy of Science Press, Washington, DC

Wisborg, P. (2013) 'Human rights against land grabbing? A reflection on norms, policies, and power', *Journal of Agricultural & Environmental Ethics,* vol 26, no 6, pp. 1199–1222

Wittman, H. (2010) 'Agrarian reform and the environment: fostering ecological citizenship in Mato Grosso, Brazil', *Canadian Journal of Development Studies/Revue Canadienne D'études du Développement*, vol 29, no 3–4, pp. 281–298

Wittman, H., Desmarais, A., and Wiebe, N. (2010) 'The origins and potential of food sovereignty', *Food Sovereignty: Reconnecting Food, Nature and Community*, pp. 1–14

Wolford, W. 2010 'Participatory democracy by default: Land reform, social movements and the state in Brazil', *Journal of Peasant Studies*, vol 37, no 1, pp. 91–109

Wood, G. (2015) *Tambora: The Eruption that Changed the World*, Princeton University Press, Princeton, US

Woods, W. I., Lehmann, J., Rebellato, L., Steiner, C., Teixeira, W. G., and Winkler-Prins, A. (eds) (2009) *Amazonian Dark Earths: Wim Sombroek's Vision*, Springer, Berlin

World Health Organization (1979) 'DDT and its derivates', *Environmental Health Criteria*, 9, WHO, Rome

World Inequality Lab (2018) *World Inequality Report: 2018, Executive Summary*: available at https://wir2018.wid.world/files/download/wir2018-summary-english.pdf, accessed 14 March 2019

World Wildlife Fund (WWF) (2018) *Living Planet Report – 2018: Aiming Higher,* M. Grooten and R. E. A. Almond (eds), WWF, Gland, Switzerland

Worster, D. (1979) *Dust Bowl: The Southern Plains in the 1930s*, Oxford University Press, Oxford, UK

Worster, D. (2016) 'The National Parks', Oxford Research Encyclopedia: American History.

Wright, A. L. (1976) 'Market, Land and Class: Southern Bahia, Brazil, 1890–1942', PhD dissertation, University of Michigan, Ann Arbor, MI

Wright, A. L. (2005) *The Death of Ramon Gonzalez*, University of Texas Press, Austin, TX

Wright, A. L. and Wolford, W. (2003) *To Inherit the Earth: The Landless Movement and the Struggle for a New Brazil*, Food First, Oakland, CA

Wright, J. (2008) *Sustainable Agriculture and Food Security in an Era of Oil Scarcity: Lessons from Cuba*, Earthscan, London

Wright, J. and Muller-Landau, H. (2006) 'The future of tropical forest species', *Biotropica*, vol 38, no 3, pp. 287–301

Wyatt, K. B., Campos, P. F., Gilbert, M. T. P., Kolokotronis, S. O., Hynes, W. H., DeSalle, R., Daszak, P., MacPhee, R. D., and Greenwood, A. D. (2008) 'Historical mammal extinction on Christmas Island (Indian Ocean) correlates with introduced infectious disease', *PloS One*, vol 3, no 11, p. e3602

Yano, K., Chu, I., and Sato, M. (1983) 'Faunal and biological studies on the insects of paddy fields in Asia. XI Records on aquatic Coleoptera from paddy water in the world', *Chinese Journal of Entomology*, vol 3, pp. 15–31

Yarnall, K. and Price, M. (2010) 'Migration, development and a new rurality in the Valle Alto, Bolivia', *Journal of Latin American Geography*, vol 9, no 1, pp. 107–124

Yuan, Z. Y. and Chen, H. Y. (2015) 'Negative effects of fertilization on plant nutrient resorption', *Ecology*, vol 96, no 2, pp. 373–380

Zanotti, L. (2016) *Radical Territories in the Brazilian Amazon: The Kayapós Fight for a Just Livelihood*, University of Arizona Press: Tucson

Zhang, W. (2011) 'Constructing ecological interaction networks by correlation analysis: hints from community sampling', *Network Biology*, vol 1, no 2, p. 81–89

Zhao, Z. H., Hui, C., He, D. H., and Li, B. L. (2015) 'Effects of agricultural intensification on ability of natural enemies to control aphids', *Scientific Reports*, vol 5, p. e8024

Zhong, W., Gu, T., Wang, W., Zhang, B., Lin, X., Huang, Q., and Shen, W. (2010) 'The effects of mineral fertilizer and organic manure on soil microbial community and diversity', *Plant and Soil*, vol 326, no 1–2, pp. 511–522

Zimmerer, K. S. and Young, K. R. (eds) (1998) *Nature's Geography: New Lessons For Conservation in Developing Countries*, University of Wisconsin Press, Madison, WI

Zomer, R. J., Neufeldt, H., Xu, J., Ahrends, A., Bossio, D., Trabucco, A., Van Noordwijk, M., and Wang, M. (2016) 'Global tree cover and biomass carbon on agricultural land: The contribution of agroforestry to global and national carbon budgets', *Scientific Reports*, vol 6, p. e29987

Index

Page numbers in *italics* denote figures.

T - #0217 - 111024 - C0 - 234/156/15 - PB - 9780367137816 - Matt Lamination